Lecture Notes in Physics

The Editorial Policy for Proceedings

The series Lecture Notes in Physics reports new developments in physical research and teaching – quickly, informally, and at a high level. The proceedings to be considered for publication in this series should be limited to only a few areas of research, and these should be closely related to each other. The contributions should be of a high standard and should avoid lengthy redraftings of papers already published or about to be published elsewhere. As a whole, the proceedings should aim for a balanced presentation of the theme of the conference including a description of the techniques used and enough motivation for a broad readership. It should not be assumed that the published proceedings must reflect the conference in its entirety. (A listing or abstracts of papers presented at the meeting but not included in the proceedings could be added as an appendix.)

When applying for publication in the series Lecture Notes in Physics the volume's editor(s) should submit sufficient material to enable the series editors and their referees to make a fairly accurate evaluation (e.g. a complete list of speakers and titles of papers to be presented and abstracts). If, based on this information, the proceedings are (tentatively) accepted, the volume's editor(s), whose name(s) will appear on the title pages, should select the papers suitable for publication and have them refereed (as for a journal) when appropriate. As a rule discussions will not be accepted. The series editors and Springer-Verlag will normally not interfere with the detailed editing except in fairly obvious cases or on technical matters.

Final acceptance is expressed by the series editor in charge, in consultation with Springer-Verlag only after receiving the complete manuscript. It might help to send a copy of the authors' manuscripts in advance to the editor in charge to discuss possible revisions with him. As a general rule, the series editor will confirm his tentative acceptance if the final manuscript corresponds to the original concept discussed, if the quality of the contribution meets the requirements of the series, and if the final size of the manuscript does not greatly exceed the number of pages originally agreed upon. The manuscript should be forwarded to Springer-Verlag shortly after the meeting. In cases of extreme delay (more than six months after the conference) the series editors will check once more the timeliness of the papers. Therefore, the volume's editor(s) should establish strict deadlines, or collect the articles during the conference and have them revised on the spot. If a delay is unavoidable, one should encourage the authors to update their contributions if appropriate. The editors of proceedings are strongly advised to inform contributors about these points at an early stage.

The final manuscript should contain a table of contents and an informative introduction accessible also to readers not particularly familiar with the topic of the conference. The contributions should be in English. The volume's editor(s) should check the contributions for the correct use of language. At Springer-Verlag only the prefaces will be checked by a copy-editor for language and style. Grave linguistic or technical shortcomings may lead to the rejection of contributions by the series editors. A conference report should not exceed a total of 500 pages. Keeping the size within this bound should be achieved by a stricter selection of articles and not by imposing an upper limit to the length of the individual papers. Editors receive jointly 30 complimentary copies of their book. They are entitled to purchase further copies of their book at a reduced rate. As a rule no reprints of individual contributions can be supplied. No royalty is paid on Lecture Notes in Physics volumes. Commitment to publish is made by letter of interest rather than by signing a formal contract. Springer-Verlag secures the copyright for each volume.

The Production Process

The books are hardbound, and the publisher will select quality paper appropriate to the needs of the author(s). Publication time is about ten weeks. More than twenty years of experience guarantee authors the best possible service. To reach the goal of rapid publication at a low price the technique of photographic reproduction from a camera-ready manuscript was chosen. This process shifts the main responsibility for the technical quality considerably from the publisher to the authors. We therefore urge all authors and editors of proceedings to observe very carefully the essentials for the preparation of camera-ready manuscripts, which we will supply on request. This applies especially to the quality of figures and halftones submitted for publication. In addition, it might be useful to look at some of the volumes already published. As a special service, we offer free of charge LATEX and TEX macro packages to format the text according to Springer-Verlag's quality requirements. We strongly recommend that you make use of this offer, since the result will be a book of considerably improved technical quality. To avoid mistakes and time-consuming correspondence during the production period the conference editors should request special instructions from the publisher well before the beginning of the conference. Manuscripts not meeting the technical standard of the series will have to be returned for improvement.

For further information please contact Springer-Verlag, Physics Editorial Department V, Tiergartenstrasse 17, W-6900 Heidelberg, FRG

J. van Paradijs   H. M. Maitzen (Eds.)

Galactic High-Energy Astrophysics High-Accuracy Timing and Positional Astronomy

Lectures Held at the
Astrophysics School IV
Organized by the European Astrophysics Doctoral Network
(EADN) in Graz, Austria, 19-31 August 1991

Springer-Verlag
Berlin Heidelberg GmbH

Editors

Jan van Paradijs
Astronomical Institute "Anton Pannekoek" and
Center for High-Energy Astrophysics
Kruislaan 403, NL-1098 SJ Amsterdam, The Netherlands

Hans Michael Maitzen
Institut für Astronomie der Universität Wien
Türkenschanzstraße 17, A-1180 Wien, Austria

ISBN 978-3-662-13948-6   ISBN 978-3-540-47767-9 (eBook)
DOI 10.1007/978-3-540-47767-9

© Springer-Verlag Berlin Heidelberg 1993
Originally published by Springer-Verlag Berlin Heidelberg New York in 1993
Softcover reprint of the hardcover 1st edition 1993

58/3140-543210 - Printed on acid-free paper

PREFACE

The 4th Predoctoral Astrophysics School of the European Astrophysics Doctoral Network (EADN) was held from August 19 - 31, 1991 at Graz-Mariatrost with the participation of 7 teachers, and 34 students from 11 European countries. With this School EADN has completed half a decade of European collaboration in the field of academic teaching in astrophysics. After the EADN Schools at Les Houches (France, 1988), Ponte de Lima (Portugal, 1989) and Dublin (Ireland, 1990) Austrian astronomy hosted the fourth School and chose Graz as its venue.

Graz is related to both subjects of the School – Galactic High-Energy Astrophysics, and High Accuracy Timing and Positional Astronomy – through historical and contemporary circumstances. It is known as one of the Kepler cities. In 1994 Graz will celebrate the 400th anniversary of Kepler's arrival there where he started both his teaching and scientific careers. Kepler worked on the most accurate and numerous positional observations available at that time through the efforts of Tycho Brahe; he can be said to have also contributed to the field of High Energy Astronomy by his book on the detection of a STELLA NOVA, which in fact was the last naked-eye supernova discovery (1604) before the recent famous supernova SN 1987A.

We would like to mention here also that the discoverer of cosmic rays – part of the first topic of the Graz School – was the Nobel Prize winner Viktor F. Hess who conducted his research at the Karl-Franzens University of Graz. Other well-known scientists, who at some time in their careers worked at this University, include Boltzmann, Schrödinger (Nobel Prize 1933) and Mach.

Like the other two classical Austrian universities of Vienna and Innsbruck the Karl-Franzens University has an Institute for Astronomy. We would like to express our deep gratitude to this institute, especially to its head Prof. Dr. Hermann Haupt and two of his students, Karin Muglach and Robert Greimel, for their support during the preparatory phase and the School weeks.

We acknowledge with grateful appreciation financial support for the School from:
- SCIENCE, the scientific stimulation programme of the Commission of the European Community;
- the Austrian Ministry of Science and Research;
- the Government of the Land Steiermark (Styria);
- the City of Graz, in addition to a truly delightful reception by its mayor Alfred Stingl and City
  Counsellor Helmut Strobl;
- the Granholm Foundation (Sweden), and
- the Oesterreichische Forschungsgemeinschaft.

Extensive cooperation and help came from the home institute of the Local Organizer, the Vienna Institut für Astronomie (head Prof. Dr. Paul Jackson), and especially from its collaborators Dr. Anneliese Schnell, Dr. Ernst Goebel and Mag. Franz Kerschbaum.

Thanks go also to the staff of the Bildungshaus Mariatrost which provided not only dormitories and meeting rooms, but also a relaxed and friendly atmosphere surrounded by magnificent natural beauty.

Last but not least, we wish to specially thank the Coordinator of EADN, Prof. Dr. Jean Heyvaerts who completed with his participation in the Graz School a period of five years at the helm of EADN.

The 4th EADN School in Graz was pronounced successful by teachers and students alike. It occurred at a very critical point in recent European history, since its beginning coincided with the coup d'etat in the former Soviet Union. Together with the scientific educational values and the charming Graz downtown atmosphere this may have contributed to a very special feeling of togetherness of young European doctoral students in astrophysics during those two weeks in August 1991.

Amsterdam/Vienna, October 1992

J. van Paradijs, H.M. Maitzen

Contents

List of participants

Anastasiadis, Anastasios	Thessaloniki (Greece)
Aparicio, Jose M.	Barcelona (Spain)
Aringer, Bernhard	Wien (Austria)
Beeharry, Girish K.	Meudon (France)
Boncheva, Theodora I.	Shoumen (Bulgaria)
Boulard, Marie-Helene	Toulouse (France)
Cognard, Ismael	Meudon (France)
Colomer, Francisco	Göteborg (Sweden)
Cuisinier, Francois	Strasbourg (France)
Del Rio, Evileo	Barcelona (Spain)
Dimitrova, Petya	Shoumen (Bulgaria)
Egonsson, Jim	Lund (Sweden)
Ferreira, Jonathan	Grenoble (France)
Greimel, Robert	Graz (Austria)
Ivanov, Milen	Sofia (Bulgaria)
Kerschbaum, Franz	Wien (Austria)
Kunz, Mathias	Tübingen (Germany)
Kuulkers, Eric	Amsterdam (Netherlands)
Manning, Rodger	Birmingham (England)
Maravelias, Sergios	Athens (Greece)
Marti-Ribas, Josep	Barcelona (Spain)
Muglach, Karin	Graz (Austria)
Peracaula, Marta	Barcelona (Spain)
Pfeiffer, Benoite	Toulouse (France)
Prins, Sacha	Amsterdam (Netherlands)
Quemerais, Eric	Verrieres le Buisson (France)
Saphonova, Margaret	Moscow (Russia)
Schmitz-Fraisse, Christine	Toulouse (France)
Schultheis, Mathias	Wien (Austria)
Siopis, Christos	Ioannina (Greece)
Torkelsson, Ulf	Lund (Sweden)
Villata, Massimo	Torino (Italy)
Wyn, Graham	Leicester (England)
Zamanov, Radoslav	Smoljan (Bulgaria)

Part I
Galactic High-Energy Astrophysics

Particle   acceleration   in   astrophysics

A.   Achterberg[1,2]

[1] Sterrekundig Instituut, Postbus 80.000, 3508TA Utrecht
[2] Centrum voor hoge-energie astrofysica, Kruislaan 403,
1098 SJ Amsterdam, The Netherlands

Abstract: I review the physical principles of particle acceleration in astrophysical objects, with an emphasis on diffusive shock acceleration.

1 Introduction

The subject of these lectures is the physics of energetic particles or photons in astrophysical plasmas. Relativistic particles play an important role in astrophysics. Historically, the first indication that very energetic particles are present in our galaxy came from the discovery of Cosmic Rays. Around 1910 it was realized that a component of the natural radioactivity measured on Earth originates in outer space. Balloon experiments by the Austrian phycisist Victor Hess in 1911 showed that the intensity of this cosmic radiation increased with height above the Earth's surface. We now know that our whole galaxy is pervaded with a tenuous gas of relativistic particles (protons, electrons, α-particles and heavier nuclei) with a power-law energy distribution $N(E)dE = \varkappa E^{-s} dE$ with slope $s \approx 2.5$ up to energies per particle of 10^{12} - 10^{14} eV. It is usually assumed that these particles are generated near the shock wave which propagates into the interstellar medium after a supernova explosion. The idea that supernovae are the source of cosmic rays was first suggested by Baade and Zwicky in 1934.

In 1953 it became clear that the optical continuum emission of the Crab nebula (a remant of the supernova of 1054 AD) is polarized, and therefore non-thermal. Measurements by Oort, Baade and Walraven confirmed that we are dealing with synchrotron radiation by relativistic electrons in a magnetic field. This radiation mechanism had already been proposed by Shklovsky as a source for the radio-emission from the Crab.

In 1954, Baade and Minkowski were able to identify the radio source Cygnus A with an optical galaxy, confirming the extra-galactic nature of many non-thermal radio sources. It is now commonly believed that the radio emission from Active Galactic Nuclei (AGN's), Quasars and the extended radiolobes and jets associated with elliptical galaxies and quasars is due to synchrotron radiation by relativistic electrons. There are many cases where the presence of relativistic particles has been inferred (directly or indirectly) in astrophysical objects. A number of examples is listed in Table 1.

4

Proces/Object	Obs. evidence	Type of particle	Energy
Solar flares	direct measurement by satellites in space	electrons , protons	MeV
Pulsars	Optical-/radio pulses	electrons. positrons?	10 MeV ?
Close binaries	Extended radio structure and bursts: SS433. Cyg X-3	electrons	≈ 10 MeV
Cosmic Rays	direct measurement	electrons,protons,nuclei	$\leq 10^{20}$ eV
Supernova Remnants	synchr. radiation	electrons	MeV-GeV
radiogalaxies, Quasars and AGN' s	synchr.radiation	electrons	MeV-GeV

Table 1.

2 Particle acceleration: general principles.

In many astronomical objects where a large amount of energy is generated part of that energy is emitted in the form of relativistic particles. These relativistic particles must be produced by some acceleration process. It is important to distinguish between *heating* and *acceleration*. One speaks of *heating* when the available energy W is distributed equally between the particle population, so that the energy per particle roughly equals E $\approx$ W/N, with N the total number of particles. A thermal (Maxwellian) distribution of particles obviously satisfies this requirement: the mean kinetic energy per particle equals $E_{th} = k_b T/2$ per degree-of-freedom. One speaks of *acceleration* when a minority of particles gets a significant fraction of the available energy , so that the energy of an individual particle in that minority satisfies E $\gg$ W/N (E $\gg$ k_bT).

2.1 Stochastic and regular Fermi acceleration

In 1949 Enrico Fermi proposed the first serious acceleration mechanism in an astrophysical context[1]. He proposed that the galactic cosmic rays are accelerated in the interstellar medium during collisions with magnetised clouds, which scatter the particles. In its simplest form, the Fermi-model can be described as a process of elastic scattering by moving scattering centers. Consider the following simple example. A particle with momentum $\mathbf{p}$ - $\gamma m \mathbf{v}$ is centrally scattered by a moving "billiard ball" which acts as a scattering center. The scattering is elastic in the frame K' moving with the billiard ball with velocity $\mathbf{V}$. In the scattering event, the momentum $p'_{\parallel}$ of the particle along the direction of

V in K′ is reversed, while the component $p'_\perp$ is unaffected (fig. 1). Denoting the various quantities before- and after scattering by the subscripts i and f one can find the energy change associated with this scattering event by two succesive Lorentz transforms: one from the laboratory frame K to K′ before scattering, and the reverse transform after scattering. The relations before- and after scattering in the frame K′ are:

$$p'_{\|i} = \Gamma(p_{\|i} - E_i V/c^2) \ , \quad p'_{\|f} = -p'_{\|i} \ , \quad p'_{\perp i} = p_{\perp i} = p'_{\perp f} \ ,$$

$$E'_i = \Gamma(E_i - Vp_{\|i}) = E'_f \ . \tag{1}$$

Here I have defined $\Gamma \equiv \left(1 - V^2/c^2\right)^{-1/2}$ the Lorentz-factor associated with frame K′. Transforming back to the lab frame one finds: $E_f = \Gamma(E'_f + p'_{\|f}V) = \Gamma(E'_i - p'_{\|i}V)$. Expressing this in the initial quantities in the lab frame one can write this as:

$$E_f = \Gamma^2\left((1 + V^2/c^2)E_i - 2(V \cdot p_i)\right) \ . \tag{2}$$

Writing $\mathbf{p} = E\mathbf{v}/c$ one finds the energy change resulting from this encounter :

$$\Delta E = E_f - E_i = 2\Gamma^2\left(\left(\frac{V}{c}\right)^2 - \frac{V \cdot v_i}{c^2}\right)E_i \ . \tag{3}$$

From now on I will drop the subscript i. In a head-on collision one has $\mathbf{V} \cdot \mathbf{v} < 0$ and the particle gains energy. In a collision where the particle overtakes the scattering center one has $\mathbf{V} \cdot \mathbf{v} > 0$ and it loses energy (assuming of course that $v > V$). I will consider mostly situations where $V \ll v \leq c$ so that $\Gamma \approx 1$.

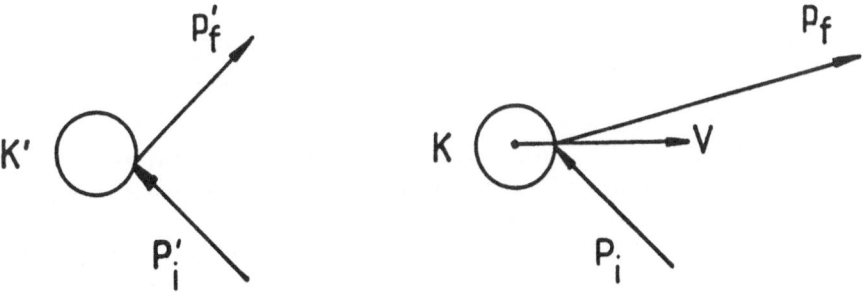

Fig. 1. *Idealised model for Fermi acceleration. A particle is scattered by a moving scattering center. In the frame where the scattering center is at rest the scattering is elastic.*

There are no two possibilities: in a *special geometry* one can fix it in such a way that only head-on collisions occur. This would be the case if a particle were trapped between an elastically reflecting wall and a stream of scattering centers moving toward that wall (fig. 2) . In that case the particle gains an amount of energy every time it goes through a cycle of reaching the wall, reflecting back into the stream and scattering back towards the wall starting a new cycle. How many scatterings there are before it reaches the wall again is unimportant. This is most easily seen by looking at the process in the frame moving with the scattering centers: in that frame each scattering is elastic, except the reflection by the wall, which is moving in that frame with velocity - V. This version of the process is called *regular Fermi acceleration*. As we will see in the next Section this forms the basis for the process of acceleration of charged particles near a shock. The energy change per cycle is of order (V ≪ v):

$$\Delta E \approx \left(Vv/c^2\right) E. \tag{4}$$

The second possibility is that of particles being scattered by a "gas" of scattering centers moving in random directions. In that case the particles again gain net energy. This is due to the fact that the number of head-on collisions (where the particle gains energy) exceeds the number of overtaking collisions (where the particle loses energy). Assuming again v ≫ V and scattering centers with density n_* and collisional cross section σ_* , the encounter rate for a given (fixed) direction of **V** equals:

$$R_* \approx n_* \sigma_* |\mathbf{v} - \mathbf{V}| \approx n_* \sigma_* v \left(1 - \frac{\mathbf{v} \cdot \mathbf{V}}{v^2}\right) . \tag{5}$$

Head-on collisions (**v**·**V** < 0) are indeed more frequent than overtaking collisions (**v**·**V** > 0). The net energy gain per unit time follows by averaging $R_* \times \Delta E$ (Eqn. 3) over all possible directions of **V**:

$$\left\langle \frac{dE}{dt} \right\rangle \approx n_* \sigma_* v \left\langle \left(1 - \frac{\mathbf{v} \cdot \mathbf{V}}{v^2}\right) \Delta E \right\rangle \approx \frac{8}{3} \frac{v}{\lambda} \frac{V^2}{c^2} E. \tag{6}$$

Here $\lambda \equiv 1/(n_* \sigma_*)$ is the mean-free-path of the particles in the gas of scattering centers, corresponding to a mean scattering rate $\langle R_* \rangle = v/\lambda = n_* \sigma_* v$. This result can be derived by writing **v**·**V** = $vV\cos\vartheta$ and using $\langle \cos\vartheta \rangle = 0$, $\langle \cos^2\vartheta \rangle = 1/3$ for an isotropic velocity distribution of scattering centers. The mean energy gain per collision equals:

$$\langle \Delta E \rangle = \left\langle \frac{dE}{dt} \right\rangle / \langle R_* \rangle \approx \frac{8}{3} \frac{V^2}{c^2} E \qquad (V \ll c) . \tag{7}$$

This version of the process is called *stochastic Fermi acceleration*.

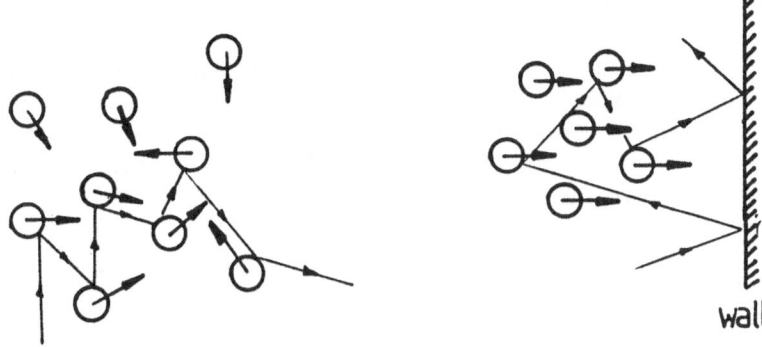

Fig. 2. *Simple model for regular and stochastic Fermi acceleration. The regular version of the proces requires a special geometry such as a unidirectional stream of scattering centers.*

If one compares the energy-gain per scattering for this proces with the gain per cycle (4) for the regular proces one sees that this process is less efficient by a factor $\approx v/V$. This is the result of the near-cancellation (to order $V/v \ll 1$) of the number of head-on and overtaking collisions.

There is a second important difference between regular and stochastic Fermi acceleration. In the regular proces all particles gain roughly the same amount of energy in each cycle, equal to the magnitude of the energy change $|\Delta E| \approx (Vv/c^2)E$ per collision. In the stochastic process however the *mean* energy increase per collision $\langle \Delta E \rangle$ is much less than the energy change ΔE per *individual* collision: $\langle \Delta E \rangle \approx (V/v)|\Delta E|$. This means that the particles in the latter process not only gain energy, but that they also *disperse* in energy. For $V \ll v$ the energy change at each scattering equals $\Delta E \approx ((\mathbf{V} \cdot \mathbf{v})/c^2)E$ which can have either sign with almost equal probability. This means that particles at an energy E go through a (slightly biased) random walk in energy with stepsize ΔE. Two particles starting out at the same energy will in general diffuse away from each other in energy.

The amount of dispersion can be calculated by considering the quantity $\delta E \equiv \Delta E - \langle \Delta E \rangle$. The average value $\langle \delta E \rangle$ is (by definition) zero. But from Eqns. 3-6 one can calculate that δE^2 increases in time. Assuming once again $V \ll v$ and using $\langle \delta E^2 \rangle = \langle \Delta E^2 \rangle - \langle \Delta E \rangle^2 \approx \langle \Delta E^2 \rangle$ one gets:

$$\frac{d}{dt} \delta E^2 \approx n_* \sigma_* v \left\langle \left(2 \frac{\mathbf{v} \cdot \mathbf{V}}{c^2} E \right)^2 \right\rangle \approx \frac{4}{3} \frac{v}{\lambda} \left(\frac{v^2 V^2}{c^4} \right) E^2. \qquad (8)$$

Here I have neglected terms of order V/v , V/c with respect to unity.

8

2.2 The spectrum due to stochastic Fermi acceleration.

The above results can be applied to calculate the energy distribution one expects as a result of the Fermi process. Stochastic acceleration will be considered first. For relativistic particles with $v \approx c$ one can define the mean velocity $\langle dE/dt \rangle$ and a diffusion coefficient D_E in energy space by:

$$\left\langle \frac{dE}{dt} \right\rangle = \frac{8V^2}{3c\lambda} E \equiv \alpha E \quad , \quad D_E \equiv \frac{1}{2}\left\langle \frac{d\,\delta E^2}{dt} \right\rangle = \frac{1}{4}\alpha E^2 \quad . \tag{9}$$

According to (9) the mean energy of an individual particle increases as $E(t) = E_o \exp(\alpha t)$, so the typical timescale of the acceleration process equals $\tau_{acc} = \alpha^{-1}$.

The flow of particles in energy under the influence of stochastic Fermi acceleration can be described by a convection-diffusion equation (known as the *confusion equation* in some circles). Let $dN = N(E,t)dE$ be the number-density of particles with energy in the range $(E, E+dE)$. If particles escape from the region where the acceleration takes place after a time T, $N(E,t)$ satisfies the following equation:

$$\frac{\partial N(E,t)}{\partial t} + \frac{\partial S(E,t)}{\partial E} = -\frac{N(E,t)}{T} + Q(E,t). \tag{10}$$

Here $Q(E,t)$ is a source term describing the injection of particles into the acceleration process. $S(E,t)$ denotes the mean particle flux in energy. It is given in terms of the mean energy gain $\langle dE/dt \rangle$ and the energy diffusion coefficient D_E by the expression:

$$S(E,t) \equiv \left\langle \frac{dE}{dt} \right\rangle N(E,t) - \frac{\partial}{\partial E}\left(D_E N(E,t) \right). \tag{11}$$

This equation can be derived using the so-called *Fokker-Planck* approximation[2]. For stochastic Fermi acceleration this equation can be written according to (9) in the form:

$$\frac{\partial N(E,t)}{\partial t} + \frac{\partial}{\partial E}\left[\alpha E N(E,t) - \frac{1}{4}\alpha \frac{\partial}{\partial E}\left(E^2 N(E,t) \right) \right] = -\frac{N(E,t)}{T} + Q(E,t). \tag{12}$$

In the steady-state ($\partial/\partial t = 0$) the solution is a power-law $N(E) \propto E^{-s}$. When particles are injected into the acceleration proces at rate R with an energy E_o one has $Q(E,t) = R\delta(E - E_o)$. The solution of (12) reads in this case[3]:

$$N(E) = \frac{4(R/\alpha E_o)}{(9 + \frac{16}{\alpha T})^{1/2}}\left(\frac{E}{E_o}\right)^{-s(E|E_o)}. \quad \text{where:} \tag{13a}$$

$$
s(E|E_o) = \begin{cases} -\dfrac{1}{2} - \dfrac{1}{2}\left(9 + \dfrac{16}{\alpha T}\right)^{1/2} & (E \leq E_o), \\[3mm] -\dfrac{1}{2} + \dfrac{1}{2}\left(9 + \dfrac{16}{\alpha T}\right)^{1/2} & (E > E_o). \end{cases} \tag{13b}
$$

If particles are retained infinitely long in the source ($T \longrightarrow \infty$) one has $N(E) \propto E^2$ for $E \leq E_o$ and $N(E) \propto E^{-1}$ for $E > E_o$, irrespective of the value of α. Generally, the distribution has a maximum at the injection energy E_o, decaying towards higher- and lower energy.

The power-law behaviour $N(E) \propto E^{-s}$ with $s \geq 1$ for $E > E_o$ is the reason that models such as this were very popular as an astrophysical acceleration mechanism until the mid-seventies. Power-law distributions are commonly observed, for instance in the case of galactic cosmic rays and the relativistic electrons in extended radiolobes associated with active galaxies and some quasars.

The problem however is that observations in many cases give a spectrum $N(E) = \kappa E^{-s}$ with $s \approx 2.5$, with only a small spread around that mean value. This means that the parameter αT (the ratio of the escape time T and the typical acceleration time $\tau_{acc} \approx \alpha^{-1} \approx c\lambda/V^2$) which determines that slope would have to be of order unity in all sources, regardless the detailed physical conditions (size, fieldstrength, luminosity) in each source. Such "fine-tuning" of physical parameters is rather unlikely in general. Ways around this have been constructed, involving "feedback loops" which adjusts the level of acceleration by turbulence in such a way that $\alpha T = O(1)$, but some special assumptions are still needed[3,4]. The process of particle acceleration near shocks, which will be considered below, does not suffer from this particular problem.

2.3 Stochastic acceleration by plasma waves and turbulence

2.3.1 Basic principles

In this section I will briefly consider a more realistic description of stochastic particle acceleration. Consider a particle subject to a *random* force. For simplicity I will consider the one-dimensional case. The dynamics of the particle is described by:

$$
\frac{dp}{dt} = F(\mathbf{x}, t) \quad , \quad \langle\langle F(\mathbf{x}, t) \rangle\rangle = 0. \tag{14}
$$

This random force could for instance be due to random electromagnetic fields associated with waves in a plasma, and is a function of the position $\mathbf{x}$ and time t. The averaging bracket $\langle\langle \ \rangle\rangle$ in this case is to be interpreted in a statistical sense as an *ensemble average* . The trajectory of the particle is $\mathbf{x} = \mathbf{X}(t)$. The change in momentum of the particle between $t = 0$ and $t = \Delta t$ can be found by integrating along the particle trajectory:

$$\Delta p = \int_0^{\Delta t} dt' \, F\left(\mathbf{X}(t'), t'\right) . \tag{15}$$

The time-integral and the ensemble average can be interchanged, and if one *assumes* that the change in the particle orbit under the influence of the random force can be neglected, one immediately finds to lowest order:

$$\langle\langle \Delta p \rangle\rangle = \int_0^{\Delta t} dt' \, \langle\langle F(\mathbf{X}', t') \rangle\rangle = 0. \tag{16}$$

This means that the particle tends to random-walk in momentum, centered around the momentum it originally started with at $t = 0$. Although the mean displacement in momentum $\langle\langle \Delta p \rangle\rangle$ vanishes to lowest order, $\langle\langle \Delta p^2 \rangle\rangle$ increases. Using (15) once again one has:

$$\langle\langle \Delta p \, \Delta p \rangle\rangle = \int_0^{\Delta t} dt' \int_0^{\Delta t} dt'' \, \langle\langle F(\mathbf{X}', t') \, F(\mathbf{X}'', t'') \rangle\rangle. \tag{17}$$

Here I have introduced the notation $\mathbf{X}' \equiv \mathbf{X}(t')$. Specifically, consider the case of acceleration by waves in a plasma. A plasma supports a veritable Zoo of plasma waves of electrostatic-, electromagnetic- and mixed types[5]. For the present discussion we will only need the fact that in that case the force $F(\mathbf{x}, t)$ can be expanded using a Fourier-integral (essentially a continuous superposition of plane waves):

$$F(\mathbf{x}, t) = \int \frac{d^3 k}{(2\pi)^3} \, \tilde{F}(\mathbf{k}) \exp\left(i\mathbf{k}\cdot\mathbf{x} - i\omega(\mathbf{k})t\right) . \tag{18}$$

Here $\omega(\mathbf{k})$ is the frequency of the wave under consideration, and $\tilde{F}(\mathbf{k})$ the Fourier-amplitude of $F(\mathbf{x}, t)$. The Fourier-amplitude generally is a complex number. It satisfies the relation $\tilde{F}(-\mathbf{k}) = \tilde{F}^*(\mathbf{k})$, which together with $\omega(-\mathbf{k}) = -\omega^*(\mathbf{k})$ ensures that $F(\mathbf{x}, t)$ is a real quantity. One can show that in a *spatially homogeneous, stationary ensemble* of waves the Fourier-amplitudes have an ensemble average which satisfies the following relation:

$$\langle\langle \tilde{F}(\mathbf{k}) \, \tilde{F}(\mathbf{k}') \rangle\rangle = |\tilde{F}(\mathbf{k})|^2 \, (2\pi)^3 \, \delta(\mathbf{k} + \mathbf{k}'). \tag{19}$$

Expanding each of the two factors $F(\mathbf{k})$ in (16) in a Fourier integral (one over $d^3 k$ and one over $d^3 k'$) and using the property (19) one can write Eqn. (17) after integration over $\mathbf{k}'$:

$$\langle\langle \Delta p^2 \rangle\rangle = \int_O^{\Delta t} dt' \int_O^{\Delta t} dt'' \int \frac{d^3 k}{(2\pi)^3} |\tilde{F}(k)|^2 \exp\left(i k \cdot (X'' - X') - i\omega(k)(t'' - t') \right). \quad (20)$$

The remaining integral over k contains an exponential term of the form e^{iS} which fluctuates strongly in time. The simplest case obtains when there is no magnetic field. The unperturbed orbit of the particle in that case is a straight line: $X(t) = X_o + vt$. This means that the phase S in the exponential term in integral (19) is given by:

$$S = \left(k \cdot v - \omega(k) \right)\left(t'' - t' \right). \quad (21)$$

Defining new variables $T = (t'' + t')/2$ and $\tau = (t'' - t')$ one can evaluate the integral using $dt'dt'' = dT d\tau$. S depends only on τ so the integral over T can be evaluated trivially:

$$\langle\langle \Delta p^2 \rangle\rangle = 2\Delta t \int \frac{d^3 k}{(2\pi)^3} |\tilde{F}(k)|^2 \int_{-\Delta t}^{\Delta t} d\tau \exp\left(i \left(k \cdot v - \omega(k) \right)\tau \right). \quad (22)$$

Using $e^{iS} \equiv \cos S + i \sin S$ we can evaluate the integral over τ:

$$\int_{-\Delta t}^{\Delta t} d\tau \exp\left(i(k \cdot v - \omega(k))\tau \right) = \frac{2\sin\left((k \cdot v - \omega(k))\Delta t \right)}{k \cdot v - \omega(k)}. \quad (23)$$

In the limit $\Delta t \longrightarrow \infty$ one can use the following theorem involving the Dirac delta-function: $\lim_{t\to\infty} \sin(xt)/x = \pi \delta(x)$. This allows one to evaluate Eqn. (22) for large Δt, defining the momentum diffusion coefficient D_P in the process:

$$\frac{\langle\langle \Delta p^2 \rangle\rangle}{2\Delta t} \equiv D_P = \int \frac{d^3 k}{(2\pi)^3} |\tilde{F}(k)|^2 2\pi \delta\left(\omega(k) - k \cdot v \right). \quad (24)$$

This result has a simple interpretation. The δ-function in (24) selects those waves for which the *resonance-condition* $\omega - k \cdot v = 0$ is satisfied. Those are exactly the waves which have a constant phase S ($\tilde{F} \propto e^{iS}$) as seen by the particle along its unperturbed orbit. As a result, the force excerted on the particle by these resonant waves does not fluctuate wildly in time and is able to change the momentum appreciably. In contrast, the action on the momentum of non-resonant waves averages out.

If the particle propagates through a magnetised plasma with a mean magnetic field B the resonance condition becomes more complicated. The reason is that the unperturbed orbit is now a spiral along the magnetic field . This spiralling motion results in the appearance of so-called *gyro-resonances* involving harmonics of the gyration frequency $\Omega_B = q|B|/\gamma mc$ of the particle. The resonance condition in this case reads:

$$\omega(\mathbf{k}) - k_\| v_\| + n\Omega_B = 0 \quad (n = \cdots -2,-1,0,1,2,\cdots). \quad (25)$$

Here $k_\|$ and $v_\|$ are the components of the wave-vector $\mathbf{k}$ and particle velocity $\mathbf{v}$ along the magnetic field, e.g. $k_\| \equiv \mathbf{k} \cdot \mathbf{B}/|\mathbf{B}|$.

The result (24) is easily generalised to more spatial dimensions. In the unmagnetised case, one defines the momentum-diffusion tensor D_{ij} by :

$$D_{ij} = \frac{\langle\langle \Delta p_i \Delta p_j \rangle\rangle}{2\Delta t} = \int \frac{d^3 k}{(2\pi)^3} \left| \tilde{F}_i(\mathbf{k}) \tilde{F}_j^*(\mathbf{k}) \right| 2\pi\delta\big(\omega(\mathbf{k}) - \mathbf{k}\cdot\mathbf{v}\big). \quad (26)$$

The change and energy $\langle dE/dt \rangle$ and the energy diffusion coefficient D_E associated with this momentum diffusion can be derived by the following argument. Let $dN = 4\pi p^2 f(\mathbf{x},t,p)dp dV$ be the number of particles in a infinitesimal volume dV with momentum in the range $(p, p + dp)$. I have assumed for simplicity that the particle momenta are distributed isotropically so that $f(\mathbf{x},t,\mathbf{p})$ depends only on the magnitude of momentum $p \equiv \sqrt{(p_x^2 + p_y^2 + p_z^2)}$. Under the influence of momentum diffusion with diffusion coefficient $D_p \equiv p_i D_{ij} p_j / p^2$, this distribution evolves in time according to:

$$\frac{\partial}{\partial t} f(\mathbf{x},t,p) = \frac{1}{p^2}\frac{\partial}{\partial p}\left(p^2 D_p \frac{\partial}{\partial p}\right) f(\mathbf{x},t,p). \quad (27)$$

The placement of the factors p^2 in this equation ensures that the number of particles is conserved : diffusion redistributes particles in momentum, it does not create or destroy them. The energy diffusion coefficient follows directly from the relation $\Delta E = (\partial E/\partial p)\Delta p = v\Delta p$, so that $D_E \equiv \langle\langle \Delta E^2 \rangle\rangle/2\Delta t = v^2 \langle\langle \Delta p^2 \rangle\rangle/2\Delta t = v^2 D_p$. The mean energy change can be defined by the relation:

$$\int dp\, 4\pi p^2 E \frac{\partial}{\partial t} f(\mathbf{x},t,p) \equiv \int dp\, 4\pi p^2 \left\langle \frac{dE}{dt} \right\rangle f(\mathbf{x},t,p). \quad (28)$$

Substituting expression (27) for $\partial f/\partial t$ into this equation, and performing two partial integrations in p one finds $\langle dE/dt \rangle$. Collecting results:

$$\left\langle \frac{dE}{dt} \right\rangle = \frac{1}{p^2}\frac{\partial}{\partial p}\left(p^2 D_p v\right) \quad , \quad D_E = \frac{\langle\langle \Delta E^2 \rangle\rangle}{2\Delta t} = v^2 D_p. \quad (29)$$

2.3.2 Acceleration by magnetosonic waves: a physical realization of Fermi acceleration

I will consider a specific example which reproduces Fermi's intuitive results. Consider a magnetic field with small fluctuations: $B = B_o + B'$ with $\langle\langle B'\rangle\rangle = 0$. The frequency of these fluctuations is small compared with particle gyrofrequency: $\omega \ll \Omega_B$. Let the waves responsible for these fluctuations be in the x-z plane with the constant (mean) field B_o along the z-axis. The magnetic field is divergence-free, which in this case implies $\nabla \cdot B = \partial B'_x/\partial x + \partial B'_z/\partial z = 0$. Particles gyrate along the magnetic field with gyro-radius $R_L = v_\perp/\Omega_B$ where $v_\perp \equiv \equiv \sqrt{(v_x^2 + v_y^2)}$. Due to this gyration, the particle "sees" a fluctuating magnetic field component $\Delta B_x \approx \Delta x(\partial B'_x/\partial x)$, with Δx the excursion of the particle along the x-axis due to its helical orbit. The z-component of the associated fluctuating Lorentz-force equals:

$$\left(F_L\right)_z = -\frac{q}{c} v_y \Delta B_x \approx -\frac{p_\perp v_\perp}{B_o} \cos^2\left(\varphi_o - \Omega_B t\right)\frac{\partial B'_z}{\partial z}. \tag{30}$$

Here I have parametrized the helical orbit by $\Delta x \approx R_L \cos\left(\varphi_o - \Omega_B t\right)$, $v_y = v_\perp \sin(\varphi_o - \Omega_B t)$ and used the $\nabla \cdot B = 0$ condition to write the result in terms of $\partial B'_z/\partial z$.
Since the magnetic field changes slowly over one gyroperiod $\Delta t_B = 2\pi/\Omega_B$ one can average over the gyromotion and get a mean force:

$$\bar{F}_z = -\frac{p_\perp v_\perp}{2B_o}\frac{\partial B'_z}{\partial z} \approx -\frac{p_\perp v_\perp}{2B_o}\frac{\partial B}{\partial z}. \tag{31}$$

Here I have used that the field strength equals : $B = \sqrt{((B_o + B'_z)^2 + B'^2_x)} \approx B_o + B'_z + + O(B'^2)$. One can easily check that the other two components of the Lorentz-force average to zero. The force given by Eqn. (31) is called the *magnetic mirror force* since it tends to deflect particles away from regions with increasing field strength. The fluctuations in the field strength are expanded as a collection of plane waves, c.f. Eqn. (18): $\delta B \equiv B - B_o = \int (d^3k/8\pi^3)B(k)\exp(ik\cdot x - i\omega t)$. The associated force follows from (31) as (with $\parallel$ denoting the z-component and using $(\partial/\partial z)\exp\left(ik\cdot x - i\omega t\right) = ik_\parallel\exp\left(ik\cdot x - i\omega t\right)$):

$$\bar{F}_z(x,t) = \int \frac{d^3k}{(2\pi)^3}\left(-\frac{ik_\parallel p_\perp v_\perp}{2B_o}\right) B(k)\exp\left(ik\cdot x - i\omega(k)t\right), \tag{32}$$

i.e. $\tilde{F}(k) = -ik_\parallel p_\perp v_\perp B(k)/2B_o$. In this case, only the parallel component of the momentum diffuses. As long as $k_\perp R_L \ll 1$ only the n = 0 resonance needs to be considered. The momentum diffusion tensor has only one component (the zz-component):

$$D_{\parallel\parallel} \equiv \frac{\langle\langle\Delta p_{\parallel}^2\rangle\rangle}{2\Delta t} = \int\frac{d^3k}{(2\pi)^3}\left(\frac{p_{\perp}v_{\perp}}{2B_o}\right)^2 k_{\parallel}^2\left|B(k)\right|^2 2\pi\delta\left(\omega(k) - k_{\parallel}v_{\parallel}\right). \quad (33)$$

The corresponding diffusion of the *magnitude* of momentum follows from the relation $\Delta p = (\partial p/\partial p_{\parallel})\Delta p_{\parallel} = (p_{\parallel}/p)\Delta p_{\parallel} = (v_{\parallel}/v)\Delta p_{\parallel}$. Using the fact that one can replace $k_{\parallel}v_{\parallel}$ by $\omega(k)$ in the integral because of the δ-function one finds:

$$D_p = (v_{\parallel}/v)^2 D_{\parallel\parallel} = \int\frac{d^3k}{(2\pi)^3}\left(\frac{p_{\perp}v_{\perp}}{2B_o v}\right)^2\omega(k)^2\left|B(k)\right|^2 2\pi\delta\left(\omega(k) - k_{\parallel}v_{\parallel}\right).$$

$$(34)$$

Low-frequency compressive waves in a magnetised plasma able to induce such fluctuations in the field strength are the so-called *fast magnetosonic waves*. They have a frequency $\omega(k) \approx kV_A$ where $V_A = B_o/\sqrt{(4\pi n_p m_p)}$ is the *Alfvén velocity*. These waves can only exist if the magnetic pressure $B_o^2/8\pi$ exceeds the thermal pressure $nk_b T$ of the plasma. Using this we can estimate from (34) the typical magnitude of D_p:

$$D_p \approx \frac{\pi}{3}\left(\frac{V_A^2}{v\lambda}\right)\left(\frac{\delta B}{B_o}\right)^2 p^2 . \quad (35)$$

Here $\lambda = 2\pi/k$ is the typical wavelength of the magnetic fluctuations, and I have assumed that about 1/3 of the energy in the fluctuations resides in the component $B'_z \approx \delta B$. The corresponding mean energy gain and energy diffusion coefficient for relativistic particles ($v \approx c$, $E \approx pc$) follows from (29) as:

$$\left\langle\frac{dE}{dt}\right\rangle = \frac{4\pi}{3}\left(\frac{V_A^2}{c\lambda}\right)\left(\frac{\delta B}{B_o}\right)^2 E \equiv \alpha E, \quad D_E \approx c^2 D_p = \frac{1}{4}\alpha E^2 . \quad (36)$$

These relations have exactly the same form as the relations (9) derived using the simple model for Fermi-acceleration of the previous Section. Note however that the physical assumptions used to derive these results are different in the two cases .

In the simple model for Fermi-acceleration of Section 2.1 it was assumed that the scattering centers deflect the particle through a large angle at each scattering. This is called *integral scattering* . Furthermore the scattering was described as a series of impulsive events rather than as the influence of a force acting continuously on the particle, as was done in this Section. This gave a value of $\alpha = \tau_{acc}^{-1}$ in terms of the mean-free-path λ and the magnitude of the velocity V of the scattering centers as $\alpha \approx 8V^2/3c\lambda$.

In the case under consideration here a particle is deflected continuously by the magnetic fluctuations. The angle of deflection Δ per "scattering" is only small: $\Delta \approx \delta B/B_o$ per wavelength λ. This is called *differential scattering*. As a result, the *effective* mean free path where these small deflections diffusively add up to a large angle of order $\pi/2$ equals $\lambda_{eff} \approx \lambda/\Delta^2 \approx \lambda B_o^2/\delta B^2$. This explains the appearance of the extra factor $(\delta B/B_o)^2$ in the expression for α in this case. The waves propagate with a velocity V_A, so the Alfvén speed is indeed the correct velocity for the scattering centers.

When the magnetic fluctuations become strong so that $\delta B \approx B_o$ one has $\lambda_{eff} \approx \lambda$ so $\alpha \approx 4\pi V_A^2/3c\lambda$, and the two approaches are equivalent up to factors of order unity. In that case the energy density of the average field $B_o^2/8\pi$ and of the fluctuations ($\approx 3\delta B^2/8\pi$) are roughly the same. In that case one speaks of *strong turbulence*. The derivation given here is then on shaky ground: in particular the assumption that the momentum change can be calculated using *unperturbed* particle orbits becomes highly questionable. It was this assumption that led to the appearance of the resonance condition.

The above method of calculating the non-linear interaction between particles and turbulent waves is known in the plasma-physics community as the *quasi-linear approximation*. It is often difficult to ascertain its validity, but as a tool for making order-of-magnitude estimates it is useful.

2.4 " Fermi deceleration": expansion losses

Consider a particle interacting with scattering centers which are passively advected by a diverging flow so that their velocity equals the local fluid velocity. For simplicity consider an one-dimensional flow with velocity $V(x)$. A scattering center at $x = x_o$ sees neighbouring scattering centers at a distance $\Delta x = x - x_o$ move with a velocity $\Delta V \approx \Delta x (\partial V/\partial x)_o$. In a diverging flow $\partial V/\partial x > 0$, and scattering centers move away from each other. A particle scattered by a scattering center at $x = x_o$ will *always* suffer an overtaking collision at its next scattering. If the next scattering occurs after a time Δt, one has $\Delta x = v_x \Delta t$. The energy loss associated with this scattering event according to (IV.3) equals, assuming $v \gg \Delta V$:

$$\Delta E \approx -2 \frac{\Delta V v_x}{c^2} E \approx -2 \frac{v_x^2 \Delta t}{c^2} \left(\frac{\partial V}{\partial x}\right)_o E. \tag{37}$$

In the simple model for which this result was derived, the particle velocity along the x-axis is reversed at the scattering: $v_x \longrightarrow -v_x + O(v/V)$. This means that the following scattering takes place near $x = x_o$ with a scattering center (nearly) at rest, and no change in energy results. Therefore the particle on average gains energy every *second* scattering. If one averages over the direction of particle velocities one has $\langle v_x^2 \rangle = v^2/3$ if particles are distributed isotropically in velocity. Using this, one gets the following expression for the mean energy loss per unit time $\langle dE/dt \rangle \approx \langle \Delta E/2\Delta t \rangle$:

$$\left\langle \frac{dE}{dt} \right\rangle = - \frac{1}{3} \frac{v^2}{c^2} \left(\frac{\partial V}{\partial x} \right)_o E \ . \tag{38}$$

This result can be generalised to a three-dimensional flow by replacing $\partial V/\partial x$ by $\nabla \cdot V = \partial V_x/\partial x + \partial V_y/\partial y + \partial V_z/\partial z$. A useful version can be derived using the relation $\Delta E = v\Delta p$ and $p = Ev/c^2$ and expressing Eqn. (38) in terms of momentum:

$$\left\langle \frac{dp}{dt} \right\rangle = - \frac{1}{3} (\nabla \cdot V) p. \tag{39}$$

This expression describes the so-called *expansion losses* that a particle suffers if it is coupled by scattering centers to a diverging flow ($\nabla \cdot V > 0$). If, on the other hand, the flow is *converging* ($\nabla \cdot V < 0$) the particle *gains* energy in what can be considered a differential version of Fermi acceleration. If the scattering centers are not passively advected, but have a velocity dispersion V_s^2 with respect to the mean fluid velocity $V(x)$ one has to add the effect of ordinary stochastic Fermi acceleration to Eqn. (38) or (39), e.g.:

$$\left\langle \frac{dp}{dt} \right\rangle = - \frac{1}{3} (\nabla \cdot V) p + \frac{8}{3} \frac{V_s^2}{v\lambda} p. \tag{40}$$

2.5. The energy balance in stochastic Fermi acceleration

So far, I have treated the acceleration process from a single-particle point of view, neglecting the back-reaction of all accelerated particles on the "scattering centers" which provide the acceleration. This is the *test-particle approximation* in which one considers the circumstances leading to the acceleration (wave turbulence) as given. In reality, the energy gained by all particles involved in the acceleration process is acquired at the expense of the energy of the scattering centers or waves responsible for the acceleration, which will lose energy. This back-reaction of the particles on the accelerating agent must be taken into account.

Let us once again consider the simple model of Sections 2.2. , allowing for energy-independent escape of the particles from the region where the acceleration takes place and injection of particles at a rate R per unit volume with energy E_o . Defining the particle energy density $U_p(t) \equiv \int dE N(E,t) E$ its time-evolution follows from integrating Eqn (12):

$$\frac{\partial U_p}{\partial t} \equiv \int dE \frac{\partial N(E,t)}{\partial t} E = RE_o + \alpha U_p - \frac{U_p}{T} \ . \tag{41}$$

The second term on the right hand side $\alpha U_p = \int dE \langle dE/dt \rangle N(E,t)$ is just the work done per unit volume by the scattering centers in accelerating the particles. By an argument of detailed balance, the energy of the scattering centers U_* must therefore change according to:

$$\frac{\partial U_*}{\partial t} = \varepsilon - \alpha U_p \qquad (42)$$

The first term on the right-hand-side is the net amount of energy fed into the scattering centers per unit time, and the second term the energy extracted by the accelerated particles. Consider the case where energy is fed continously into the scattering centers (waves), and the system reaches a steady-state with $\partial/\partial t = 0$. The energy-balance equations (41) and (42) then are solved simultaneously by:

$$U_p = \frac{RE_o T}{1 - \alpha T} \quad , \; \varepsilon = \alpha U_p \quad \longleftrightarrow \quad \alpha T = \frac{\varepsilon}{\varepsilon + RE_o} \quad . \qquad (43)$$

In that case the level of waves and particles in the source adjusts in such a way that ratio of the escape- and acceleration time $\alpha T = T/\tau_{acc}$ depends only on the ratio ε/RE_o of energy fed per unit time into the wave turbulence and the relativistic particles. This is the feedback mechanism referred to in Section II.2. The observed energy spectra $N(E) \propto E^{-2.5}$ correspond according to (13b) to a value of $\alpha T \approx 0.6$. This would be realized provided $\varepsilon/RE_o \approx 2$, i.e. twice as much energy is injected into the region in the form of turbulent waves than in the form of particles. In principle this is an attractive idea, since it reduces the problem of the observed spectra to a one-parameter problem, which except for the parameter ε/RE_o does not directly involve the other physical conditions of the source. However, there is no a-priori reason why ε/RE_o would take roughly the same value in all sources, as seems to be implied by observations.

3. Particle acceleration near astrophysical shocks

About fifteen years ago, several authors[6-9] independently proposed the mechanism of diffusive shock-acceleration as a possible astrophysical source of energetic particles in our own galaxy (e.g. cosmic rays) or in active galaxies and quasars. In this mechanism, particles gain energy during repeated scattering across a shock front. This energy gain is due to the fact that "scattering centers" advected with the bulk flow in the up- and downstream region near the shock converge relative to each other due to the compression created at the shockfront, see fig. 3. As such this mechanism is a physical realization of the regular Fermi mechanism of Section II.1. The mean energy gain per shock crossing is of order $\Delta E \approx \left(\Delta u \, v/c\right)E$ (c.f. Eqn. 4), corresponding to a momentum gain $\Delta p \approx \left(\Delta u/v\right)p$. Here $\Delta u = (u_- - u_+) \cdot n$ the velocity difference across the shock, $\mathbf{u}_-$ ($\mathbf{u}_+$) the fluid velocity ahead of (behind) the shock, $\mathbf{n}$ is the shock normal and v the particle velocity.

The necessary scattering can be provided by the fluid particles themselves in the case of Thomson scattering of photons by electrons in an ionized plasma. For charged particles it is due to collective effects, in particular gyro-resonant scattering of charged particles by short-wavelength MHD waves (wrinkles in the magnetic field).

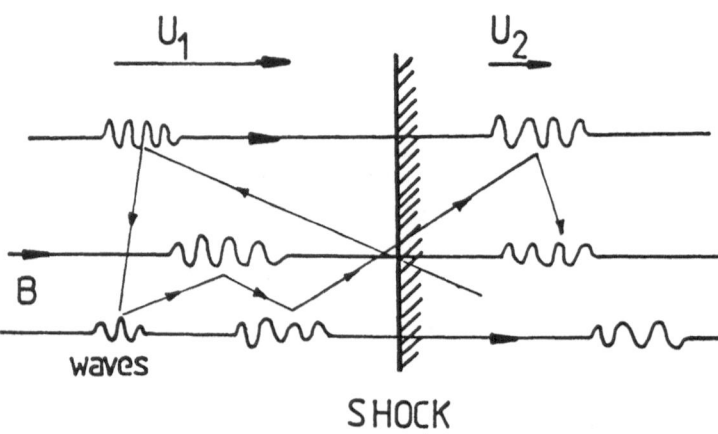

Fig. 3. *Physical picture of shock-acceleration. Scattering centers advected with a flow converge relative to each other at a shock front. The role of the shock is to create a compression in the gas. For a normal shock as shown here this compression is given by $r = u_-/u_+ = \rho_+/\rho_-$,where ρ is the density of the gas. Particles gain energy every time they complete a cycle of crossing the shock from the downstream to the upstream region and back.*

The rate at which particles cross the shock is proportional to the scattering rate. The scattering maintains a good coupling between the bulk fluid and the accelerated particles. This results in diffusive propagation which traps particles near the shock front in a "boundary layer". This in turn allows a small fraction of particles to gain energies far in excess of the mean kinetic energy $mu_s^2/2$ per particle of mass m in the upstream flow as seen from a shock propagating with velocity u_s.

The simplest version of the mechanism assumes that the scattering mean-free-path $\lambda_{mfp} > d$, with d the thickness of the shock. The shock is then treated as a discontinuity. If this ordering not satisfied, particles or photons no longer see the shock as a discontinuity, but as a continuous transition in the flow velocity. Particles then gain energy at rate determined by the local compression rate $\nabla \cdot u$ in the flow, c.f. Eqn (39).

In what follows , I will review some of the features of the shock-acceleration process, with an emphasis on the various astrophysical applications. For more detailed derivations of various equations I refer to the review by R. D. Blandford and D. Eichler[10].

3.1. A simple statistical model

Consider a simple case of a planar, infinitely thin and steady shock. A scattering mechanism operating on both sides of the shock front ensures that particles are scattered repeatedly across the shock. I will define a *cycle* as an event in which a particle with momentum p crosses the shock from downstream to upstream and (after some time) back again. One can define the following (suitably averaged) quantities:

momentum change/cycle: $\quad \Delta(p) \ll p$,

probability of escape/cycle: $\quad P(p) \ll 1$, $\qquad$ (44)

probability of destruction/cycle: $\quad A(p) \ll 1$.

Let $dN \equiv n(p)dp$ be the number of particles in the momentum interval $\left(p, \ p + dp \right)$ *currently involved* in the acceleration process. After completing one cycle, the various quantities have undergone an average change according to:

$$p \longrightarrow \bar{p} = p + \Delta(p) \,, \qquad (45a)$$

$$dp \longrightarrow \bar{dp} \approx \left(1 + \frac{\partial \Delta(p)}{\partial p} \right)dp \,, \qquad (45b)$$

$$dN \equiv n(p)dp \longrightarrow \overline{dN} \equiv n(\bar{p})\bar{dp} = \left(1 - P(p) - A(p) \right)dN \,. \ (45c)$$

Equation (45c) assumes that no particles are added to the pool of accelerating particles at momentum p. Substituting Eqns. (45a) and (45b) into Eqn. (45c) and making a Taylor expansion to first order yields a differential equation for n(p). This shows how the distribution n(p) results from the competition between momentum gain and particle losses:

$$\frac{\partial}{\partial p} \left(\Delta(p)n(p) \right) = - \left(P(p) + A(p) \right)n(p) \equiv - W(p)n(p). \qquad (46)$$

Here I have defined $W(p) \equiv P(p) + A(p)$ which is the total probability for a particle to be removed from the acceleration process during one cycle. Integrating Eqn. (46) gives the differential number density of particles participating in the acceleration:

$$n(p) = n(p_0) \frac{\Delta(p_0)}{\Delta(p)} \exp\left\{ - \int_{p_0}^{p} dp' \left(W(p')/\Delta(p') \right) \right\}. \tag{47}$$

The integration constant $n(p_0)$ is determined by the injection process. If the escape of particles from the acceleration process is due to advection into the downstream medium, as is usually the case, an observer behind the shock will measure a particle distribution *emerging* into the downstream region proportional to $P(p)n(p)$:

$$F_{\bullet}(p) = x \left(P(p)/\Delta(p) \right) \exp\left\{ - \int_{p_0}^{p} dp' \left(W(p')/\Delta(p') \right) \right\}. \tag{48}$$

Here $F_{\bullet}(p)dp$ is the number-density of particles in the momentum interval $(p, p + dp)$ in the downstream region. The constant x must be calculated from a detailed consideration of the boundary conditions at the shock.

3.2. Diffusive shock acceleration of charged particles

The most straightforward example of this process is the acceleration of charged particles near a normal shock propagating in a magnetised plasma. I will assume that the scattering process is elastic in the rest frame of the plasma. When scattering is due to the interaction of the charged particles with low-frequency MHD waves it can be considered elastic provided the wave phase velocity in the plasma rest frame $|\omega/k| \ll u_s$ with u_s the shock velocity. Scattering keeps the particle distribution nearly isotropic in the rest frame of the plasma on both sides of the shock. No particles are destroyed in this case so one has $A(p) = 0$.

Let the lab frame K be the frame where the shock is at rest. In that frame the fluid streams into the shock with velocity $u_- = u_s$ and leaves the shock with velocity $u_- = u_s/r$ where $r \geq 1$ is the *compression ratio*. I will look at the process in terms of the energy change in the frame $K_{\bullet}$ in which the downstream medium is at rest. In that frame the shock moves with velocity u_s/r towards positive x.

A particle will cross the shock into the upstream medium provided $v_x > u_{\bullet} = u_s/r$, corresponding to an angle ϑ_i with respect to the x-axis in the range $u_s/rv_i^{\bullet} \leq \cos\vartheta_i \leq 1$. After several scatterings upstream it crosses back into the downstream region at an angle ϑ_f where $u_s/rv_f^{\bullet} > \cos\vartheta_f \geq -1$, completing one cycle. All these quantities are defined in the rest-frame $K_{\bullet}$ of the downstream fluid. Seen from that frame, the upstream scattering centers approach at a velocity Δu given by the usual law for velocity addition in relativity:

$$\Delta u = \frac{u_- - u_+}{1 - \frac{u_- u_+}{c^2}} = \frac{(r - 1)u_s}{r - (u_s/c)^2} . \tag{49}$$

Conversely, an observer at rest in the rest frame K_- of the upstream fluid sees the downstream scattering centers approach with velocity $-\Delta u$.

The energy with which it enters the upstream flow as seen from the rest frame K_- of the upstream flow equals:

$$E_i^- = \Gamma\left(E_i^+ + \Delta u p_x^+\right) = \Gamma E_i^+\left(1 + (v_i^+ \Delta u \cos\vartheta_i /c^2)\right). \tag{50}$$

Here $\Gamma \equiv \left(1 - \Delta u^2/c^2\right)^{-1/2}$. When it re-enters the downstream flow across the shock its energy in the frame K_- is unchanged: the scattering which has taken place in between is elastic in that frame. So one can use the the same transformation:

$$E_f^- = \Gamma E_f^+\left(1 + (v_f^+ \Delta u \cos\vartheta_f/c^2)\right) = E_i^- = \Gamma E_i^+\left(1 + (v_i^+ \Delta u \cos\vartheta_i/c^2)\right).$$

From this one immediately finds the ratio of the initial and final energy in the downstream rest frame K_+ after the completion of one cycle:

$$\frac{E_f^+}{E_i^+} = \frac{\left(1 + (v_i^+ \Delta u \cos\vartheta_i /c^2)\right)}{\left(1 + (v_f^+ \Delta u \cos\vartheta_f /c^2)\right)} . \tag{51}$$

Note that the limits on the values of $\cos\vartheta_i$ and $\cos\vartheta_f$ given above ensure that $E_f^+ \geq E_i^+$ so the particle always gains energy during one cycle.

This is an implicit equation which one can solve using $v/c = (\gamma^2 - 1)/\gamma^2$. However, in the limiting case $\Delta u/v^+ \ll 1$ (and therefore $u_s \ll c$) and in the case of highly relativistic particles or photons ($v_i^+ = v_f^+ \approx c$) one can solve (V.8)) immediately. In the first case $v_f^+ = v_i^+ + O\left(\Delta u/v^+\right)$. The change in energy per cycle is small and equals, neglecting terms of order $(\Delta u/v)^2$ (compare Eqn. 3):

$$\Delta E^+ = E_f^+ - E_i^+ \approx \left(\frac{v^+ \Delta u}{c^2}\right)\left(\cos\vartheta_i - \cos\vartheta_f\right)E_i. \tag{52}$$

In the second case with $v \approx c$ one has:

$$E_f^+ = E_i^+ \times \left(\frac{1 + (\Delta u/c)\cos\vartheta_i}{1 + (\Delta u/c)\cos\vartheta_f}\right). \tag{53}$$

If $\Delta u \ll v$, c both expressions are equivalent. I will consider that case first.

It is convenient to rewrite (53) in terms of the particle momentum $p = Ev/c^2$, dropping the subscript i in the process:

$$\Delta p \approx \frac{\Delta u}{v} \left(\cos\vartheta_i - \cos\vartheta_f \right) p. \tag{54}$$

The flux of particles crossing the shock from downstream to upstream per unit area with an angle between ϑ_i and $\vartheta_i + d\vartheta_i$ is proportional to[8]:

$$\delta n \propto \cos\vartheta_i \sin\vartheta_i d\vartheta_i. \tag{55}$$

A similar expression is valid for the particles crossing back into the downstream flow with ϑ_i replaced by ϑ_f. Averaging the momentum change (54) over all allowed values of these angles yields the average momentum gain per cycle. Neglecting terms of order $\Delta u/v$ with respect to unity and defining a new variable $\mu \equiv \cos\vartheta$:

$$\frac{\langle \Delta p \rangle}{p} = \frac{\Delta u}{v} \int_0^1 d\mu_i \int_{-1}^0 d\mu_f \; \mu_i \, \mu_f \times \left(\mu_i - \mu_f \right) = \frac{4}{3} \frac{\Delta u}{v}. \tag{56}$$

A (small) fraction of particles escape from the shock by advection into the downstream region. The scattering in the downstream region tends to make the particle distribution isotropic. If there are no boundaries downstream allowing free escape, the particles are dragged along with the downstream fluid, with an average velocity equal to the streaming velocity u_* behind the shock. Seen from the lab-frame, where the shock is stationary, this corresponds to a net flux of particles with momentum p into the downstream region equal to:

$$\Sigma_*(p) \approx u_* F_*(p) + O(u_*/v). \tag{57}$$

Here $F_*(p)dp$ is the number density of particles with momentum in the range $(p, p + dp)$ behind the shock. The net flux back across the shock into the upstream region equals:

$$\Sigma_b(p) \approx \frac{1}{2} \int_0^1 d\mu_i \; v\mu_i \, F_*(p) \approx \frac{1}{4} v F_*(p) + O(u_*/v). \tag{58}$$

The chance of escape per cycle in the steady state must equal the ratio of these fluxes: $P(p) = \Sigma_* / \Sigma_b = 4u_*/v = 4u_s/rv \ll 1$. Collecting results, one finds for the shock-acceleration process of charged particles in the limit $\Delta u = (r - 1)u_s/r \ll v, c$:

$$A(p) = \frac{4(r - 1)}{3r} \frac{u_s}{v} p \quad , \quad A(p) = 0 \quad , \quad P(p) = \frac{4u_s}{rv} \quad . \tag{59}$$

Substituting this in (48) yields a power-law momentum distribution in the downstream region, with a slope depending *only* on the compression ratio of the shock:

$$F_+(p) = F_0\left(p/p_0\right)^{-\frac{r+2}{r-1}} . \tag{60}$$

The integration constant F_0 follows from an elementary consideration of particle conservation. Let us assume that a seed-distribution of particles is present far upstream of the form $F_-(p) = n_-\delta(p - p_0)$, isotropic in the rest-frame of the plasma. The corresponding flux of particles advected into the shock equals $\Sigma_-(p) = u_s F_-(p)$. The flux into the downstream region equals $\Sigma_+(p) = u_s F_+(p) = u_s F_+(p)/r$. Since no particles are destroyed, particle conservation in the steady state implies $\int dp\,\Sigma_-(p) =$ $=\int dp\,\Sigma_+(p)$. All particles are accelerated, so $F_+(p) = 0$ for $p < p_0$. Performing the integrals yields the relation $F_0 = 3r p_0^{-1} n_-/(r - 1)$. So we arrive at the final result, valid in the *steady state* for a *mono energetic* population of seed particles with number-density n_- in the upstream region:

$$F_+(p) = \frac{3r}{r - 1}\frac{n_-}{p_0}\left(\frac{p}{p_0}\right)^{-\frac{r+2}{r-1}} \qquad (p \geq p_0) . \tag{61}$$

What is remarkable about this result is that this steady-state spectrum does *not* depend on the details of the scattering process. In contrast, the spectrum for stochastic Fermi acceleration (Eqn. 13) contains the parameter αT = escape time/acceleration time which sensitively depends on the scattering rate v/λ. The reason for this behaviour is that in shock acceleration the particle only gains net energy every time it completes a *cycle* and not per scattering event. Only when it returns to the downstream region and is scattered there does the energy change become irreversible. This situation is completely analogous to the example used in Section 2.1 of a particle being scattered by a uniform stream of scattering centers streaming towards an elastically reflecting wall. The only difference is that the wall is replaced by a second set of scattering centers. The function of the shock is to create a velocity difference between the two sets of scattering centers.

Since the relative gain in momentum per cycle $\Delta(p)/p$ and the chance of escape per cycle $P(p)$ both scale as u_s/v, the downstream distribution at high momentum is a power-law in momentum depending *only* on the compression r at the shock. A power-law does not contain any *intrinsic* momentum scale. This is in contrast for example with an exponential distribution $F(p) \propto e^{-\sigma p}$ which contains the intrinsic momentum scale $p \approx \sigma^{-1}$.

When the particles far upstream ("seed particles") are not all at the same momentum, but in a distribution $F_-(p)$ so that $dN = F_-(p)dp$ is the number-density of particles in the range $(p, p+dp)$ far ahead of the shock, the downstream distribution of accelerated particles is given by:

$$F_+(p) = \frac{3r}{r - 1}\int_0^P dp_0\, p_0^{-1} F_-(p_0)\left(\frac{p}{p_0}\right)^{-\frac{r+2}{r-1}} . \tag{62}$$

In a perfect classical fluid with specific-heat-ratio $\Gamma = 5/3$ the maximum compression in a shock equals $r = (\Gamma + 1)/(\Gamma - 1) = 4$ (ref. 11). The slope of the distribution therefore has a minimum value $(r + 2)/(r - 1) = (3\Gamma - 1)/2 = 2$.

3.3. Cycle time and momentum gain for particles in shock acceleration

The maximum energy particles can gain in the process of shock acceleration is determined by energy losses or the finite age and size of the shock. Here the influence of the scattering becomes important. The scattering rate ν_s, or equivalently the mean-free-path $\lambda \approx v/\nu_s$ of the particles, determines how many shock crossings occur per unit time . This crossing rate in turn determines the rate at which particles gain momentum.

Consider a particle crossing the shock into the upstream region. After its first upstream scattering it starts diffusing with respect to the upstream flow. In most cases that diffusion is one-dimensional, guided by the magnetic field. The reason is that particles are tied to the field in their spiralling motion by the Lorentz force. They can slide freely along the field like beads along a wire. Diffusion *across* the field may be neglected in most circumstances.

I consider again the case of a normal shock (fig. 3) where the magnetic field is along the x-axis. Seen from the rest-frame K_- of the upstream flow the particle starts at a distance $x_o \approx \lambda_-$ in front of the shock. Given the scattering frequency ν_s it travels a distance $\delta x \approx v_x/\nu_s$ between two scatterings. In a time t the particle experiences $N \approx \nu_s t$ scatterings. As with any diffusion process the *net* distance $|\Delta x|$ travelled equals:

$$\Delta x^2 \approx N \langle \delta x^2 \rangle \;=\; t \langle v_x^2 \rangle / \nu_s \;=\; \frac{1}{3} \frac{v^2}{\nu_s} t \equiv D_- t. \tag{63}$$

Here I have defined the upstream spatial diffusion coefficient by $D_- = v_-/3\nu_s = \lambda_- v/3$. The probability of finding a particle between x and $x + dx$ at time t if it was at x_o at $t = 0$ equals [12]:

$$dW(x|x_o) \;=\; \frac{\exp\!\left(-(x - x_o)^2/(2D_- t)\right)}{\sqrt{(2\pi D_- t)}}\, dx. \tag{64}$$

At a time t the shock has progressed a distance $u_s t$ into the upstream medium, and only the particles with $x \geq u_s t$ have not yet been overtaken by the shock. Putting $x_o \approx \lambda_-$ the probability the particle has *not* been overtaken at time t equals:

$$W(t) = \int_{u_s t}^{\infty} \frac{\exp\!\left(-(x - \lambda_-)^2/(2D_- t)\right)}{\sqrt{(2\pi D_- t)}}\, dx = \frac{1}{2}\,\mathrm{erfc}\!\left(\sqrt{\left(\frac{u_s^2 t}{2D_-}\right)} - \frac{\lambda_-}{\sqrt{(2D_- t)}}\right). \tag{65}$$

Here erfc(z) $\equiv$ $(2/\sqrt{\pi})\int_{z}^{\infty}dx\ \exp(-x^2)$ is the complementary error function[13] which satisfies erfc(-∞) = 2, erfc(0) = 1 and erfc(z) $\approx$ exp(-z)/$\sqrt{\pi}$ z for z >> 1. For t $\longrightarrow$ ∞ we have W(t) $\longrightarrow$ 0: every particle will eventually be overrun by the shock. Physically that is obvious: the distance it can diffuse $\Delta x \approx \sqrt{(D\ t)}$ increases more slowly in time than the distance $\Delta x_s = u_s t$ the shock travels in the same period. For t $\approx$ 0 one has W(t) $\approx$ erfc(-∞)/2 = 1: there has not yet been time for the shock to catch up with the particle. Finally, at t = λ_-/u_s, W(t) = erfc(0)/2 = 1/2, and about half of the particles which crossed the shock at t - 0 will have crossed the shock again into the downstream medium. Therefore a particle typically spends a time $\Delta t_- \approx \lambda_-/u_s$ in the upstream medium. In a similar way one can show that a particle typically spends a time $\Delta t_+ \approx \lambda_+/u_+ = r\lambda_+/u_s$ downstream if it crosses the shock again. Note that during that time it undergoes many scatterings. The number of scatterings in the upstream medium is of order $N_- = v_s \Delta t_- \approx (v/u_s)$ >> 1. This gives a typical cycle time $t_{cy} = \Delta t_- + \Delta t_+$ [14]:

$$ t_{cy} \approx (\lambda_- + r\lambda_+)/u_s . \qquad (66) $$

The mean momentum gain per cycle equals $\Delta p = (4\Delta u/3v)p$. So, while being accelerated by the shock, the momentum of a particle increases in time according to:

$$ \left\langle \frac{dp}{dt} \right\rangle \approx \frac{\Delta p}{t_{cy}} = \frac{4}{3}\frac{(r-1)u_s^2}{r(\lambda_- + r\lambda_+)v}p \equiv \frac{u_s^2}{v\bar{\lambda}}p. \qquad (67) $$

Here I have defined an "effective mean-free-path" by $\bar{\lambda} \equiv 3(r\lambda_- + r^2\lambda_+)/4(r-1)$.

This result looks quite similar to the result (6) for stochastic Fermi acceleration. The difference is that the relevant velocity here is the shock velocity u_s which in general is much larger than the velocity of the scattering centers ($\approx$ the Alfvén-velocity V_A for scattering by magnetic irregularities). Therefore the acceleration proceeds on a much shorter timescale. In most cases the mean-free-path λ on either side of the shock depends on the momentum p, i.e. $\bar{\lambda} = \bar{\lambda}(p)$. This momentum dependence reflects the details of the scattering process.

3.4. The maximum energy attainable in shock acceleration

The energy-gain due to shock acceleration for relativistic particles with Lorentz-factor γ >> 1 follows from (68) by using v $\approx$ c , E $\approx$ pc. Writing this equation in terms of γ, introducing a mean energy-loss per unit time corresponding to - $\langle d\gamma/dt\rangle_{loss}$, the particle experiences a net energy gain according to:

$$\frac{d\gamma}{dt} = \frac{u_s^2}{c\,\overline{\lambda}(\gamma)}\,\gamma - \left(\frac{d\gamma}{dt}\right)_{loss} \qquad (69)$$

For a strong hydrodynamical shock in an ideal gas with compression ratio $r = 4$. In one has $\overline{\lambda} = \lambda_- + 4\lambda_+$. The maximum Lorentz-factor γ_{max} obtains when the energy gain per cycle equals the mean loss incurred in a cycle time so the two terms on the right-hand- side of Eqn. (69) balance.

In a situation where losses can be neglected the maximum energy achieved by particles depends on the time T (e.g. age of the shock) available for acceleration , and the injection Lorentz factor γ_0. This follows from the implicit equation:

$$T = \int_{\gamma_0}^{\gamma_{max}} d\gamma \left(c\overline{\lambda}(\gamma)/u_s^2\gamma\right). \qquad (70)$$

Usually this time can be related to the typical size L of a shock (e.g. a supernova shock) and its velocity by $u_s T \approx L$. This shows that, although the scattering mechanism is not important for the shape of the distribution of accelerated particles for $\gamma \ll \gamma_{max}$, it is important in that it determines the cycle-time t_c and the maximum energy that can be achieved. I will consider gyro-resonant scattering by magnetic irregularities as an example.

3.5. Gyro-resonant scattering by Alfvén waves

The scattering mechanism most often invoked in the context of diffusive shock acceleration is gyro-resonant pitch-angle scattering on low-frequency MHD waves, such as Alfvén waves . Particles with charge Ze gyrating around a magnetic field $\boldsymbol{B}_0$ with gyro-radius of order $R_L = pc/Ze|\boldsymbol{B}_0|$ are scattered by waves with frequency ω and wavevector $\boldsymbol{k}$ satisfying the resonance condition $\omega - k_\parallel v_\parallel \pm \Omega_B = 0$. Here $k_\parallel \equiv \boldsymbol{k}\cdot\boldsymbol{B}_0/|\boldsymbol{B}_0|$, $v_\parallel = \boldsymbol{v}\cdot\boldsymbol{B}_0/|\boldsymbol{B}_0|$ and $\Omega_B \equiv Ze|\boldsymbol{B}_0|/\gamma mc$ is the cyclotron frequency . I will consider Alfvén waves propagating along the magnetic field, so that $\boldsymbol{k} = k_\parallel \boldsymbol{B}_0/|\boldsymbol{B}_0|$. If we put the magnetic field along the z-axis the Alfvén waves have a magnetic field perturbation associated with them given by (compare Eqn. 18) :

$$\delta\boldsymbol{B} = \int \frac{dk_\parallel}{2\pi}\, B(k_\parallel)\left(1 , i , 0\right)\exp\left(ik_\parallel z - i\omega_A t\right) . \qquad (71)$$

Here $\omega_A = k_\parallel V_A$ is the Alfvén frequency, with $V_A = B_0/\sqrt{\left(4\pi n_p m_p\right)}$. The Lorentz force associated with $\delta\boldsymbol{B}$ changes the pitch angle ϑ between the magnetic field and the particle momentum. Defining $\mu \equiv p_\parallel/p = \cos\vartheta$, this change is given by:

$$\frac{d\mu}{dt} = -\frac{Ze\,v_\perp}{pc} \int \frac{dk_\parallel}{2\pi} \; B(k_\parallel) \exp\left(i\Omega_B t + ik_\parallel v_\parallel t - i\omega t - i\varphi_o\right). \quad (72)$$

This expression is completely analogous to the equation for the acceleration for a particle by a random force treated in Section II.3. One can therefore immediately calculate the diffusion in pitch angle ϑ by analogy with Eqn. (24):

$$\frac{\langle\langle \Delta\mu^2 \rangle\rangle}{2\Delta t} \equiv D_\mu = \left(\frac{Ze}{\gamma m}\right)^2 (1 - \mu^2) \int \frac{dk_\parallel}{2\pi} \; \left|B(k_\parallel)\right|^2 2\pi \delta(k_\parallel v_\parallel + \Omega_B - \omega). \quad (73)$$

Performing the integration over $k_\parallel$, using the property $\delta(ax) = |a|^{-1}\delta(x)$ of the δ-function, yields in the limit of low-frequency waves ($\omega \ll \Omega_B$):

$$D_\mu = \Omega_B(1 - \mu^2) \left. \frac{|k_\parallel| |B^2(k_\parallel)|}{B_o^2} \right|_{k_\parallel = -\Omega_B/v\mu}. \quad (74)$$

The resonance $k_\parallel = +\Omega_B/v\mu$ also occurs , but for waves with a different polarization, in particular waves with $\delta B_y = -i\delta B_x$ (see 71). This resonance condition has a simple physical interpretation. The wave given by (71) is a *circularly polarized* wave where the magnetic field δB rotates through 360° around B_o over one wavelength $\lambda = 2\pi/k_\parallel$. The particle completes one rotation in one gyroperiod $\Delta t_B = 2\pi/\Omega_B$, and in that time it travels a distance $\Delta s = v_\parallel \Delta t = 2\pi v\mu/\Omega_B$ along the magnetic field. The particle will preserve the relative orientation between the field-perturbation δB and its own velocity component $v_\perp$ provided it travels one wavelength in one gyroperiod. So (up to a sign) one must have:

$$\Delta s = 2\pi v\mu/\Omega_B = \pm \lambda = \pm 2\pi/k_\parallel \quad \longleftrightarrow \quad k_\parallel = \pm \Omega_B/v\mu. \quad (75)$$

In that case the wave-induced Lorentz force $\mathbf{F} = (q/c)\mathbf{v}_\perp \times \delta\mathbf{B}$ has a constant value along the unperturbed particle orbit.

This diffusion in pitch-angle means that after a time Δt the pitch angle diffuses by an amount $\Delta\mu \approx \sqrt{(D_\mu \Delta t)}$. A particle will diffuse through an angle of the order of $90^O = \pi/2$ ($\Delta\mu \approx 1$) in a time $\tau_{90} \approx 1/D_\mu$, reversing its direction of motion along the magnetic field. This corresponds to a typical mean-free-path $\lambda \approx v\tau_{90}$. This mechanism supplies the scattering needed to confine particles near the shock. The effective scattering frequency due to this proces equals $\nu_s \approx 1/\tau_{90} \approx D_\mu$. The resonance condition gives a wave number of the resonant waves of order $k_\parallel = \Omega_B/v\mu \approx 1/R_L$.

This quasi-linear treatment, (presumably) valid for small amplitudes $|\delta B| \ll B_o$ of the wave magnetic field, gives a spatial diffusion coefficient for relativistic particles of order:

$$D(p) \equiv \frac{v^2 \tau_{90}}{3} \approx \frac{vR_L}{3}\left(|\delta B(k)|/B_o\right)^{-2}\Big|_{k \approx 1/R_L}. \qquad (76)$$

Here τ_{90} is the time it takes to scatter a particle through a 90 degree angle. Here I have defined $\delta B(k) = k_\parallel |B(k_\parallel)|/\pi$ so that $\delta B(k)^2/8\pi$ is the magnetic energy density carried by the waves in a bandwidth $\Delta k_\parallel \approx k_\parallel$ of the fluctuation spectrum.

An often made (but unsubstantiated) approximation puts $|\delta B(k)|/|B_o| \approx 1$ at all wavelengths. In that case $\tau_{90} \approx 1/\Omega_B$ so $D(p) \approx$ $\approx vR_L/3 \approx pcv/3Ze|B_o|$ and the scattering mean-free-path equals the gyro radius. The rationale behind this estimate is that saturation of the wave turbulence will occur when the magnetic field perturbations reach an amplitude comparable to the mean field. However, recent computer simulations[31] of the wave-particle interactions show that the scaling with $|\delta B(k)|^{-2}$ predicted by Eqn. (76) breaks down for fairly small wave amplitudes . The pitch-angle scattering is strongly affected by non-linear effects, resulting in a scattering mean-free-path of about 5 - 30 gyro radii. In the estimates below , valid for strong shocks, I parametrize the uncertainty of the details of the scattering process by putting the effective mean-free-path equal to $\bar{\lambda} = R_L/\varepsilon = R_o(\gamma\beta/\varepsilon)$, where $\varepsilon \equiv 1/\Omega_c \tau_{90} < 1$, and $R_o \equiv mc^2/Ze|B|$ is the gyro-radius of a $\gamma \approx 1$, $\beta = v/c = O(1)$ particle.

3.6   Maximum energy for specific loss-mechanisms

Using the results of the previous Section, adopting a scaling $\bar{\lambda} \approx \varepsilon^{-1}R_o(p/mc)$ ($\varepsilon \leq 1$) for the effective mean-free-path one can write the following relation for the net energy gain of a particle undergoing shock acceleration, valid for arbitrary $\gamma \geq 1$:

$$\frac{d\gamma}{dt} = \varepsilon\frac{u_s^2}{cR_o} - \left(\frac{d\gamma}{dt}\right)_{loss}. \qquad (77)$$

Without losses $\gamma(t) = \gamma_o + \varepsilon u_s^2 t/cR_o$ increases *linearly* in time, reflecting the fact that the cycle time becomes longer with increasing energy : $t_{cy} \approx \bar{\lambda}/u_s \propto p$. The maximum energy follows from equating the right-hand-side of this equation to zero. For $\gamma \ll \gamma_{max}$ the losses can be neglected and the particle distribution will be the power-law in momentum given by Eqn. (61). This power law will cut off exponentially at $p \approx \gamma_{max}mc$. In table 2, the most important energy limiting processes invoked in an astrophysical context are shown with the associated value of γ_{max} , as can be calculated from Eqn. (77).

limiting process	$(d\gamma/dt)_{loss}$	γ_{max}
synchr. /Compton losses electrons : $\tau_s \equiv \frac{3}{4} m_e c / \sigma_T U$. $U \equiv \frac{B^2}{8\pi} + U_{rad}$	γ^2/τ_{es}	$\left(\epsilon u_s^2 \tau_{es}/cR_e\right)^{1/2}$
expansion losses ions and electrons: $t_{exp} \approx L/u_s$	γ/t_{exp}	$\epsilon L u_s/cR_o$
relativistic bremsstrahlung electrons: $\tau_{eB} \approx 2\pi/n_i \alpha \sigma_T c \langle g \rangle$	γ/τ_{eB}	$\epsilon u_s^2 \tau_{eB}/cR_e$
proton-proton collisions: $t_{loss} \approx 1/n_p \langle \sigma_{pp} \rangle c$ proton-photon collisions : $t_{loss} \approx 1/n_{ph} \langle \sigma_{p\gamma} \rangle c$	γ/t_{loss}	$\epsilon u_s^2 t_{loss}/cR_p$
geometry, e.g. trapping of particles in a magnetic loop of size L : $T < L/u_s$		$\epsilon L u_s/cr_o$

Table 2 *In this table* $\sigma_T \equiv (8\pi/3)(e^2/m_e c^2) \approx 6.65 \times 10^{-25} cm^2$ *is the Thomson cross section,* L *is the size of the source.* $\alpha \equiv e^2/\hbar c \approx 1/137$ *is the fine-structure constant,* n_i *the ion density,* $\langle g \rangle$ *a Gaunt factor of order unity weakly dependent on energy.* n_p *the non-relativistic proton density and* n_{ph} *the photon density at* $h\nu > m_\pi c^2/\gamma$, *with* m_π *the pion mass. Proton-proton collisions* $p + p \longrightarrow p's + \pi's$ *and proton-photon collisions* $p + \gamma \longrightarrow p + \pi's$ *have a cross section of order* $\langle \sigma_{pp} \rangle \approx \langle \sigma_{p-ph} \rangle \approx 10^{-27} cm^2$. $R_e \equiv m_e c^2/eB$ *and* $R_p \equiv m_p c^2/eB$.

A special case is the limitation placed on the acceleration process by the *geometry* or *time-dependence* of the shock. So far, it was assumed that the shock is infinitely large and infinitely old. The geometry of the shock places a limitation on the maximum energy for the following reason. In front of the shock accelerated particles form a so-called *precursor*. This precursor has a scale of order $L \approx u_s/D_-$. This is the scale at which diffusion of particles away from the shock is exactly compensated by advection of particles into the shock by the upstream flow. Putting the distance Δx that a particle can diffuse away from the shock in a time t equal to the distance the fluid travels towards the shock in the same period one has (c.f. Eqn. 63):

$$\Delta x = \sqrt{(D_- t)} \approx u_s t \longleftrightarrow t \approx D_-/u_s^2, \quad \Delta x \approx D_-/u_s = L. \tag{78}$$

For the scaling of the diffusion coefficient adopted here relativistic particles have a precursor length $L \approx (c\gamma/3\varepsilon u_s)R_o$. Particles with higher energy are able to get further ahead of the shock. If the shock itself has size L, the particles will "notice" the finite size of the shock if $L \gtrsim$ L. At higher energies particles get so far ahead of the shock that they are likely to miss it when they are scattered back. The coupling between the fluid and the particles has become too weak to make efficient acceleration possible. This will occur at a value of γ of the order:

$$\gamma_{max} \approx \frac{\varepsilon u_s L}{cR_o} = \varepsilon\left(\frac{u_s}{c}\right)\frac{|Ze|BL}{mc^2} . \tag{79}$$

This limitation due to the size of the shock will be important for young shocks. For very old (and therefore very large) shocks the limitation on particle energy will generally be due to radiative or collisional losses of the particles.

3.7. Astrophysical sites of shock acceleration.

Shock acceleration is expected to operate at a variety of astrophysical objects. These include blast waves which result from explosive phenomena such as Solar or stellar flares and supernovae , the bow shock of Earth and other planets , the outer edge of shock-bounded bubbles blown into the interstellar medium by stellar winds , the so-called super-bubbles which result from sequential supernovae in OB-star associations in the galactic plane .

The collimated supersonic outflows (jets) associated with Young Stellar Objects , galactic X-ray sources such as SS433, and Cygnus X-3 , and the nuclei of active galaxies and quasars are terminated by a strong shock where they impinge on the interstellar or intergalactic medium. In the case of active galaxies and radio-loud quasars one, sees extended radio lobes. In the stronger radio galaxies (the so-called Fanaroff-Riley II sources with a radio luminosity exceeding $L_\nu \approx 10^{34}$ erg/s per Hz around $\nu \approx 1.4$ Ghz) one sees the so-called "hot spots" where the jets terminate. One usually assumes that these hot spots are the sites of strong *in situ* particle acceleration associated with the terminating shocks. For a review see Meisenheimer and Röser[15].

Further possible sites of shock acceleration are accretion flows. In accretion disks mass slowly spirals towards a compact object such as a white dwarf, neutron star or black hole under the influence of frictional forces and possibly dissipation in shocks. The gravitational binding energy liberated in this fashion is thought to be the primary energy source powering non-thermal emission and the (collimated) outflows from Young Stellar Objects, galactic X-ray sources , active galaxies and quasars. In accretion flows with little or no angular momentum standing accretion shocks have been proposed as a site of efficient particle acceleration[16].

The acceleration process requires a strong coupling through scattering between the energetic particles and the bulk flow. Therefore there is the danger of particles being "dragged" with the flow to the compact object, none reaching an observer at infinity. In particular this is true for quasi-spherical accretion flows, which do not have an "edge" like an accretion disk, across which particles can escape to infinity. This has been discussed recently in some detail for spherical accretion by Schneider and Bogdan[17].

It is impossible, within the limited confines of this review, to do justice to all the fine points and outstanding problems of the theory of shock acceleration or particle acceleration in general in all these various objects. In table 3 I have summarized the typical parameters for the shock acceleration process in a number of astrophysical objects. These include the size, age and shock velocity u_s, the mechanism limiting the energy gain and the maximum value of γ that can be reached, according to the expressions given in table 2 of the previous Section.

Table 3.

Object	size L	age t	u_s	B	limiting mechanism	γ_{max}
Solar Flare	10^{11} cm	1000 s	1000 km/s	10G	protons: geometry [1]	$10^3 \varepsilon$
Type II shock					electrons: synchr. loss.	$20 \sqrt{\varepsilon}$
Earth bowshock	10^{10} cm	----	300 km/s	$30 \mu G$	protons : geometry	kin. energy ≤ 100 keV
Supernova	10 pc	1000 yr	1000 km/s	1 mG	protons: exp. loss.	$10^6 \varepsilon$
Remnants					electrons: synchr. loss.	$3 \times 10^3 \sqrt{\varepsilon}$
Superbubbles	5 kpc	10^7 yr	10 km/s	0.1 mG	protons: exp. loss.	$10^7 \varepsilon$
					electrons: synchr. loss .	$10^2 \sqrt{\varepsilon}$
Active Galactic Nuclei [2]	$\geq R_g \equiv \dfrac{2GM}{c^2}$ $\approx 10^{13} M_8$ cm	---	$\alpha \sqrt{(R_g/r)} c$	???	protons : photon/pion production	$10^{10} \varepsilon \alpha B(G) t_6$

[1] *I have labeled with "geometry" all limiting mechanisms where the size of the shock limits the energy gain . This will happen when $D(p)/u_s L \geq 1$, with L the typical size of the shock, and D(p) the diffusion coefficient.*

[2] *These estimates assume that the accelerating region is close to the Schwarzschild radius of a black hole with a mass of $10^8 M_8$ solar masses. The shock velocity is taken to be a fraction α of the free-fall velocity $v_{ff} \approx c \sqrt{(R_g/r)}$. The typical loss-time t_6 in units of 10^6 seconds varies between 10^{-3} and 1 for different models [14,15].*

3.8 Observational evidence for shock acceleration.

3.8.1   Interplanetary shocks

The best *direct* evidence for shock acceleration comes from observations from spacecraft. In a number of quasi-parallel interplanetary shocks (i.e. **B** and **n** aligned within 45°) energetic protons and an associated enhancement of MHD wave activity has been recorded more than an hour before the shock passes the spacecraft. The observations can be explained by the theory of diffusive shock acceleration. Accelerated particles "boiling off " the shock form a precursor in the upstream region with a typical dimension $L \approx D/u_s$, the typical scale on which the diffusion of particles away from the shock, and the advection with the flow into the shock balance each other. The angular distribution of the accelerated particles in the precursor region is anisotropic. For super-Alfvénic shocks with $u_s > V_A \equiv |\mathbf{B}|/\sqrt{(4\pi\rho)}$ this anisotropy is sufficient to generate Alfvén waves by a gyro-resonant two-stream instability. These waves provide the scattering needed to trap particles near the shock. Similar observations have been made at the bow-shocks of Earth and Jupiter, but the interpretation of these measurements is difficult.

3.8.2 Cosmic-ray acceleration in the Galaxy

In all other cases, the evidence is more or less circumstantial. The best studied case is cosmic-ray acceleration by supernova remnants (SNR) in our galaxy. The observed spectrum of galactic cosmic rays, when corrected for the (energy dependent) shielding influence of the heliosphere, is a power-law in kinetic energy $T \equiv (\gamma - 1)mc^2$: $dN/dT = xT^{-2.7}$ for T in the range of 3 - 10^5 GeV/nucleon[19]. Above that range, the spectrum seems to flatten, and then steepen again. This power-law behaviour is followed by protons and other primary nuclei, but the elemental abundances at a given energy/nucleon differ significantly from Solar values.

The global energetics of cosmic rays in the Galaxy seems to support the idea of SNR as the primary sources[20,21]. A supernova typically releases an energy $E_{snr} \approx 10^{51}$ erg into the interstellar medium, mostly in the form of kinetic energy of the ejecta and swept-up interstellar matter. The supernova rate Q_{snr} is estimated at 1 per 30-100 years . Cosmic-ray propagation in the Galaxy is usually described in terms of the so-called "leaky box model" [20,21] in which a typical cosmic ray is retained in the Galaxy for about $t_r \approx 3 \times 10^6$ yr in a "confinement volume" of about $V \approx 10^{65}$ cm^3. This corresponds to a dilute halo with a thickness at least 5 times that of the stellar disk i.e. a scale height of about $H \approx 1$ kpc above the galactic plane. This idea is confirmed by observations of diffuse synchrotron emission from cosmic-ray electrons in neighbouring spiral galaxies.

The retention time is derived by measuring the amount of secondary cosmic rays produced in collisions between primary cosmic rays an nuclei in the interstellar medium. Calculations show that a typical primary cosmic ray sees a "grammage" $\langle\rho\rangle L \approx 14\,g/cm^2$. Here $\langle\rho\rangle$ is the mean density of matter in the confinement volume, and $L \approx ct_r$ the typical total pathlength. One of products of these collisions is the radioactive isotope ^{10}Be with a lifetime of about 4×10^6 yr. From measuring its abundance one can also estimate the retention time t_r. Since the corresponding path-length $L \approx 3 \times 10^6$ ly ≈ 1 Mpc is much larger that the thickness of this diffuse halo surrounding our galaxy ($H \approx 1$ kpc) the propagation of a cosmic ray must be diffusive. From the properties of random walks one can estimate the typical cosmic ray mean-free- path λ from the number of steps a cosmic ray needs to reach the edge of the flattened halo if it starts in the middle of the galactic disk:

$$L = N\lambda \,, \quad H = \sqrt{N}\lambda \quad \longleftrightarrow \quad \lambda \approx H^2/L \approx 1 \text{ pc}. \qquad (80)$$

The cosmic-ray energy density measured at Earth equals $\varepsilon_{cr} \approx 1$ eV/cm$^3 \approx \approx 2 \times 10^{-12}$ erg/cm^3. Taking this to be a representative value, one can calculate the fraction of the energy from supernovae which has to be channelled into cosmic rays in order to maintain their current energy density in the galaxy:

$$E_{cr}/E_{snr} \approx \varepsilon_{cr}V/Q_{snr}E_{snr}t_r \approx 10^{-2}. \qquad (81)$$

Stellar winds have been proposed as a source of cosmic rays, mainly on the ground that the isotopic abundances of cosmic rays resemble those observed in the coronae of the Sun and other active stars . However, the power dissipated by stellar winds in the interstellar medium is about a factor 5 less than that put in by supernovae. The efficiency with which this power would have to be converted into cosmic rays is correspondingly larger, which seems unlikely.

It is also suggestive that the break in the cosmic-ray spectrum at about 10^{14} eV/nucleon (which corresponds to $\gamma \approx 10^6$) is intriguingly close to the value of γ_{max} estimated for the process of shock acceleration by SNR blastwaves (table 3). This suggests that primary cosmic rays below 10^{14} eV/nucleon originate at SNR, but that particles at higher energies have a different origin. Acceleration at superbubbles could give $\gamma_{max} \approx 10^7$ if scattering is very efficient so that $\varepsilon \approx 1$. But this is not nearly enough to explain the Ultra-High-Energy cosmic rays above 10^{18} eV [22]. When expansion losses dominate, the maximum attainable energy scales as $\gamma_{max} \approx \varepsilon L u_s/cr_o$. So if one believes that shock acceleration is the process responsible for the production of these UHE particles, one needs either a very large, or a very fast shock. This has lead to the suggestion that these particles are accelerated at the termination shock of a galactic wind with $L \approx 300$ kpc, $u_s \approx 400$ km/s and $B \approx 1$ μG [23] . For protons this gives a maximum Lorentz factor equal to $\gamma_{max} \approx 5 \times 10^8 \varepsilon (u_s/400 \text{ km s}^{-1}) \times (L/300 \text{ kpc}) B_{\mu G}$. Other nuclei with a mass $m = A m_p$ and charge Ze can boost γ_{max} by a factor Z/A and the energy by a factor Z. It is therefore very important to measure the chemical composition of the UHE cosmic rays. The occurence of high-Z nuclei above 10^{15} eV such as members of the Iron group ($Z = 26$, $A \approx 54$) would be much easier to account for than, say, ^{4}He with $Z = 2$ $A = 4$. The low particle flux arriving at Earth in this energy range has made composition measurements difficult until now. For a review on these UHE cosmic rays see: Wdowczyk and Wolfendale[24].

3.9 "Realistic" shock acceleration: non-linear effects.

Most of the estimates derived in this paper have been based on a test-particle approach in which particles are accelerated by a shock of prescribed strength in a simple, one-dimensional flow. This test-particle approximation, although useful for making estimates, is not likely to apply in practice if shock acceleration is efficient in the sense that a significant fraction of the incoming momentum flux $\rho u_s^2 + P$ is converted into energy density of accelerated (i.e. non-thermal) particles at the shock. It is important to realize that shock acceleration is by its very nature a very non-linear process: the energy gained by the particles at the shock is extracted from the kinetic energy of the incoming flow. This will influence the strength of the shock by slowing down the upstream flow.

The strong coupling between the accelerated particles (or photons) and the bulk flow mediated by scattering is a necessary ingredient of the theory, even though the *details* of the scattering do not enter directly into the resulting distribution of accelerated particles. This coupling allows the bulk fluid to feel the pressure-gradients in the gas of accelerated particles ("radiation pressure"). The precursor formed by the accelerated particles in the upstream region slows down the incoming fluid due to the associated pressure gradient $\nabla \Pi$, with the pressure of the accelerated particles given by $\Pi(\mathbf{x},t) = \int dp F(p,\mathbf{x},t)(pv/3)$.

The effect of an increasing particle pressure towards the shock is twofold. First of all, the deceleration of the incoming flow leads to a weaker (viscous) subshock with compression $r < 4$. If enough energy is put into the accelerated particles, the viscous subshock is erased altogether, and one is left with a "cosmic-ray dominated" or "radiation dominated" shock in which the velocity changes smoothly over several precursor scales. Secondly, for flat spectra one expects the cosmic-ray gas to behave as an ideal relativistic gas with $\Gamma_{cr} \approx 4/3$. This makes the system more compressible, so that the *overall* compression across the precursor + viscous subshock (if present) may exceed the value $r = (\Gamma + 1)/(\Gamma-1) = 4$ (for $\Gamma = 5/3$) for a pure hydrodynamic shock. The value of the total compression depends sensitively on the efficiency of the acceleration process.[25,26]

This non-linear modification of the shock structure will affect the momentum distribution of the accelerated particles. The mean-free-path $\lambda(\gamma)$ increases with particle energy. High-energy particles diffuse further into the decelerating precursor flow and sample a larger effective compression across the shock, leading to a larger energy gain per cycle and a flattening of the distribution $F(p)$ at higher energies. This breaks the scale-free power-law behaviour of Eqn. (60). The deviation from the simple power law becomes noticible if the velocity change $\Delta u = L(\partial u / \partial x)$ across the precursor becomes of the same order as u. This defines a momentum scale p_s in terms of the velocity gradient in the precursor and the momentum-dependent spatial diffusion coefficient $D(p)$ as $\left(D(p_s)/u^2\right)\left(\partial u / \partial x\right) \approx 1$.

To calculate this one must consider the full equation for the momentum distribution $F(p,x,t)$ of the accelerated particles, integrating over momentum to get the particle pressure $\Pi(x,t)$. In the simple case of elastic scattering centers advected with the fluid one has [27,32]:

$$\frac{\partial}{\partial t} F(p,x,t) + \nabla \cdot \Sigma(p,x,t) = \frac{\partial}{\partial p}\left(\frac{1}{3}(\nabla \cdot u) \, pF(p,x,t)\right) \qquad (82a)$$

$$\text{where:} \quad \Sigma(p,x,t) \equiv uF(p,x,t) - (D \cdot \nabla)F(p,x,t). \qquad (82b)$$

The right-hand-side of (82a) shows the effect of expansion losses on the distribution as defined in Eqn. (39). The expression for the flux Σ simply reflects the competition between dragging of energetic particles by the fluid with velocity u and diffusion with respect to the fluid.

Results of numerical solutions of the full set of equations for the fluid and Eqns. (16) have confirmed this intuitive picture [28,29]. When the total compression is large, the spectrum $F(p) = xp^{-s}$ becomes quite flat at higher energies ($s < 2$). Most of the particle pressure then resides at the high-energy tail of the distribution, depending quite sensitively on the momentum cut-off $p_{max} \approx \gamma_{max}mc$. The problem then becomes time-dependent, since γ_{max} (in absense of losses) depends on the age T of the shock, cf. Eqn. (70). For instance, for a scaling $\lambda(\gamma) \approx R_L/\epsilon = R_o\gamma/\epsilon$ one has $\gamma_{max} \approx \epsilon u_s T/cR_o$. This means that shocks in which a significant fraction of the available mechanical momentum flux ρu^2 is converted into pressure of energetic particles are probably intrinsically time-dependent structures.

3.10. Outstanding problems

Although the basic principles of diffusive shock acceleration are well understood, there are still a number of unsolved problems in the theory which, at least in this authors view, stand in the way of a fully quantitative comparison of theory and observations. The most pressing of these problems are:

A. The injection problem.

Most astrophysical shocks are collisionless in the sense that the conversion of directed kinetic energy of the upstream flow into randomized motion ("heat") in the downstream flow proceeds through collective (plasma) effects rather than by ordinary Coulomb collisions. The typical random kinetic energy per proton downstream equals $m_p u_s^2/2$. Shock acceleration takes a few of these particles, and accelerates them to energies much larger than this value. How many particles are picked up from the "thermal" pool at the shock to be accelerated further is not really known.

This question is probably intimately connected with the detailed microscopic structure of quasi-parallel collisionless shocks in a magnetised plasma. The overall efficiency of the acceleration process can not really be calculated without some knowledge of this injection process. The momentum- and energy conservation laws for cosmic-ray mediated shocks (generalised Rankine-Hugoniot relations) in the steady state usually allow (at least) two solutions : one with low efficiency and a strong viscous subshock, and a high-efficiency solution with a weak subshock, or no viscous subshock at all [26,30]. Which solution is chosen will depend on the injection process, unless enough seed particles are already present in the upstream flow to force the system into forming a cosmic-ray dominated shock . One might hope that numerical simulations will answer this question.

B. Electron acceleration .

Non-thermal continuous radio emission is almost always interpreted as synchrotron radiation of relativistic electrons in a magnetic field. Relativistic protons remain invisible due to their small radiation losses. It is not known how much of the energy put into accelerated particles at a shock goes into protons and ions, and how much goes into electrons. This makes it very difficult to interpret the data in terms of the energy requirements needed to power the source.

Electrons have to be accelerated to some threshold momentum before they can be picked up by the shock acceleration process. This threshold momentum usually exceeds the typical momentum $\sqrt{(m_e m_p)}u_s$ of a thermalised electron in the post-shock flow. An important threshold is associated with the requirement of efficient scattering. Protons boiling of the shock generate low-frequency MHD waves . These waves have frequencies below the proton cyclotron frequency $\Omega_p = eB/m_p c$, and wavenumbers below $k_{max} \approx \Omega_p/V_A$ ($V_A \equiv B/\sqrt{(4\pi\rho)}$ is the Alfvén velocity) . Electrons do not interact resonantly with these waves until their gyroradius $R_e \approx pc/eB > k_{max}^{-1} \approx m_p V_A c/eB$. So in order to be scattered efficiently, the electron momentum must exceed a value $p > m_p V_A$. Another possibility would be to employ other waves (e.g. so-called Whistler waves) with shorter wavelengths to scatter the electrons below the threshold momentum. This aspect of the theory is as yet not fully explored.

A further problem with electron acceleration in Active Galactic Nuclei or Quasars is that the radiation fields in the nuclei of these objects are very intense. This means that radiation losses (Compton losses) are strong, and shock acceleration may not be efficient enough to overcome these losses. Another source of relativistic electrons therefore seems likely, such as a pair cascade induced by decay of pions into γ-rays. The pions are generated by proton-proton/proton-photon collisions of shock-accelerated protons, for which radiation losses do not become very important . Usually the losses due to photon-proton collisions have already terminated the spectrum of protons accelerated at the shock. Such models have been proposed in various incarnations[33-35]

3.11   Relativistic shocks

I now briefly consider the case of relativistic shocks. From Eqn. (53) one sees that the energy jump per cycle can become quite large if the velocity jump Δu across the shock seen from the downstream rest frame becomes large in the sense that $\Delta u \approx c$. For very strong shocks in a relativistic gas (equation of state $P = \varepsilon/3$, with ε the energy density) one can derive a relation $u_- u_+ = c/3$. For $u_- = u_s \approx c$ one has $u_+ \approx c/3$. This gives a value of $\Delta u \approx c$ (see Eqn. 49). The energy of a relativistic particle ($E \approx pc$) in the downstream rest frame changes in a cycle according to:

$$\frac{E_f}{E_i} \approx \frac{p_f}{p_i} = \frac{1 + \mu_i}{1 + \mu_f} . \tag{83}$$

Here I have assumed $v \approx c$ and defined $\mu = \cos\vartheta$ with ϑ the angle between the particle velocity and the shock normal. This energy gain can be very large if $\mu_f \approx -1$. One can derive the spectrum of the accelerated particles by a simple argument if one assumes that the downstream distribution function is kept isotropic by frequent scattering. Since the escape probability and the momentum gain are independent of momentum for very relativistic particles ($v = c = $ constant) there is no *intrinsic* momentum scale in the problem. The spectrum must therefore be a power-law in momentum, which is scale-free. In the downstream rest-frame the microscopic distribution function (defined by $dN =$ $= f(\mathbf{x}, t, \mathbf{p}) d^3 p$ with dN the number density of particles in a momentum-volume $d^3 p$ around $\mathbf{p}$) a little bit downstream of the shock ($\mathbf{x} = \mathbf{x}_s^+ \equiv \mathbf{x}_s + \varepsilon$, with $\varepsilon \longrightarrow 0$) is:

$$f_+(\mathbf{x} = \mathbf{x}_s^+, t) = \frac{dN}{(d^3 x\, d^3 p)_+} = \frac{F_+(p)}{4\pi p_+^2} = x p_+^{-q} \tag{84}$$

The distribution function $f(\mathbf{x},\mathbf{p},t)$ is a Lorentz invariant. Since we treat the shock as a discontinuity, it is impossible to "hide" or "store" particles in the shock. That means the distribution function must be *continuous* across the shock. First of all I express the downstream momentum in terms of the corresponding momentum measured in the upstream rest-frame K_-. Using $E \approx pc$ and the fact that the relative velocity between K_- and K_+ = $-\Delta u$ one immediately finds from the Lorentz transformation of energy, defining $\beta = \Delta u/c$ and $\Gamma = (1 - \beta^2)^{-1/2}$:

$$p_+ = \Gamma p_- (1 + \beta\mu_-) . \tag{85}$$

The invariance of $f(\mathbf{x},\mathbf{p},t)$ together with the continuity of f across the shock then immediately gives the distribution just in front of the shock in the upstream rest-frame, defining $\mathbf{x}_s^- = \mathbf{x}_s - \varepsilon$ with $\varepsilon \longrightarrow 0$:

$$f_-(\mathbf{x}_s^-,\mathbf{p}_-) = f_+(\mathbf{x}_s^+,\mathbf{p}_+) = \kappa p_+^{-q} = \kappa \Gamma^{-q} p_-^{-q}(1 + \beta\mu_-)^{-q} . \tag{86}$$

Note that if $\beta \approx 1$ ($u_s \approx c$), the distribution f_- is very sharply peaked towards the direction $\mu_- = -1$ pointing away from the shock. This is an example of relativistic beaming: seen from the upstream rest-frame, the downstream distribution is sharply peaked even though it is isotropic in the downstream rest-frame.

The slope q can be determined as follows: in the steady state the flux $\mathbf{S}(\mathbf{x},p)$ of particles at a given momentum must satisfy $\nabla \cdot \mathbf{S} = 0$. If all quantities only depend on the coordinate x along the shock normal this reduces to $\partial S_x(x,p)/\partial x = 0$. This implies:

$$S_x^-(x) = \int_{-1}^{+1} d\mu_- (c\mu_- + u_s) f_-(x,p_-,\mu_-) = \text{constant.} \tag{87}$$

The term $c\mu_- + u_s$ is just the relative velocity Δv_x between the shock and the particle in frame K_- . If no particles are incident far ahead of the shock so that all accelerated particles are injected at the shock front one must have $S_x^- = 0$. Substituting Eqn. (86) yields an implicit integral equation for q. Defining $\beta_s \equiv u_s/c$:

$$\int_{-1}^{+1} d\mu_- (\mu_- + \beta_s)(1 + \beta\mu_-)^{-q} = 0. \tag{88}$$

For β and $\beta_s \ll 1$ one can expand $(1 + \beta\mu_-)^{-q} \approx 1 - q\beta\mu_- + O(\beta^2)$ and evaluate the integral. This yields, using Eqn. (49) in the form $\beta \approx$ $\approx (r-1)\beta_s/r$

$$2\beta_s - \frac{2}{3} q\beta = 0 + O(\beta_s^2) \quad \longleftrightarrow \quad q = 3\beta_s/\beta = \frac{3r}{r - 1} . \tag{89}$$

As can be seen from definition (84) this is completely equivalent to our result (60) which was derived in a quite different manner using a statistical argument. So this procedure reproduces the correct slope of the

$$F_*(p) \equiv 4\pi p^2 f_*(p) \propto p^{(2-q)} \propto p^{-\frac{r+2}{r-1}} \,. \tag{90}$$

momentum distribution for a non-relativistic shock.

In the relativistic case one has to perform the full integration. The resulting expression can be written (assuming $q > 2$):

$$(1 + \beta)^{q-2} - (1 - \beta)^{q-2} - \frac{q-2}{q-1} \frac{1-\beta\beta_s}{1-\beta^2} \left\{ (1 + \beta)^{q-1} - (1 - \beta)^{q-1} \right\} = 0. \tag{91}$$

This equation can be solved analytically by assuming $(1 - \beta)/(1 + \beta) \ll 1$, which turns out to be a good approximation. The equation then becomes, neglecting the $(1 - \beta)^{q-2}$, $(1 - \beta)^{q-1}$ terms:

$$\frac{q-2}{q-1} = \frac{1-\beta}{1-\beta\beta_s} \,. \tag{92}$$

Using the definition of $\beta = \Delta u/c$:

$$\beta = \frac{(r-1)\beta_s}{r - \beta_s^2} \quad \text{with } r \equiv u_s/u_* \,. \tag{93}$$

one finds after some straightforward algebra:

$$q = 1 + \frac{r(1 + \beta_s)}{\beta_s(r-1)} \,. \tag{94}$$

For those shocks in extremely relativistic fluids which satisfy $u_s u_* = c^2/3$ one has $r = 3\beta_s^2$ and this equation reduces to:

$$q = 1 + \frac{3\beta_s(1 + \beta_s)}{3\beta_s^2 - 1} \,. \tag{95}$$

This expression is valid for $\beta_s > 1/\sqrt{3} \approx 0.57$ ($r > 1$). The upstream sound speed in this case equals $c\sqrt{(\partial P/\partial\varepsilon)} = c/\sqrt{3}$, so this condition corresponds to the usual condition of supersonic flow in the upstream region.

This analytical formula correctly reproduces numerical results of Kirk and Schneider[36]. For $\beta_s \approx 0.8$ they find $q \approx 5.6$, whereas Eqn. (94) gives $q = 5.69$. For $\beta_s \longrightarrow 1$ we both find $q \approx 4$. This last case corresponds to $F(p) \propto p^{-2}$, c.f. Eqn.(90). A more precise calculation in this case has to take account of the fact that relativistic beaming becomes so strong that the downstream distribution becomes anisotropic. One then finds a somewhat steeper slope: $q \approx 4.2$. So strangely enough, although the physics is rather different, a strong relativistic shock with

$\beta_s \approx 1$ and compression ratio $r = 3$ produces a particle spectrum with almost the same slope as a ordinary non-relativistic hydrodynamical shock with compression ratio 4, the maximum compression allowed for a shock in an ideal monoatomic fluid.

3.12   Shock-drift acceleration

There is an acceleration mechanism associated with oblique shocks (magnetic field not along the shock normal) which does not rely on repeated scattering across the shock. This mechanism is called *shock drift acceleration*[37,38]. Consider a shock propagating along the x-axis with velocity $V_{x-} = U_s$, with the plane shock front in the y-z plane, and the magnetic field in the x-z plane. Let the shock compression ratio equal $r = \rho_+/\rho_- = V_{x-}/V_{x+}$, and the angle between the shock normal (x-axis) and the magnetic field be given by ψ_- and ψ_+ upstream and downstream respectively. In a frame where the shock is stationary at $x = 0$ (the *shock frame*), there is an electric field along the shock front which is induced by the bulk plasma motion: $\mathbf{E} = -(\mathbf{V} \times \mathbf{B})/c = (V_x B_z /c)\mathbf{e}_y$. The boundary-conditions at the shock are the continuity of the normal magnetic field B_x and the tangential electric field $E_y = V_x B_z /c$ across the shock. This leads to:

$$B_+\cos\psi_+ = B_-\cos\psi_- \; ; \quad B_+\sin\psi_+ = rB_-\sin\psi_-. \tag{96}$$

This jump in the *tangential* component of the magnetic field is important. If one considers particles with a gyro-radius R_L larger than the shock thickness, the jump in the magnetic field strength for $\psi_- \neq 0$ has an important consequence. If a particle crosses the shock from upstream to downstream, it suddenly finds itself in a region with a stronger magnetic field. Its gyro-radius $R_L = pc/|Z|eB$ becomes smaller. For some particles, this leads to an orbit, which allows them to cross the shock again into the upstream region: in some sense they have been reflected by the shock (or rather, the discontinuity in the magnetic field created by the shock). This reflection is most easily described in the so-called de Hoffmann-Teller frame (dHTF) where the induced electric field vanishes. This frame moves with respect to the shock frame with a velocity $\mathbf{V}_{dHTF}$ along the shock, given by:

$$\mathbf{E}_{dHTF} = \mathbf{E} + (\mathbf{V}_{dHTF} \times \mathbf{B}) = 0 ,$$

$$\longrightarrow \quad \mathbf{V}_{dHTF} = (V_x \tan\psi_-)\mathbf{e}_z = (V_x \tan\psi_+)\mathbf{e}_z. \tag{97}$$

In this frame, the plasma velocity $\mathbf{V}$ and the magnetic field $\mathbf{B}$ are parallel on both sides of the shock. This transformation only exists if $V_{dHTF} < c$, so $\tan\psi_- < c/V_{x-} = c/U_s$. Shocks which satisfy this requirement are called *sub-luminal*. In the dHTF there is no electric field, so the energy of a particle is conserved (magnetic component of the Lorentz force does no work). In this frame the reflection by the shock is elastic. However, when one transforms back to the shock-frame , there is an energy change given by the direct analogue of Eqn. (51):

$$\frac{E_f}{E_i} = \frac{\left(c^2 + V_{dHTF} v_{zi}\right)}{\left(c^2 + V_{dHTF} v_{zf}\right)} . \tag{98}$$

In order to calculate this change, one has to follow the detailed particle orbits to find (a) which particles are reflected, and (b) the value of v_{zf} given the initial conditions, among them v_{zi} . It turns out that most particles are either transmitted by the shock, or reflected only once.

For shocks with $V_{dHTF} \ll c$ it turns out that -on average- the particles traverse the shock conserving their so-called magnetic moment $\mu \propto \sin^2 \vartheta / B$ in the dHTF. Here ϑ -as before- is the pitch-angle between the momentum vector and the magnetic field. Particles are reflected provided $\vartheta > \vartheta_c$ with $\sin \vartheta_c = \sqrt{(B_-/B_+)}$, all evaluated in the dHTF. Those particles which have a velocity along the field of order $v_\parallel \approx V_{dHTF}$ so that they have a small velocity with respect to the shock can gain a significant amount of energy[39]. This is especially true for nearly-luminal shocks with $V_{dHTF} \approx c$. On the other hand, in order not to be transmitted by the shock the initial particle velocity must be of order $v \approx V_{dHTF}$, which for nearly-luminal shocks means that they have to be relativistic already.

3.13 Conclusions

The theory of diffusive shock acceleration has firmly established itself in the astrophysical community as an important mechanism for the generation of energetic particles. In its simplest form, it is deceivingly straightforward. Practical application to astrophysical objects however is hindered by the fact that some of the details of the process such as injection, and the acceleration of electrons remain ill-understood. It seems therefore premature to use the proces as a panacea whenever efficient generation of energetic particles is required by the observations. Some of these problems are closely linked to our lack of understanding of all the details of quasi-parallel shock structure in astrophysical environments.

References

1) Fermi, E. : 1949, *Phys. Rev.* **75**, 1169.
2) Schmidt, G.: 1979, *Physics of high temperature plasma*, Academic Press, 1979, p. 374.

3) Achterberg, A.: 1979, *Astron. Astrophys.* **76**, 276.

4) Burn, B.J.: 1975, *Astron. Astrophys.* **45**, 435.

5) e.g. Ichimaru, S.: 1973, *Basic principles of plasma physics*, W.A. Benjamin Inc.

6) Axford, W.I., Leer, E., Skadron, G.: 1977, *Proc. 15th Int. Cosmic Ray Conf.*

7) Krimsky, G.E.: 1977, *Dokladay Acad. Nauk. SSR* **242**, 1306.

8) Bell, A. R.: 1978, *Mon. Not. R. astr. Soc.* **182**, 147.

9) Blandford, R. D., Ostriker, J. P.: 1978, *Astrophys. J. Lett.* **221**, L29.

10) Blandford, R. D., Eichler, D.: 1987, *Physics Reports* **154**, 1.

11) Landau, L. D., Lifshitz, E. M.: 1959, *Fluid Mechanics*, Pergamon Press, p. 329.

12) e.g.: Liboff, R.L.: 1990, *Kinetic Theory*, Prentice-Hall International, p. 53.

13) Abramovitz, M., Stegun, I.A.: 1972, *Handbook of mathematical functions*, Dover, NY, p. 297.

14) Lagage, P. O., Cesarsky, C. J.: 1981, *Plasma Astrophysics*, ESA SP-161, p. 317.

15) Meisenheimer, K., Röser, H-J.: 1989, *Hot Spots in Extragalactic Radio Sources*, Lecture Notes in Physics, Springer Verlag, Berlin.

16) Spruit, H. J.: 1988: *Astron. Astrophys* **194**, 319.

17) Schneider, P., Bogdan, T. J.: 1989, *Astrophys. J.* **347**, 496.

18) Scholer, M.: 1985, in: *Collisionless Shocks in the Heliosphere: Reviews of Current Research*, B. T. Tsurutani & R.G. Stone, eds., Geophysical Monograph 35, American Geoph. Union, p. 287.

19) Wefel, J.P.: 1988, in *Genesis and propagation of Cosmic Rays*, M.M. Shapiro, J.P. Wefel eds., NATO ASI Vol. 220, p. 1. D. Reidel Publ., Dordrecht, Holland.

20) Longair, M. S.: 1981, *High energy astrophysics*, Cambr. Univ. Press, Ch. 23.

21) Cesarky, C. J.: 1980, *Ann. Rev. Astron. Astrophys.* **18**, 289.

22) e.g.: Hillas, A. M.: 1984: *Ann. Rev. Astron. Astrophys.* **22**, 425.

23) Jokipii, J. P., Morfill, G.: 1987, *Astrophys. J.* **312**, 170.

24) Wdowczyk, J., Wolfendale, A. W.: 1989, *Ann. Rev. Nucl. Part. Sci.* **39**, 43.

25) Drury, L. O'C., Volk, H. J.: 1981, *Astrophys. J.* **248**, 344. .

26) Achterberg, A., Blandford, R. D., Periwal, V.: 1984: *Astron. Astrophys.* **132**, 97.

27) e.g.: Skilling, J.: 1975, *Mon. Not. R. astr. Soc.* **172**, 557.

28) Achterberg, A.: 1987, *Astron. Astrophys.* **174**, 329, 1987.

29) Falle, S. A. E. G., Giddings, J. R.: 1987, *Mon. Not. R. astr. Soc.* **225**, 399.

30) Drury, L. O'C., Volk, H. J.: 1981, *Astrophys. J.* **248**, 344.

31) Max, C.E., Zachary, A. L., Arons, J.: 1988, *Proc. Joint Varenna-Abatsumani International School & Workshop on Plasma Astrophysics*, ESA SP-285, p. 45.

32) Schlickeiser, R.: 1989, *Astrophys. J.* **336**, 243 & 264.

33) Kazanas, D., Ellison, D. C.: 1986, *Astrophys. J.* **304**, 178.

34) Sikora, M., Shlosman, I.: 1989, *Astrophys. J.* **336**,

35) Sikora M. et al.: 1987, *Astrophys, J. Lett.* **320**, 593. L81.

36) Kirk, J. G., Schneider, P.: 1987, *Astrophys. J.* **315**, 425.

37) Armstrong, T.P. et al.: 1985, in *Collisionless Shocks in the Heliosphere, Reviews of Current Research*, B. T. Tsurutani & R.G. Stone, eds., Geophysical Monograph 35, American Geoph. Union, p. 271

38) Chiueh, T.: 1988, *Astrophys. J.* **333**, 366.

39) Lieu, R., Quenby, J. J.: 1990, *Astrophys. J.* **350**, 692.

Interstellar Medium and Supernova Remnants

E. A. Dorfi

Institut für Astronomie der Universität Wien, Türkenschanzstr. 17,
A-1180 Wien, Austria

Abstract: The current status of theories of the interstellar medium as well as of supernova remnants is reviewed. Some shortcomings and the importance of including magnetic fields and cosmic rays are discussed in detail and at the end this additional physical input is compared to observations in the X- and γ-range.

Introduction

There is much evidence both from observational facts as well as from theoretical grounds that large fractions of the interstellar medium (ISM) are controlled by energetic phenomena like supernova explosions and stellar winds from young OB-associations. Therefore and in accordance with the title of this fourth European Astrophysical Doctoral Network (EADN)-school I will mainly concentrate on the high energy aspects of the interstellar medium and in the second part discuss the evolution of supernova remnants in more detail. In doing so I want to show that within the framework of present ISM theories which do not incorporate the various essentially non-linear interactions of cosmic rays and magnetic fields into a global description of the ISM, several observational aspects cannot be covered. Up to now it seems to be unclear whether a simple modification of the existing ISM theories (e.g. by fixing parameters) or whether the inclusion of new physical phenomena is needed to obtain a more consistent picture of the overall properties of the ISM. Keeping in mind the richness of the mutual interactions I tend to the opinion that all existing ISM theories can be regarded only as guidelines and are deficient by not taking into account the non-linear interactions of cosmic rays and magnetic fields. However, being faraway from giving a consistent theory of the ISM I also focus at different chapters of these lectures on the basic modifications caused by cosmic rays and magnetic fields. Starting from a simple analytical description of the SNR evolution (in spherical symmetry) the last two paragraphs describe some observational consequences at higher energies (X- and γ-rays) and theoretical X- and γ-ray luminosities are calculated for a SNR evolution including the non-linear modification by particle acceleration in shock waves.

In dealing with the overall properties of the ISM I will briefly consider the effects of superbubbles and galactic winds. Superbubbles or supershell as observed in HI are likely to evolve to scale heights which cannot be solely related to the galactic disk but can provide a 'vent' to the galactic halo. In this context the ISM is not a 'closed system' and has various connections to the sourrounding galactic halo. Again I want to point out the importance of cosmic rays for these processes. I am encouraged by the fact that in the year 1912 Victor Hess discovered the so-called *Höhenstrahlung* (nowadays cosmic rays) during several balloon flights in Graz, the city where this year's EADN-school takes place.

1 Theories of the Interstellar Medium

This chapter reviews some present theories of the interstellar medium. Starting with the two-phase model based on equilibrium cooling the major shortcomings are discussed leading to a model of the 'violent' interstellar medium where a third hot phase is controlled by supernova explosions (see e.g. the recent review by Spitzer 1990). Some consequences and predictions of the three-phase model are explained and contrasted to observations. The influence of cosmic rays on several processes is emphasized which may lead to a better agreement with the observational facts if the nonlinear modifications of shock acceleration are taken into account. At the end more recent developments on the so-called disk-halo connection and on galactic winds are presented.

1.1 Constituents of the ISM

To develop a 'theory' of the interstellar medium it is essential to introduce briefly the main constituents and summarize their properties and various interactions. However, in this short overview I have to ignore some properties of the interstellar medium like changes in the chemical abundances or a more detailed discussion of interstellar clouds. The main emphasis lies on several high energy aspects and in dealing with the diversity and frequency of the energizing phenomena it becomes clear that almost all interstellar material is overrun several times by (weak) shock waves. Turbulent as well as systematic motions like rotation are observed on quite different scales. Therefore a static decription cannot be used to model this environment, however it may be possible to reach a steady state situation or at least a slowly evolving medium.

Beside the energies contained in turbulent or ordered motions basically four components and their 'lively' interactions have to be included in a consistent description of the ISM.

a) **Interstellar plasmas.** They consist of electrons, ions, neutrals and dust particles with ranges in temperature and density of about

$$10 \lesssim \ T/[\mathrm{K}] \ \lesssim 10^6$$
$$10^{-3} \lesssim \rho/[\mathrm{g\,cm}^{-3}] \lesssim 10^7 \tag{1}$$

where the low temperature and high density values are typical for the dense cold cores of molecular clouds (see also Table 1). On the other hand the high temperatures and low densities correspond to the interiors of SNR's. The mass fraction of dust m_d to gas m_g will be taken to be roughly $m_d/m_g \simeq 0.01$.

b) **Radiation fields.** The radiation field in the ISM displays large intensity variations ranging from the 2.7 K background radiation up to the very high energy γ-rays produced by inverse Compton scattering or neutral pion decay. The matter is either exposed to the radiation field or the photons are generated themselves in the ISM.

c) **Cosmic rays.** These high energy particles are in almost all cases charged and well above the thermal energies of the background plasma with the main energy contribution coming from GeV-particles. The cosmic rays are closely linked to the magnetic field and at rare occasions single particles are observed carring energies of up to 10^{21}eV/nuc.

d) **Magnetic fields.** From observations it is obvious that a large scale ordered ($L \simeq 1\,\mathrm{kpc}$) as well as smaller scale ($1 \lesssim L \lesssim 100\,\mathrm{pc}$) 'turbulent' magnetic fields exist. However, the orientation of these fields depends on the observing direction but a clear tendency of field lines aligned with the spiral arms can be inferred from the data.

An important observational fact concerning these constituents of the ISM is shown by their energy densities (including also turbulence and ordered motions) (e.g. Axford 1981a)

$$u_{th} \simeq u_{\mathrm{mag}} \simeq u_{\mathrm{rad}} \simeq u_{\mathrm{CR}} \simeq u_{\mathrm{turb}} \simeq u_{\mathrm{kin}} \simeq$$
$$\simeq 1\,\mathrm{eV\,cm}^{-3} = 1.6\,10^{-12n}\mathrm{erg\,cm}^{-3} \tag{2}$$

which are all of the same order of magnitude. Comparing these almost equal energy densities contained in the different components the ISM seems to have reached a certain equipartition. The large number of processes involved makes such a conclusion plausible although a detailed balance is a priori not clear. Figure 1 summarizes some (not all) processes interchanging energies between various components as well as some observational consequences.

Interstellar Medium Dynamics

Intercloud Medium	H II–regions, stellar winds	SNR's	radiation	cosmic rays	magnetic fields	
evaporation, 2–phases, condensation, conduction	ionisation, champagne flows, dust, star formation	shocks, cloud crushing, 3–phases, triggered star formation	21 cm line, cooling, H_2, CO, molecules, extinction	γ–rays, π^0, heating, ionisation	Alfvén–waves, stability, heating, braking	interstellar clouds
	ionisation, mass and energy input, dust	ionisation, mass and energy input, cloud formation	heating, ionisation, UV, O VI, X–rays	CR–transport, dissipation, 2.–order–Fermi, spallation	$\langle n_e \rangle$, $B_\parallel$ turbulence, dynamo, Parker–instability	intercloud medium
		explosion in wind cavities, Type II–SNR's	Ly α, thermal bremsstrahlung, radio recombination	particle acceleration, 1.–order–Fermi, Ne^{22}	B–compression, conduction, braking	H II–regions, stellar winds
			synchrotron, radio, UV, X– and γ–rays, cooling	particle acceleration, 1.–order–Fermi, galactic CR's	B–compression, conduction, relativistic electrons	SNR's
				γ–rays, π^0, inverse Compton	reconnection, polarisation $(B_\perp)$, Zeeman and Faraday $(B_\parallel)$	radiation
					gyration, confinement, Alfvén–waves, galactic wind	cosmic rays

2–phase model

3–phase model

... and more processes

Fig. 1. Overview of several energy interchanging processes between the constituents of the ISM and some of their observational consequences

For simplicity reasons I do not distinguish between molecular and diffuse clouds (cf. Table 1). The interactions with dust particles as well as the formation of dust grains are neglected although these 'heavy' particles may have a significant influence on dynamical interstellar processes (like dust driven stellar winds, ambipolar diffusion inside dense clouds, depletion of metals on grain surfaces, formation of H_2-molecules, etc.). The cloud chemistry and shock chemistry are not considered in this article and therefore I will assume the plasmas to be chemically homogeneous throughout the following sections.

1.2 Interstellar plasmas

The interstellar plasmas cover a broad range in density and temperature (cf. Sect. 1.1) and consist of electrons, neutrals, ions and dust particles. The plasma is assumed to be at least quasineutral but in most applications I will use a magnetohydrodynamical approximation which does not distinguish between the different species. During the following sections I will assume pressure equilibrium which is plausible because no large scale motions driven by pressure gradients are observed on galactic scales. The existence of different phases at pressure equilibrium is possible only in the case of complex thermodynamics and is given by a detailed balance between heating and cooling mechanisms.

1.2.1 Two-Phase-Model

This description of the ISM is based on the work of Field, Goldsmith and Habing (1969) which starts with a equilibrium configuration, i.e. a balance between cooling $\mathcal{L}$ and heating $\mathcal{G}$ with a density ρ and a temperature T. Since the plasma is assumed to be optically thin no radiation transport is necessary and we get the simple relation

$$\mathcal{L}(\rho, T) = \mathcal{G}(\rho, T). \tag{3}$$

More useful in the subsequent considerations the net effect is given by

$$L(\rho, T) = \mathcal{L}(\rho, T) - \mathcal{G}(\rho, T). \tag{4}$$

Following Field's (1965) ideas such a configuration is unstable if disturbances exist that can decrease the entropy S of the system which is written like

$$\left(\frac{\partial L}{\partial S} \right)_X < 0, \tag{5}$$

where X denotes a thermodynamical quantity. Assuming a perfect gas and an isobaric perturbation Equation (5) yields

$$\left(\frac{\partial L}{\partial T}\right)_P = \left(\frac{\partial L}{\partial T}\right)_\rho - \frac{\rho}{T}\left(\frac{\partial L}{\partial \rho}\right)_T < 0. \tag{6}$$

At this point a more detailed discussion of individual heating and cooling processes is needed.

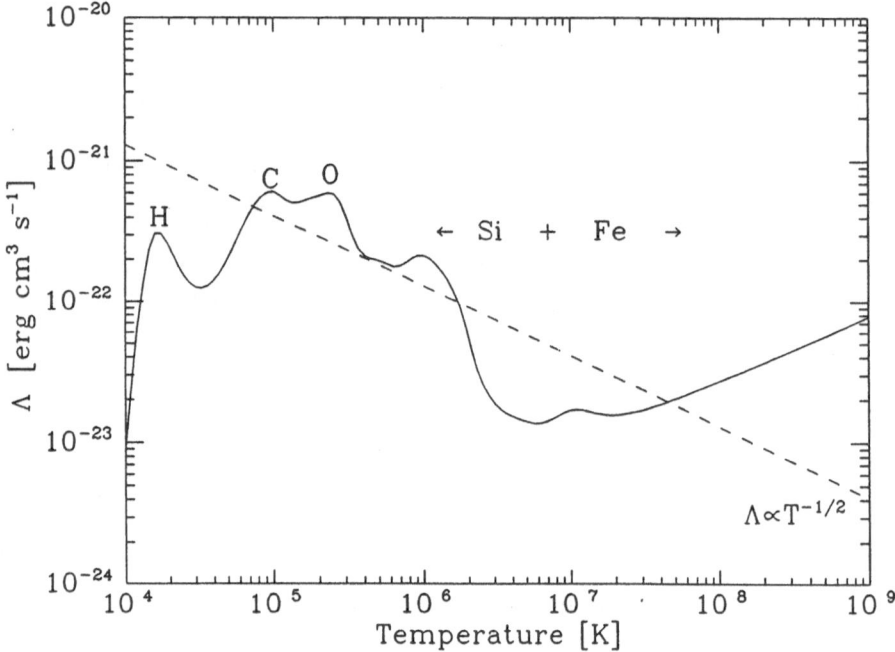

Fig. 2. The total cooling in $\mathrm{erg\,cm^3 s^{-1}}$ from a hot plasma as a function of temperature

Cosmic rays are the main source of ionization in the denser interiors of clouds and even energetic particles with a few tens of MeV/nuc can penetrate dense clouds (Cesarsky & Völk 1978) because they are not hindered by diffusion. Such diffusion would be caused by resonant small scale MHD-waves which are strongly damped in a weakly ionized medium due to ion-neutral friction (Kulsrud & Pearce 1969). The ionization rate by cosmic rays ζ_{CR} is measured by the abundance of certain molecules and typical values of ζ_{CR} are within the range of $10^{-17}\,\mathrm{s^{-1}}$ up to $10^{-15}\,\mathrm{s^{-1}}$. A mean value of $10^{-16}\,\mathrm{s^{-1}}$ is in agreement with UV-measurements of the [HD/H$_2$]-ratio (Barsuhn & Wamsley 1977). From the [DCO$^+$/HCO$^+$]-ratio Guelin et al. (1982) derive a value of $\zeta_{CR} = 3.5\,10^{-16}\,\mathrm{s^{-1}}$. The heating rate corresponding to this cosmic ray ionization rate depends now on many following secondary processes as well as on the degree of ionization and chemical composition where the ionized particles and electrons lose their energy. Nevertheless in this case of general considerations the mean heating by cosmic rays in an atomic cloud of low ionization is around $\epsilon_h \simeq 6\,\mathrm{eV}$ per primary ionization and the total heating rate for a hydrogen density n_H can be written according to Dalgarno & McCray (1972)

$$\mathcal{G}_{CR} \simeq 10^{-28} \left(\frac{\zeta_{CR}}{10^{-17}} \right) n_H \; \mathrm{erg\,cm^{-3}s^{-1}}. \qquad (7)$$

Other heating mechanisms operate in the ISM like heating by starlight, soft X-rays, photoelectrons from grains, collisions of dust grains (heated by the near-infrared radiation field, cf. Sect. 1.3) with molecules. Beside these direct heating mechanisms the ISM is also heated in collective hydrodynamical processes due to the compression by stellar winds, shocks, SNRs, MHD-waves and dissipation of turbulence. Just to quantify one other mechanism, heating by photoelectrons from dust grains is considered where de Jong et al. (1980) find

$$\mathcal{G}_{ge} \simeq 4\,10^{-26}\, n_H \; \mathrm{erg\,cm^{-3}s^{-1}}. \qquad (8)$$

These two examples show that the basic form of the direct heating mechanisms can be stated like

$$\mathcal{G} = nG \propto n \qquad (9)$$

where G is the sum of all relevant heating contributions.

The cooling of a tenuous plasma proceeds through the conversion of kinetic energy into radiation by various processes and for cooling to be efficient the plasma has to be optically thin towards the produced photons. Cooling clearly depends on the chemical composition and the degree of ionization. Line cooling and recombination radiation are the most important cooling mechanisms at temperatures $T < 10^7$. At higher temperatures bremsstrahlung dominates the energy loss of a hot plasma, i.e. the cooling rate $\propto T^{1/2}$. All effects together are building the so-called cooling function $\Lambda(T)$ defined through

$$\mathcal{L} = n^2 \Lambda(T). \qquad (10)$$

Detailed calculations on the ionization structure and emission properties of a hot optically thin plasma have been carried out by Raymond & Smith (1977) where more information can be found also on the radiative mechanisms, on the ionization and recombinations rates as well as on the collisional cross sections. A similar cooling function is displayed in Fig. 2 where the main contributions of different chemical elements to radiative cooling are indicated by the corresponding chemical symbols. Note again that these relative cooling contributions depend on the chemical composition of the hot plasma (e.g. Böhringer & Hensler 1989). For many applications I will adopt Kahn's approximation of the cooling function in the range between $2\,10^5 \leq T/[\mathrm{K}] \leq 10^7$ (Kahn 1976) which can be stated as a simple power law of the temperature

$$\Lambda(T) = 1.3 \, 10^{-19} T^{-1/2} \text{ erg cm}^3 \text{s}^{-1} \qquad (11)$$

shown in Fig. 2 by the dashed straight line. The exact numerical value of Eq. (11) changes with different authors and is meant only as an order of magnitude estimate. In the high temperature range ($T > 2 \, 10^7$K) the plasma is totally ionized and can therefore radiate only through free-free transitions (i.e. bremsstrahlung) leading to a cooling law of $\Lambda(T) \propto T^{1/2}$.

Collecting the expressions for heating (9) and cooling (10) the equilibrium condition (3) as well as pressure equilibrium are given by

$$L = n^2 \Lambda(T) - n \, G \qquad \text{and} \qquad n \, T = const. \qquad (12)$$

or put in a more suitable form

$$\frac{\Lambda(T)}{T} = \frac{G}{nT}. \qquad (13)$$

Applying Eqs. (12) the instability criterium (5) reads now

$$\begin{aligned}
\left(\frac{\partial L}{\partial T}\right)_P &= \frac{\partial \Lambda}{\partial T} n^2 - \frac{n}{T} \left(\frac{\partial L}{\partial n}\right)_T \\
&= \frac{\partial \Lambda}{\partial T} n^2 - \frac{n}{T} (2n\Lambda - G) \\
&= n^2 \left[\frac{\partial \Lambda}{\partial T} - \frac{2\Lambda}{T} + \frac{G}{nT}\right] \\
&= n^2 \left[\frac{\partial \Lambda}{\partial T} - \frac{\Lambda}{T}\right] < 0.
\end{aligned} \qquad (14)$$

In summary this instability will occur in regimes where the cooling function $\Lambda(T)$ rises moderately with temperature, i.e.

$$\frac{\partial \ln \Lambda}{\partial \ln T} < 1. \qquad (15)$$

The equilibrium situation can be further analysed if $\Lambda(T)/T$ is plotted versus T as done in Fig. 3. The depicted curve marks all possible equilibrium states for a given cooling law $\Lambda(T)$. Above this curve we have a net heating of the medium and below a net cooling. Fixing a certain heating rate (i.e. a certain value of G in Eq. (9) which depends on the local conditions of the ISM) a horizontal line can intersect at most at four values our equilibrium curve determining the possible equilibrium states. It is also clear that only a limited zone of heating rates is permitted to generate a multi-phase ISM indicated by the two horizontal dashed lines in Fig. 3 and called 'multi-phase-zone'. Each of these intersections gives rise to a 'phase' of the ISM which could be thermally stable or could last long enough to be observed. However, only two of these phases are stable according to the stability criterium of Eq. (15) because the other two intersections exhibit

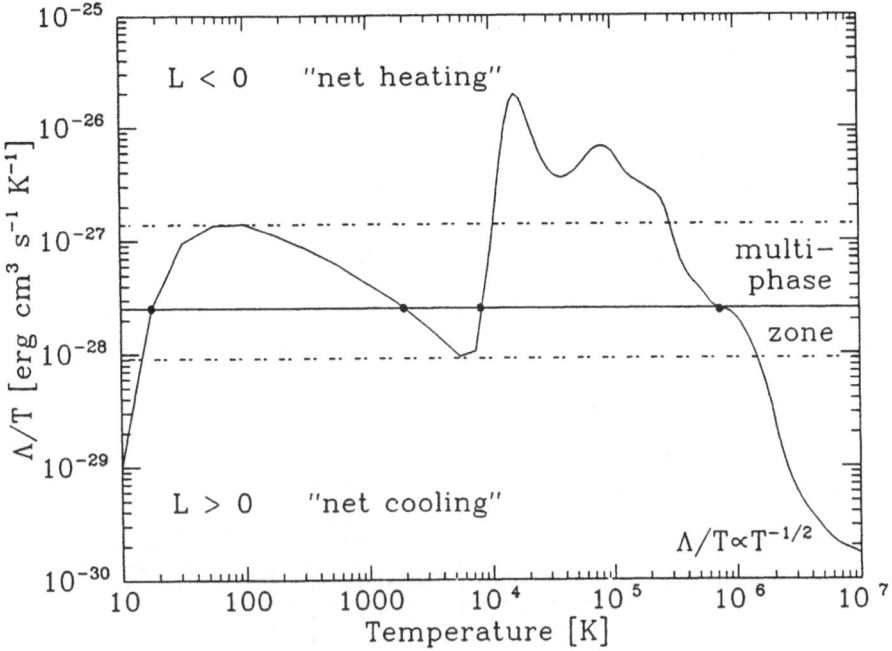

Fig. 3. Equilibrium states of the two-phase model based on a constant heating rate and on constant pressure as a function of temperature

negative slopes at the $\Lambda(T)$ curve. The equilibrium values at the low temperature and high density around $T \simeq 100\,\mathrm{K}$ and $n \simeq 10\,\mathrm{cm}^{-3}$ are called 'clouds' whereas the values of $T \simeq 10^4\,\mathrm{K}$ and $n \simeq 10^{-1}\,\mathrm{cm}^{-3}$ characterize the so-called 'intercloud medium'. Note that this discussion is based on the assumption of pressure equilibrium between the two phases.

Within the framework of the two-phase model some observational facts cannot be explained. The thermal gas is unstable at high temperatures but several X-ray observations (e.g. McCammon & Sanders 1990) indicate large portions of hot plasmas. To maintain a hot medium either an additional heating mechanism is needed (like dissipation of turbulent motions or damping of Alfvén waves) or the ISM is far from an equilibrium state. In the latter case the cooling time t_c must be large compared to the typical evolution time scale of the ISM. The two-phase model predicts a warm partially ionized medium around a temperature of $T \simeq 10^4\,\mathrm{K}$ which is not observed in this manner. The estimated mean electron density $\langle n_e \rangle$ in this warm medium disagrees with dispersion measurements of pulsar signals (cf. Sect. 1.5.1) which gives the mean electron density integrated along the line of sight. The interstellar OVI absorption lines make the presence of a widespread hot gas very likely (Jenkins & Meloy 1974). At first these lines have been attributed to this warm medium but nowadays it is clear that these lines are due to the conductive interfaces between cold clouds and gas at X-ray emitting temperatures. The conductive energy transport is mainly provided by the motions of electrons and therefore depends on the orientation of the magnetic field in the interface region which connects the magnetic field to other properties of the ISM. However, the local nature of this process makes an estimate of the overall influence very difficult.

Table 1. Average properties of molecular regions in the ISM

	L[pc]	n[cm^{-3}]	$M[M_\odot]$	Δu[km/s]	T[K]
giant molecular cloud					
complex	$20 - 80$	$100 - 300$	$8\,10^4 - 10^6$	$6 - 15$	$8 - 15$
member	$3 - 20$	$10^3 - 10^4$	$10^3 - 10^5$	$4 - 12$	$15 - 40$
core	$0.5 - 3$	$10^4 - 10^6$	$10 - 10^3$	$1 - 3$	$30 - 100$
clump	< 0.5	$> 10^6$	< 10	$4 - 15$	$30 - 200$
dark cloud					
complex	$6 - 20$	$10^2 - 10^3$	$10^3 - 10^4$	$1 - 3$	$\simeq 10$
member	$0.2 - 4$	$10^2 - 10^4$	$5 - 500$	$0.5 - 1.5$	$8 - 15$
core	$0.1 - 0.4$	$10^4 - 10^5$	$0.3 - 10$	$0.2 - 0.4$	$\simeq 10$

1.2.2 Three-Phase-Model

Beginning with the work of Cox & Smith (1974) the so-called three-phase model of the ISM has been developed by McKee & Ostriker (1977) which introduces a third hot phase controlled by supernova explosions and stellar winds from young OB-associations. Depending on several parameters the supernova remnants (SNR's) can overlap and the hot and tenuous interiors of the SNR's can only cool slowly forming the third hot phase. Note that this model does not refer to an equilibrium situation but is better characterized by a so-called 'violent' ISM. Since SNe are responsible for the main source of hot gas in our Galaxy a quantitative theory of SNR evolution is essential for an accurate description of the hot interstellar plasmas.

This particular model suggests that the thermal plasma is distributed among four phases. a) Most of the galactic volume is filled with the hot low-density ionized gas of old SNRs called HIM whose parameters are given in Table 2. b) Embedded in that hot phase is a large number of cold neutral and relatively dense clouds (CNM) containing most of the mass and having a small filling factor. c) Each of these clouds is surrounded by a warm envelope occupying large volumes with small masses. For simplicity reasons these envelopes consist of the warm neutral medium (WNM) where the low degree of ionization is maintained by the very soft X-rays emitted from SNRs. d) Further out the envelopes of the clouds show the warm but ionized medium (WIM) whose ionization balance is controlled by the UV photons from hot stars (OB, white dwarfs, planetary nebulae) and by conduction from the hot phase.

The values for the HIM,CNM, WIM, WNM are now calculated under the assumption of pressure equilibrium and various interchange processes (evaporation, fragmentation and condensation of clouds, thermal instabilities, shock waves) also mentioned in the overview (Fig. 1). The result of such a balance (McKee & Ostriker 1977) is summarized in Table 2 and the work of Ikeuchi et al. (1984) illuminates in more detail the possible states and the time-dependence of the ISM including the role of star formation, cloud-cloud collisions as well as of gravitational instabilities occurring in cloud complexes.

Since the fraction of ionization in the WIM and WNM is largely determined by the interstellar radiation field (cf. Sect. 1.3) we see again a vital interaction between the various constituents of the ISM. Note that the volume filling factor of the cloud envelopes (WNM+WIM) covers more than 30% of the total galactic space.

Table 2. Properties of the 3-phase-model according to McKee & Ostriker (1977)

	CNM	WNM*	WIM†	HIM
f	0.025	0.15	0.23	0.595
$\langle T \rangle$ [K]	80	8000	8000	$4.5\,10^5$
$\langle n \rangle$ [cm^{-3}]	42	0.37	0.25	$3.5\,10^{-3}$
$X = n_e/\langle n \rangle$	10^{-3}	0.15	0.68	1.0

* (produced by the soft X-ray background)
† (produced by the ionizing stellar UV-field)

To develop the theory of the third hot phase a few estimates concerning the SNR evolution have to be supplied. It is convenient to define the dimensionless 'porosity' parameter

$$Q = r_{SN} V_{SNR} \tau_{SNR} \tag{16}$$

where r_{SN} is the SN-rate per unit time and volume, V_{SNR} denotes the final SNR volume and τ_{SNR} the mean SNR lifetime. The SN-rate per galaxy r as well as the mean SNR lifetime can be written as

$$r_{SN} = \frac{r}{A_{disk} H_{SNR}}, \qquad \tau_{SNR} = \frac{R_{SNR}}{\bar{v}_{ISM}} \tag{17}$$

and $A_{\text{disk}} H_{\text{SNR}}$ defines the galactic volume available for SNR's, i.e. the disk area $A_{\text{disk}} \simeq \pi(15\,\text{kpc})^2$ times the typical scale height of stars $H_{\text{SNR}} \simeq \pm 100\,\text{pc}$ able to evolve into a SN. The mean SNR lifetime is given by the final SNR radius R_{SNR} divided by the mean velocity in the ISM $\bar{v}_{\text{ISM}}$. If the porosity parameter Q remains constant over several SNR lifetimes τ_{SNR} there is a simple relation to the volume filling factor

$$f = \frac{Q}{1+Q}, \qquad Q < 1. \tag{18}$$

To get further insight into the three-phase model estimates for the values of R_{SNR}, V_{SNR} and τ_{SNR} are needed.

As will be shown in Sect. 2.3 the radius and time where a radiative shell is formed can be written

$$t_{\text{shell}} \simeq 2.13\,10^4\ E_{51}^{3/14} n_0^{-4/7}\ \text{yr}$$
$$R_{\text{shell}} \simeq 20\ E_{51}^{2/7} n_0^{-3/7}\ \text{pc} \tag{19}$$
$$v_{\text{shell}} \simeq 368\ E_{51}^{1/14} n_0^{1/7}\ \text{km s}^{-1}.$$

The quantity E_{51} denotes the SN-explosion energy in units of 10^{51}erg and n_0 the number density of the external medium in cm^{-3}. Assuming that the further evolution is characterized by a pressure modified 'snowplow' solution (cf. Sect. 2.3) we get

$$R \propto t^{2/7}, \qquad \dot{R} = v \propto t^{-5/7}. \tag{20}$$

The constants of proportionality have to be fixed at $t = t_{\text{shell}}$ by $R = R_{\text{shell}}$ and $v = v_{\text{shell}}$. Therefore the velocity scales like

$$v = v_{\text{shell}} \left(\frac{t}{t_{\text{shell}}}\right)^{-5/7} \tag{21}$$

which can be solved for the time $t = \tau_{\text{SNR}}$ where the expansion velocity of the remnant has decreased to some mean velocity in the ISM, i.e. $v = \bar{v}_{\text{ISM}} = 20\,\text{km s}^{-1}$. This velocity can be obtained from the typical Alfvén velocity v_A (27) if a mean magnetic field of $B = 5\,\mu\text{G}$ and a mean density of $n = 0.3\,\text{cm}^{-3}$ is assumed.

This results in an estimate of the final SNR-radius and the final SNR-age of

$$\tau_{\text{SNR}} \simeq 1.26\,10^6\ E_{51}^{11/35} n_0^{-13/35}\ \text{yr},$$
$$R_{\text{SNR}} \simeq 64\ E_{51}^{11/35} n_0^{-13/35}\ \text{pc}. \tag{22}$$

Inserting these last estimates into the equation for the porosity parameter (16) the following expression is derived

$$Q \simeq 0.098\ N_{100}\ E_{51}^{44/35} n_0^{-52/35}, \tag{23}$$

where N_{100} stands for the number of SN-explosions per century distributed randomly over the galactic disk. This value of $Q = 0.098$ implies a filling factor (18) of $f = 0.089$.

Applying this 3-phase model to the ISM leads to several correct predictions concerning e.g. the mean interstellar pressure of about $nT \simeq 3700\,\mathrm{cm}^{-3}\mathrm{K}$ or the mean electron density of $\langle n_e \rangle \simeq 0.04\,\mathrm{cm}^{-3}$ in the WIM cloud envelopes with a typical dispersion of $\langle n_e^2 \rangle \simeq 0.08\,\mathrm{cm}^{-3}$ which is in agreement with dispersion measurements based on pulsar signals (cf. Sect. 1.5.1). The expanding SNRs overrun many clouds and can accelerate them. The energy balance maintained by this input and the losses due to inelastic cloud collision can be used to calculate the mean velocity dispersion of interstellar clouds of $\langle v_{cl} \rangle \simeq 8\,\mathrm{km\,s}^{-1}$ (McKee & Ostriker 1977), a value in agreement with the observations (e.g. Jenkins 1978ab). However, a shock passing a cloud leads to the development of several instabilities on the surface (Woodward 1976) which makes the simple picture of almost spherical clouds being accelerated rather doubtful. But due to the complicated cloud shapes, the essentially non-spherical geometry, their formation and evolution the actual value of f_{CNM} is very difficult to estimate. Within the frame of the 3-phase model the X-ray intensity of the local bubble can also be explained (see e.g. the review by Cox & Reynolds 1987) which describes the local ISM within a radius of 100 pc having a typical temperature of $10^6\mathrm{K}$ and a density of $5\,10^{-3}\mathrm{cm}^{-3}$.

Some observational facts cannot be modelled by the 3-phase model in its present versions. The HI-distribution in our galaxy is rather smooth as inferred from Lyα and 21cm observations which leads to an increase of the filling factor f_{CNM} of the cold neutral medium (Lockman et al. 1986). Concerning other uncertainties due to geometry and structure of clouds (e.g. more sheet-like than spherical) it is unclear whether some disagreement with the observations can be avoided by introducing more realistic cloud models. The HI-data of Heiles (1987) go along the same direction which favour a rather smooth distribution of the neutral gas in our Galaxy. These observations at 21cm show that a substantial fraction up to 40% of HI is warm whereas only about 2% are predicted by the current 3-phase model. These observations suggest two different HI components. On the other hand these data reveal the existence of much larger structures ($L \simeq 1\,\mathrm{kpc}$) in HI pointing again to a connection with the galactic halo; they are called supershells, worms or chimneys (cf. Sect. 1.6.1). Hence the SNR cavities appear to fill much smaller volumes than predicted by the standard 3-phase model which neglects the so-called disk-halo connection including the formation of superbubbles and galactic fountains and providing a 'vent' for the galactic interstellar medium. As concluded by Heiles (1991) the interpretation of the HI-data and the corresponding filling factors depends also on the topological structure of the ISM and it may be that the interstellar HI covers the galactic plane like a bedsheet covers a bed but does not occupy much volume, i.e. HI may have a large 2-dimensional filling factor but a very small 3-dimensional volume filling factor.

At this point I want to mention that all these arguments do not include the influence of cosmic ray acceleration in the SNR evolution although the production of energetic particles can strongly modify the expansion by non-linear effects in shock waves (cf. Sect. 2.5). The observed properties of cosmic rays indicate that the volume factor f_{HIM} of the hot phase is also constrained due to a mean density of about $\bar{n} \simeq 0.1\,\mathrm{cm}^{-3}$ assuming a path lenght ('grammage') of $X \simeq 5\,\mathrm{g\,cm}^{-3}$ (cf. Sect. 1.4 and Cesarsky (1980) for the transport and confinement of cosmic rays within our Galaxy). However, the uncertainties of these estimates do not permit an actual value for f_{HIM} but favour not too small numbers. The mentioned OVI-observations require a high value of f_{HIM} to explain the almost ubiquitous presence of hot gas. A possible way to join these different observational facts is to consider the ISM not as a closed system but to include the galactic halo as an additional energy sink (cf. Sect. 1.6).

The SNe are not uniformly distributed over the galactic disk. The regions with galactocentric radii of $R \lesssim 3\,\mathrm{kpc}$ are dominated by old Type II SNe whereas the Type I SNe are rather concentrated in OB-associations within $4 \lesssim R \lesssim 7\,\mathrm{kpc}$ and at $R \gtrsim 7\,\mathrm{kpc}$ both types are common. Therefore we expect less overlapping SNRs decreasing the porosity parameter as defined by eq. (16) to $Q \ll 1$.

1.3 Radiation fields

At the location of the sun within our galaxy ($R \simeq 10\,\mathrm{kpc}$) a diffuse interstellar radiation field is observed where a comparable energy flux exists at several frequency ranges shown in Fig. 4 extending from microwaves up to γ-rays. A large number of properties of this background radiation can be found in the IAU Symposium No. 139 edited by Bowyer & Leinert (1989). At this point I intend to show only the general behaviour of the radiation field and point out that the errors in the fluxes of Fig. 4 can differ from one frequency range to another resulting from rather different observational techniques with different intrinsic accuracies.

Starting at the lower end of frequencies we can infer the black body radiation at a temperature $T = 2.7\,\mathrm{K}$ followed by the very pronounced radiation peak in the infrared ($\lambda \simeq 1\,\mu\mathrm{m}$, $\log \nu \simeq 14.5$) produced by the large number of cool stars. Accordingly, it is not possible to talk about the interstellar radiation field without mentioning the importance of dust particles which obscure all regions at visual and UV wavelengths. The absorbed photons heat the dust grains leading to a far-infrared emission which provides a major part ($\sim 30\%$) of the total luminosity of the Galaxy (cf. also Fig. 4). Several more aspects of dust emission and absorption like extinction laws, scattering and polarization of star light can be found e.g. in a recent review by Mathis (1990). Apart from the dominance of dust emission in the range around $100\,\mu$m dust particles play a vital role also for the chemistry in interstellar clouds, in particular through the formation of H_2 or through cooling and heating by photoelectrons from grain surfaces and I refer e.g. to the IAU Symposium 135 'Interstellar dust' edited by Tielens & Allamandola (1989) where much more information is collected concerning the different aspects of interstellar dust particles.

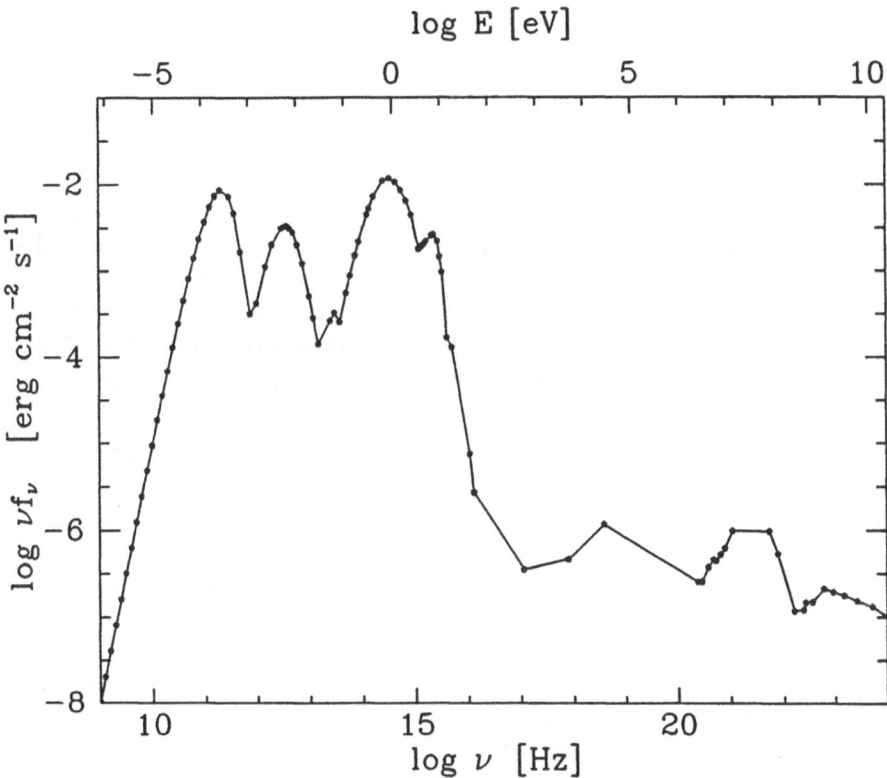

Fig. 4. The interstellar radiation field in the vicinity of the sun as a function of frequency

The values between $\lambda = 1\,$mm, $\log \nu = 11.5$ and $\lambda = 910\,$Å, $\log \nu = 15.5$ have been taken from the model of Mathis et al. (1983) where contributions from four stellar components distributed differently throughout the galaxy are included. However, IRAS observations at $\lambda = 12\,\mu$m and $\lambda = 25\,\mu$m show significantly higher fluxes which can be explained by emission from small dust grains or very small PAH grains which are not considered in the model (e.g. Mezger 1990).

The Lyman-α-edge located at $13.6\,$eV or $\log \nu = 15.517$ is clearly visible through a drop of the radiative energy density by more than two orders of magnitude. The EUV flux data in the solar vicinity are taken from the collection of Reynolds (1990). A theoretical modelling of the diffuse interstellar radiation field around the Lyα-edge has been done by Weyman (1967). These numbers are strongly dependent on the hydrogen column density and on the position within the Galaxy. The ionizing flux in the galactic halo can be much higher due to less absorbing hydrogen. In this sense the values given in Fig. 4 can be upper as well as lower limits to the actual UV-flux at a certain location in the Galaxy. Together with the X-ray flux penetrating deeper into dense clouds the UV-flux is responsible for the amount of the warm and partly ionized medium (WIM or WNM) as seen in Sect. 1.2.2. Studies of the photoionization in the halo of the Galaxy can give an upper limit for the amount of ionizing radiation in this highly obscured energy range (Bregman & Harrington 1986). According to these authors the range between 13.6 eV and 45 eV is dominated by light of young OB-stars, between 45 eV and 2 keV planetary nebulae provide their major contribution whereas the extragalactic X-ray background becomes prominent at the higher photon energies. However, the ionizing flux of OB-stars is the most important contributor to the ionizing radiation field provided that the stopping length of the photons is greater than a few parsecs. Observations of diffuse X-rays and ionized atoms suggest that the filling factor of the hot and tenuous gas is not too small (cf. Sect.1.2.2) resulting also in a low absorption probability.

For this purpose it is enough to approximate the huge number of X-ray observations by simple fits showing the general trend. The compilation of X-ray data in the range from 1 keV or $\log \nu = 17.4$ up to 100 keV or $\log \nu = 19.4$ due to Schwartz & Gursky (1974) has been used where the differential X-ray intensity is represented by two power laws with spectral indices of -0.40 vs. -1.38 joined together at 21 keV or $\log \nu = 18.7$. Note that this sharp break in the slopes has no physical significance. The spectral complexity in the X-ray range indicates that several different mechanisms are contributing to the diffuse background (bremsstrahlung, Compton scattering, discrete X-ray sources, thermal emission of hot plasmas, recombination radiation). A power law extrapolation with the same spectral index of -1.38 up to 300 keV is used to join the γ-ray data. This diffuse X-ray component consists probably of contributions from extragalactic sources (QSOs or nearby galaxies) and from galactic hot plasmas produced in SNRs and stellar winds.

The flux in the γ-range is represented by several measurements and not by a simple power law. The data included range from 0.3 MeV to 27 MeV (Trombka et al. 1973) and from 0.7 MeV to 4.5 MeV (Vedrenne et al. 1971). The two measurements at 22 MeV and 32 MeV are taken from Mayer-Hasselwander et al. (1972). All these γ-ray fluxes indicate an excess of diffuse γ-emission if the X-ray flux at 100 keV (Metzger et al. 1964) is extrapolated to the OSO-III data at 100 MeV (Kraushaar 1970). This feature has no confirmed explanation but is clearly visible in Fig. 4 around the frequency of 10^{21} Hz.

The diffuse galactic γ-ray emission has recently been reviewed by Bloe-men (1989) and the γ-ray data ranging from 50 MeV up to 5 GeV shown in Fig. 4 are taken from this publication. Gamma-rays in the energy range above 100 MeV are produced by decay of π^0-mesons generated through colli-sions of cosmic rays around 1 GeV with the thermal plasma. The lower γ-ray energies are due to bremsstrahlung of cosmic ray electrons $E_{e,CR} \lesssim 100$ MeV and due to inverse Compton scattering of high energy cosmic ray electrons $E_{e,CR} \lesssim 10$ GeV on soft photons (e.g. 2.7 K background radiation with $h\nu \simeq 6\,10^{-4}$ eV). All these three processes add up to the observed diffuse galactic emission but the π^0-decay exhibits a characteristic bump (Stecker 1973) whereas the bremsstrahlung and inverse Compton emission are basi-cally given by power laws. However, the observed γ-ray intensity samples the matter or photon density along the line of sight and since our Galaxy is almost transparent to γ-ray photons up to 10^{14} eV the observations can be used to trace the interstellar matter and/or cosmic ray particles (see also Sect. 1.4.3). Although the relative contributions of these processes are expected to be a function of the position in the Galaxy the γ-ray intensity has been used in correlation with the observed (or inferred) cosmic ray in-tensity to determine the spatial distribution of H_2 in our Galaxy. H_2 is not observed directly but estimated from the observations of the CO $(J = 1 \rightarrow 0)$-transition assuming a known conversion factor from the CO abundance to H_2. The results, problems and different approaches are discussed e.g. in Bloemen (1989) but the γ-ray observations show that the total H_2-mass in our Galaxy does not exceed $1.2\,10^9 M_\odot$ compared to the total HI-mass of $4.8\,10^9 M_\odot$ (Henderson et al. 1982).

Table 3. Maxima of the interstellar radiation energy flux observed at the location of the sun

range	$\log \nu$	source
submm	11.3	2.7 K background
far IR	12.5	dust emission
optical	14.5	average starlight
UV	15.3	OB-stars, scattering on dust
X-rays	18.7	dilute hot plasmas, lines
γ-rays	21.7	π^0-decay from CR-collisions

From the above discussion it should be clear that the radiation fluxes compiled in Fig. 4 can only exhibit the overall properties of the interstellar radiation field, because some regions in frequency are subject to observational uncertainties which could alter the presented values by factor of three (or more).

Integrating over frequency the average specific intensity of the interstellar radiation field as displayed in Fig. (4) yields

$$4\pi J_{\mathrm{rad}} = \int_0^\infty f_\nu d\nu = 3.88\,10^{-2} \mathrm{erg\,cm^{-2}s^{-1}}. \tag{24}$$

The corresponding energy density is then given by

$$u_{\mathrm{rad}} = 4\pi J_{\mathrm{rad}}/c = 0.81\,\mathrm{eV\,cm^{-3}}. \tag{25}$$

Summarizing Fig. 4 very different processes operate in the ISM to produce the diffuse background radiation. Note that locally large deviations from this averaged radiation field can exist. I emphasize again that the radiative energy density (Eq. 25) provides an important contribution to the overall energy balance of the ISM.

1.4 Cosmic Rays

1.4.1 General Remarks on Cosmic Rays

Comprehensive and up-to-date information on all subjects related to cosmic rays can be obtained from the proceedings of the biannual *International Cosmic Ray Conference*, the last one in Dublin (1991). Hence, I will mention only some basic facts of cosmic rays related to their various influences on the interstellar medium. Since the general process of particle acceleration in shock waves is discussed by Bram Achterbergs lecture (this volume) in great detail I can concentrate on the case of acceleration at the shock waves of an expanding SNR where in a nonlinear development the shock structure is modified by the reaction of the energetic particles on the thermal plasma flow.

Cosmic rays are high energy particles with energies well above the thermal velocities of the interstellar plasma. With respect to 2-body collisions cosmic rays are almost collisionless particles except for very low energies ($\lesssim 1\,\mathrm{MeV}$). Nevertheless, these charged particles are closely linked to the magnetic field because they gyrate around the magnetic fields lines. For a proton the gyroradius in a homogeneous magnetic field B is given by

$$r_{\mathrm{g}} = 10^{-6} \left[\frac{E}{[\mathrm{GeV}]}\right] \left[\frac{B}{[1\,\mu\mathrm{G}]}\right]^{-1} \quad \mathrm{pc}. \tag{26}$$

The interaction of cosmic rays with magnetic fields is considerably more complicated because in the presence of magnetic irregularities with a scale comparable to the gyroradius the particles are resonantly scattered in pitch angle (Jokipii 1966). Moreover, in the case of a cosmic ray gradient such Alfvén waves are generated by the streaming instability (Lerche 1967,

Wentzel 1968, Kulsrud & Pearce 1969) and the energetic particles are then scattered resonantly by circularly polarized Alfvén waves propagating parallel to the magnetic field B. Hence, cosmic rays can generate their own wave field which is scattering itself the particles. Since the particle speed is usually much larger than the Alfvén velocity in a medium of density ρ

$$v_A^2 = \frac{B^2}{4\pi\rho} \tag{27}$$

the resonance condition requires that the wavelength of the Alfvén waves is equal to the gyro radius of the charged particle. If the damping of these waves is small (usually the case in the hot ISM) the waves can grow rapidly and the bulk cosmic ray velocity is reduced to the Alfvén speed.

The original idea to accelerate particles on magnetic fluctuations has been proposed by Fermi (1949) to explain cosmic rays. At that time he had to postulate an interstellar magnetic field of the order of μ G to scatter the particles on magnetic clouds moving at random velocities. However, such an acceleration mechanism proceeds very slowly in time but the situation changes radically if an ordered motion of the fluctuations can be used occurring in a hydromagnetic shock wave. This process is nowadays called diffusive shock acceleration (cf. Sect. 1.4.3) because the energetic particles can be scattered (i.e. diffuse) across the shock front separating the downstream and upstream media with different velocities. This concept appears to have been 'in the air' and has been invented by several people (Krymsky 1977, Axford et al. 1978, Bell 1978ab, Blandford & Ostriker 1978). This process of shock acceleration provides a natural explanation of the power law energy distribution of cosmic rays. Depending on the physical parameters in shock waves the cosmic rays can gain energies up to about 10^{15}eV/nuc (e.g. Axford 1981a, Völk & Biermann 1988). The nonlinear modifications of the basic acceleration mechanism have been considered by Drury & Völk (1981) or Axford et al. (1982). These authors show that particle acceleration can be very efficient converting almost all incoming kinetic energy into cosmic rays leaving the thermal plasma cold although the gas has passed across the shock (cf. also Sect. 1.4.3).

The chemical composition of the primary component of cosmic rays is in general rather similar to the solar abundances (except for the underabundance of hydrogen) if corrected for spallation products (so-called secondaries) produced by passage through the interstellar medium (e.g. Simpson 1983). Most measurements are done in an energy range 100 MeV − 100 GeV/nuc and at higher energies the chemical information is still limited. However, there are some chemical anomalies which can not be attributed to interactions with the interstellar plasma. I want to mention the ^{22}Ne/^{20}Ne isotope ratio which is increased by a factor of ~ 4 and following Meyer (1981) this overabundance can be explained if a small fraction (~ 0.02) of cosmic rays originates from material of quiescent He-burning stellar layers. Such a suggestion leads to the question of how efficient terminal shocks of stellar winds can accelerate high energy particles. This topic has been reviewed by Cesarsky & Montmerle (1983) but according to them only a small fraction of the overall galactic cosmic rays can be generated by these stellar winds.

The grammage is energy dependent (due to the energy dependent collisional cross sections) and at an energy of 5 GeV typical values are $X \simeq 7\,\mathrm{g\,cm^{-2}}$ for the traversed interstellar material. However, Prishchep & Ptuskin (1975) propose a galactic halo to reconcile the cosmic ray age determinations via measurements of unstable nuclei such as ^{10}Be with the observed grammage derived from the observed primary to secondary ratio.

The spectrum for all particles are similar power laws in the energy range between about 10^{10}eV/nuc and 10^{15}eV/nuc and from observations at the earth the differential spectrum of primary Galactic cosmic rays is given by

$$dN(E) \propto E^{-2.75 \pm 0.1} dE \tag{28}$$

(e.g. Simpson 1983) where E denotes the particle energy. A clear break is visible in the spectrum at an energy of about 10^{15}eV/nuc and at higher energies the power law spectrum steepens indicating an increasing contribution from extragalactic sources. Up to the mentioned particle energy the cosmic rays obey a large degree of isotropy which decreases for higher energies. This observational fact suggests again that extragalactic particles become progressively more prominent at energies above 10^6 GeV. On the other hand from the small observed anisotropy for the galactic energies we can infer that a large number of sources generates and accelerates cosmic rays and/or that the ISM provides a considerable amount of scattering.

Following the above statements I will restrict the subsequent discussion to the so-called galactic component of cosmic rays characterized by particles energies less than 10^{15}eV/nuc. Note that as already stated in Eq. (2) the energy density of these particles is of the order of

$$U_{\mathrm{CR}} \simeq 1\,\mathrm{eV\,cm^{-3}}. \tag{29}$$

The mean particle age is again energy dependent and can be inferred from the radioactive decay of the secondary nuclei, e.g. ^{10}Be. For low energy cosmic rays ($\lesssim 1$ Gev/nuc) the average age is about $\tau_{\mathrm{CR}} \simeq 2\,10^7$yr (Garcia-Muñoz et al. 1977). Writing the grammage like $X \simeq c\bar{\rho}\tau_{\mathrm{CR}}$ an estimate can be obtained for the mean galactic density of $\bar{\rho} \simeq 0.22$ H-atoms cm^{-3} seen by these energetic particles. Since the observed mean density in our galaxy within a vertical scale height of ± 100 pc is about 0.4 H-atoms cm^{-3} (Burton & Liszt 1981) the cosmic ray storage (or confinement) volume $V_{\mathrm{gal,CR}}$ will be roughly given by $H_{\mathrm{CR}} \simeq \pm 200$ pc times the galactic disc radius of 15 kpc.

The chemical analysis of terrestrial deep sea sediments and meteorites suggests that the cosmic rays intensity near the solar system has not varied more than a factor of 2 to 3 over the last few 10^8 to 10^9 years (e.g. Honda 1979). Due to the finite life time of cosmic rays in our Galaxy a certain power is needed to maintain the observed cosmic ray energy density. This can be estimated by taking the mean storage volume and the mean life time of cosmic rays in our galaxy, i.e.

$$H_{\mathrm{CR}} \simeq 400\,\mathrm{pc}, \quad D_{\mathrm{CR}} \simeq 15\,\mathrm{kpc}, \quad V_{\mathrm{gal,CR}} = \pi H_{\mathrm{CR}} D_{\mathrm{CR}}^2 \simeq 10^{67}\mathrm{cm}^{-3} \quad (30)$$

where H_{CR} and D_{CR} denote the mean galactic scale height and the mean disk area of cosmic rays. Combining the last two equations a simple estimate leads to the cosmic ray luminosity of our Galaxy

$$\bar{L}_{\mathrm{gal,CR}} \simeq \int_{V_{\mathrm{gal,CR}}} \frac{U_{\mathrm{CR}}}{\tau_{\mathrm{CR}}}\, dV \simeq 4\,10^{40}\,\mathrm{erg\,s}^{-1} \quad (31)$$

Since SN-explosions are the most powerful events in our Galaxy it is worth to compare the mean energy release by SNe to this number. Assuming a SN-rate of $\tau_{\mathrm{SN}} = 1/30\,\mathrm{yr}^{-1}$ and a typical explosion energy of $E_{\mathrm{SN}} = 10^{51}\mathrm{erg}$ the cosmic ray luminosity relates to the SN-luminosity by

$$\bar{L}_{\mathrm{gal,CR}} \simeq 0.04\,\bar{L}_{\mathrm{SN}} = 0.04\,E_{\mathrm{SN}}\tau_{\mathrm{SN}}. \quad (32)$$

Following this conservative estimate (e.g. Blandford 1988) it is clear that if only $0.04\,E_{\mathrm{SN}}$ is transferred to high energy particles, this amount of energy is sufficient to explain the observed cosmic ray energy density. However, the mean energy released per SN-explosion can be smaller than $10^{51}\mathrm{erg}$. The observed energy dependence of the grammage, i.e. $X(p) \propto p^\alpha$ and $\alpha \approx 0.6$ (Engelmann et al. 1985) derived from the primary to secondary ratio increases the estimate of Eq. (32) by a factor up to about 10 depending on assumptions on the spectral index of the primary cosmic rays and on α (Drury et al. 1989). Thus if SNRs are the main source of galactic cosmic rays both effects enhance the value of $0.04\,E_{\mathrm{SN}}$ and consequently, the process of diffusive shock acceleration has to transfer as an upper limit about

$$\bar{L}_{\mathrm{gal,CR}} \simeq 0.3\,\bar{L}_{\mathrm{SN}}. \quad (33)$$

into high energy particles. Employing these numbers it becomes clear that cosmic rays have to influence the dynamics of SNR evolution because the shock structures are modified by the nonlinear acceleration process (cf. Sect. 2.5).

1.4.2 Influence of Cosmic Rays on the ISM Dynamics

We can distinguish two main aspects of cosmic ray influence on dynamical effects in the ISM. First, changes which are related to a modification of the shock structure itself if particles are significantly accelerated, and secondly, a different behaviour of large scale flows due to the existence of a relativistic fluid exerting an additional pressure. The first part is discussed in Sect. 1.4.3

and hence, I briefly mention several other aspects of cosmic ray interactions occurring in the ISM (see also Völk 1987).

From the above considerations it should be clear that particle acceleration can modify the evolution of SNR's since the kinetic energy dissipated in the shock waves is shared between the thermal plasma and the cosmic rays. As a consequence the amount of X-ray emitting gas is different compared to the case where no particles are accelerated (cf. Sect.2.7).

As discussed in Sect. 1.2 cosmic rays provide the most important source of heating and ionizing the dense parts of interstellar clouds. Beside this energy deposition they play a crucial role in driving the chemistry of diffuse clouds which is responsible for the formation of a large number of complex molecules (see e.g. the review by Dalgarno 1987).

Since the work of Parker (1966) it is well established that cosmic rays and magnetic fields play an important role in a possible hydrostatic equilibrium perpendicular to the galactic plane where the gravitational acceleration is balanced by the gas pressure, the magnetic pressure and the cosmic ray pressure. The magnetic field is parallel to the galactic plane (cf. also the observations of the magnetic field in Sect. 1.5.3) but such a configuration is unstable if the scale of the perturbation exceeds a certain limit (Parker 1966). The field lines are compressed in one region where the matter can slide down along the magnetic field reaching a new equilibrium and forming clouds and subsequently OB-associations separated by regions of field lines extending to larger galactic latitudes (Mouschovias et al. 1974). The final states of this Parker instability have been calculated by Mouschovias (1976). The problems of maintaining a stable vertical configuration in the presence of numerous energy releasing phenomena has been reviewed e.g. by Stephens (1991).

Recently a number of models driving a galactic wind by cosmic rays and the dissipation of Alfvén waves have been constructed (Breitschwerdt et al. 1990) showing the importance of this relativistic fluid with $\gamma_c = 4/3$ acting on the thermally lifted gas. A more detailed description is placed at the end of this chapter in Sect. 1.6.2.

The cosmic rays collide with the thermal plasma producing neutral pions which decay into observable γ-rays above $h\nu \gtrsim 100\,\mathrm{MeV}$. This process provides a natural explanation for the good correlation of the gas density with the diffuse γ-rays (Kniffen et al. 1977) if the cosmic ray flux at any point is proportional to the gas density. This can be justified by arguing that the cosmic rays are constrained by the magnetic fields which themselves are linked to the interstellar gas. The diffuse γ-emission is then proportional to the line of sight integral of the mass density squared (i.e. the column density squared) and consequently, the accumulation of matter in clouds yields a higher γ-luminosity. However, different groups arrive at different conclusions about the radial distribution of molecular hydrogen in our Galaxy, i.e. the so-called molecular ring at $4-7\,\mathrm{kpc}$ from the galactic center (see e.g. the recent review of Bloemen 1989). Nevertheless it seems clear that γ-rays produced by cosmic ray interactions are a very useful tool to probe the large scale structure of the galactic H_2-distribution.

All radio observations of SNR's are based on the synchrotron emission of relativistic electrons gyrating around magnetic field lines (see Sect. 1.5.2 for actual magnetic field values). Since the acceleration mechanism in shock waves presented above works also in the case of electrons this large number of data is a very strong although indirect indication that also protons or ions can be accelerated in SNR's. At this point I can mention measurements on the shock acceleration of particles in a much less energetic environment, namely the earth's bow shock (e.g. Kennel et al. 1985) which are in nice agreement with the predictions from diffusive particle acceleration (Lee 1982).

This short list of cosmic ray interactions (also included in the overview, Fig. 1) stresses again the fact that any realistic modelling of the ISM has to include cosmic rays as well as magnetic fields. On the other hand the subsequent considerations will show that such a task requires rather cumbersome numerical simulations to account for the nonlinearities involved in all these interactions.

1.4.3 Diffusive Shock Acceleration and Cosmic Ray Hydrodynamcis

In the so-called test particle picture where the reaction of the energetic particles on flow is neglected the acceleration process leads to a simple power law (e.g. Drury 1983, or Achterberg this Volume). A schematic picture of a shock wave is shown in Fig. 5 where the particles are scattered between the downstream and the upstream regions gaining energy at each passage of the shock front. The pitch-angle scattering occurs at magnetic fluctuations moving with the Alfvén velocities (27) $v_{A,1}$ and $v_{A,2}$, respectively. Note that $v_{A,2} = \sqrt{r}v_{A,1}$ where r denotes the compression ratio. In a shock wave with a downstream velocity u_1 and an upstream velocity u_2 we have

$$r = \frac{u_1}{u_2}, \quad q = \frac{3r}{r-1} \tag{34}$$

and the solution depends only on the compression ratio r and the particle distribution function $f(p)$ in momentum p is given by

$$f(p) \propto p^{-q}. \tag{35}$$

A detailed discussion including solutions with injection of particles, wave fields and different upstream particle distributions can be found e.g. in Drury (1983) and therefore will be omitted here.

Because of their relevance to the subsequent SNR evolution I mention the nonlinear modifications occurring in diffusive particle acceleration if the backreactions of the energetic particles on the thermal plasma and on the

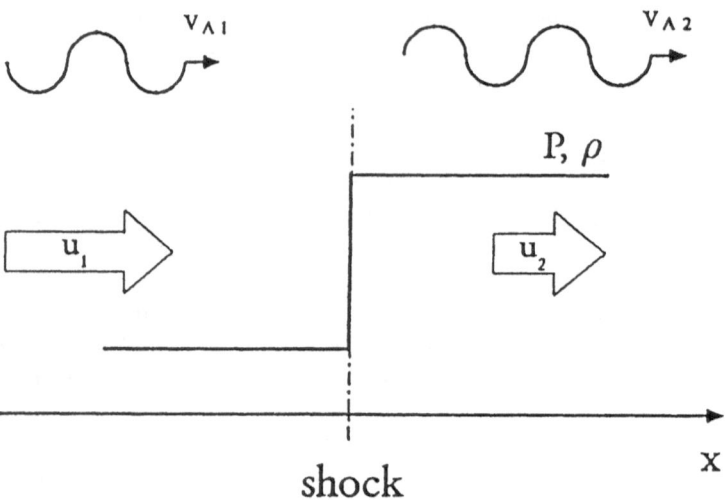

Fig. 5. Particle acceleration in shock waves without backreaction, i.e. the test-particle limit

magnetic field are included. We start with a description of a time independent shock structure including cosmic ray acceleration (Drury & Völk 1981). The Mach-number M and the ratio N of the cosmic ray pressure P_c to the total pressure in a flow of velocity u and density ρ can be defined by

$$M^2 = \frac{\rho u^2}{\gamma_c P_c + \gamma_g P_g}, \quad N = \frac{P_c}{P_c + P_g}. \tag{36}$$

The energetic particles accelerated at the shock develop a gradient into the upstream region through diffusion slowing down the incoming plasma already before the shock front is encountered. This so-called precursor has a length scale associated with the mean cosmic ray diffusion coefficient $\bar{\kappa}$ and the plasma inflow velocity u_1, i.e. $\bar{\kappa}/u_1$ (cf. Fig. 6). For moderate Mach numbers $M \lesssim 6$ a pure plasma subshock is still embedded in the structure where the transitions of the particle pressure P_c as well as of the cosmic ray flux F_c remain continuous across the whole shock front. In the case of strong shocks ($M \gtrsim 6$) even smooth transitions are possible where the thermal plasma is only compressed adiabatically. Taking fully relativistic particles, i.e. $\gamma_c = 4/3$ the compression ratio increases to $r = 7$ and in a totally smoothed shock front the amout of kinetic energy dissipated to the thermal plasma is reduced to $r^{-2} = 7^{-2} = 0.02$ meaning that 98% of the shock energy is dumped into cosmic rays. However, in a certain parameter regime of the M, N-plane (N, M defined by Eq. 36) the downstream solution is not uniquely determined by the upstream values and the Mach number (e.g. Drury 1983). Nonlinear shock structures in the presence of the scattering field and accelerated particles have been constructed by Völk et al. (1984) showing that the waves in strong shocks grow to large amplitudes already in the precursor region of the shock but this (formal) conclusion is based on a linear perturbation analysis of the wave field (McKenzie & Völk 1984). Up to now it is not clear how the system will react in such a situation but nonlinear damping mechanisms like Landau damping (Völk et al. 1984) may limit the amplitudes of the waves.

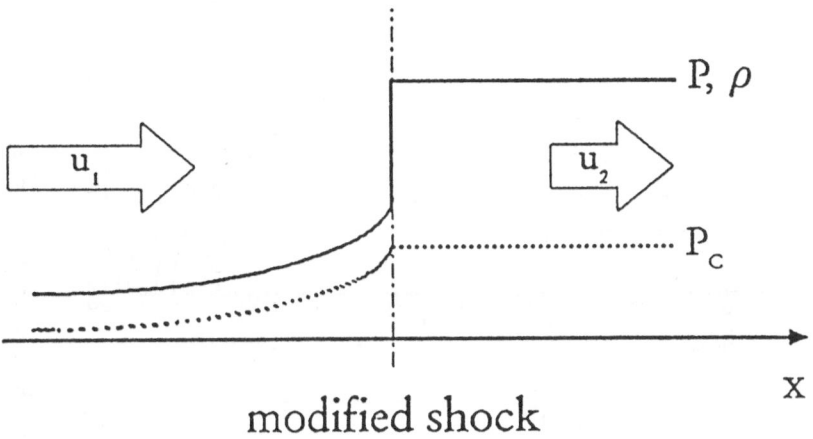

Fig. 6. Schematic view of nonlinear diffusive particle acceleration in a shock wave including the precursor region and a pure gas subshock

The question how the energetic particles modify the shock structure can be approached by treating the accelerated particles like a hydrodynamical fluid. In this case the cosmic ray pressure P_c or the energy density E_c are incorporated in the hydrodynamical description of the thermal plasma and these quantities are then defined by appropriate moments of the particle distribution function $f(p)$. The resulting equations are presented in Sect. 2.5.

Such a description of *cosmic ray hydrodynamics* (see Sect. 2.5.2) contains only hydrodynamical quantities which may be difficult to determine (e.g. Drury et al. 1989). However, I will use this set of equations throughout the following sections to take into account the nonlinear modifications in shock waves. The propagation of small disturbances in this two-fluid system has been investigated by Ptuskin (1981) who shows that very long wavelength waves couple to the cosmic rays and travel at an enhanced speed of $[(\gamma_c P_c + \gamma_g P_g)/\rho]^{1/2}$ used also in the definition of the Mach-number through Eq. (36). Short wavelength disturbances decouple from the cosmic rays and propagate at the ordinary gas sound speed of $(\gamma_g P_g/\rho)^{1/2}$.

The time-dependent evolution of a plane shock wave accelerating particles is displayed in Fig. 7 where the flow is reflected at the left boundary

68

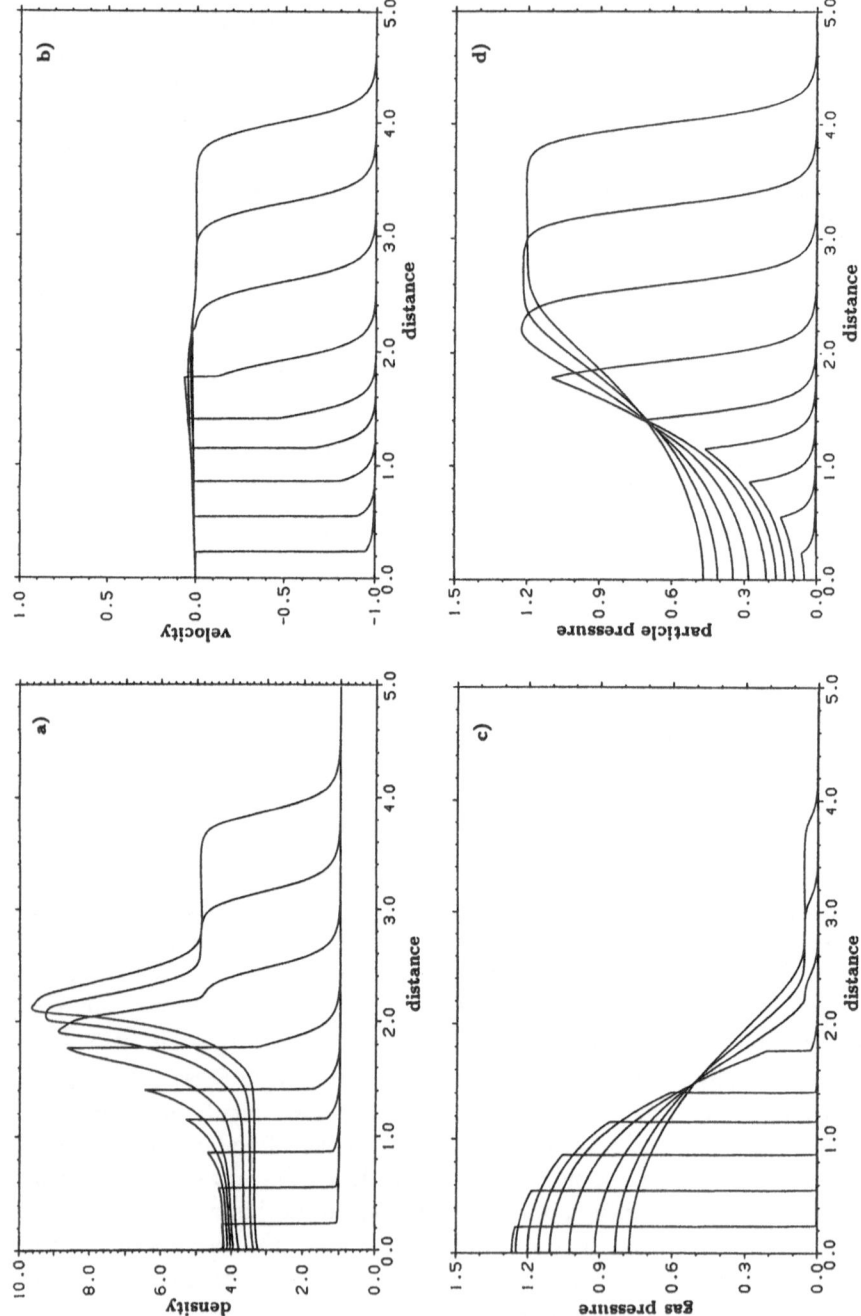

Fig. 7. Time dependent particle acceleration in a plane shock wave with $N = 0.5$ and $M = 11.2$ travelling from left to right.

sending a shock wave to the right. The parameters used fix the Mach-number at $M = 11.2$, the upstream pressure ratio of $N = 0.5$ and the adiabatic index of cosmic rays of $\gamma_c = 3/2$. The time is given in units of the diffusion time scale $t_d = \bar{\kappa}/u_s^2$ and u_s denotes the shock speed. After about $8\,t_d$ the time asymptotic values are reached showing a smoothed shock propagating to the left and compressing the thermal gas only adiabatically (cf. Fig. 7c). Almost all kinetic energy dissipated at the shock goes into cosmic rays which build up the pressure P_c in downstream region. Note that during the transition from a gas dominated shock wave to a cosmic ray dominated one the overall compression ratio can become rather large (up to 10 seen in the density, cf. Fig. 7a) because we have a precursor region where the incoming gas is decelerated and compressed and on top of that flow a subshock enhances the total compression. In addition we see motions towards the shock wave in the downstream region caused by the diffusive decrease of the cosmic ray gradient which is compensated by a flow pushed by the remaining gradient of the thermal gas. These small positive velocities (cf. Fig. 7b in the region $1 \lesssim x \lesssim 2$) are maintaining the pressure equilibrium in the downstream region. The late evolution can be compared to the time asymptotic analytical solutions of Drury & Völk (1981). The final compression ratio is $r = 6$ in agreement for strong shocks with $\gamma_c = 3/2$.

We emphasize that such transitions should also occur during the evolution of SNRs if particle acceleration becomes so efficient that most of the downstream pressure is provided by the energetic particles and not by the thermal plasma. In Sect. 2.5 we will discuss such remnants in more detail where most of the SN-explosion energy is transferred to cosmic rays leading to significantly lower gas temperatures compared to the cases without important particle acceleration.

1.5 Magnetic fields

As mentioned in the previous section cosmic rays and magnetic fields are linked together and I start this chapter by summarizing some observational facts. Basic physical effects exist to measure either the magnetic field strength $B_\parallel$ parallel to the line of sight or in the perpendicular direction, i.e. $B_\perp$.

1.5.1 Measurements of $B_\parallel$

Without going too much into details magnetic fields lead to a splitting of atomic levels and this pattern can be observed if the transition has a sufficiently large Landé factor g. This is the case for the hydrogen atom where Zeeman splitting in the 21 cm line has been found first in different galactic directions by Verschuur (1969). The other case of detectable Zeeman-splitting is the OH-molecule which is well suited to probe the magnetic field strength in dense clouds by maser emission. A magnetic field component $B_\parallel$ parallel to the line of sight leads to a wavelength displacement $\Delta\lambda$ given by the relation

$$\Delta\lambda = \frac{ge}{2\pi m_e c^2}\lambda^2 B_\parallel. \tag{37}$$

Note that the Zeeman splitting samples the magnetic fields in the cold matter of our galaxy.

A linear polarized wave changes its angle of polarization due to Faraday rotation. The different propagation speed for left- vs. right-hand side circular polarized radiation is related to the direction of the gyration of electrons causing a different index of refraction. Thus the angle of rotation Φ of linear polarized radiation is a function of the wavelength λ defining the so-called rotation measure RM, i.e.

$$\Phi = \text{RM}\lambda^2 + \Phi_0, \quad \text{RM} = 0.81 \int n_e B_\parallel dl \tag{38}$$

where n_e is given in cm^{-3}, $B_\parallel$ in μG and the integral along the distance to the source in parsecs. In the case of pulsed emission the time a signal reaches the detector again depends on frequency leading to a so-called dispersion measure

$$\text{DM} = \int n_e dl \tag{39}$$

which averages the electron density n_e along the line of sight. The ratio RM/DM provides a direct measurement of the parallel field strength $B_\parallel$ along the line of sight averaged over the electron density. In the direction of the Crab pulsar this quantity is $\text{DM} = 56.9\,\text{pc}\,\text{cm}^{-3}$ (Manchester 1972) which for a source distance of $d = 2\,\text{kpc}$ gives a mean electron density of $n_e = 0.028\,\text{cm}^{-3}$. Taking the average electron density to be $\langle n_e \rangle \simeq 0.03\,\text{cm}^{-3}$ leads to a rule of tumb for the mean pulsar distance (e.g. Lyne 1990)

$$d \simeq \frac{\text{DM}}{30}\,\text{kpc}. \tag{40}$$

A large number of RM/DM measurements have been used to determine the structure of the magnetic field (Manchester 1974, Hamilton & Lyne 1987) indicating a local field in the spiral arm of about $B_\parallel \simeq 2 - 3\,\mu\text{G}$ directed towards $\ell = 90°$ as well as a field reversal around $\ell = 50°$. In addition there exists a random field of the same magnitude having a scale of about $100\,\text{pc}$. Note that the RM and DM is produced in the medium with an important contributions of the electron density n_e along the line of sight, i.e. mainly in the warm ionized medium, the WIM already introduced in Sect. 1.2.

1.5.2 Measurements of $B_\perp$

Historically the polarization of starlight is the first observational fact which has been used to deduce the existence of a magnetic field revealed by an alignment of non-spherical dust grains along the magnetic field lines. The physical mechanism of alignment of dust grains against the bombardment of gas particles can be found in Purcell (1979) and is based on the rare ejection of H_2-molecules which carry large kinetic energies and which are formed on the surface of these dust grains. The observed linear polarization leads to an estimate of the strength or at least of the orientation of the perpendicular component $B_\perp$. An example of the overall magnetic field structure of the dark cloud near Rho Ophiuchus is given in Vrba et al. (1976) showing a nice alignment of the magnetic field with the filaments. These filaments are directed towards more dense blobs with a star forming region inside and the vectors in these blobs tend to be perpendicular to the filaments. The magnetic field strength in these blobs is of the order of about $10\,\mu G$.

Synchrotron emission of relativistic electrons gyrating around magnetic field lines produces linear polarized radiation parallel to the direction of the acceleration (Ginzburg & Syrovatskii 1965). Thus the polarization is perpendicular to the magnetic field leading to a polarization perpendicular to $B_\perp$. In a uniform field the degree of polarization can be rather high, i.e. up to 75% but in all astrophysical situations we have to add a random component to the magnetic field decreasing the observable polarization.

Table 4. Magnetic field properties of spiral galaxies as obtained from radio polarization observations

Galaxy	Typ	d [Mpc]	$\langle B \rangle$ [μG]	structure
M31	SbI-II	0.7	4 ± 1	ASS
M33	ScdII	0.7	5 ± 2	BSS?
NGC253	Sc(p)	2.5	13 ± 4	?
M81	SabI-II	3.2	8 ± 2	BSS
IC342	ScdI-II	4.5	7 ± 2	ASS
M51	SBcI	6.0	11 ± 3	BSS

ASS: axisymmetic spiral structure
BSS: bisymmetric spiral structure

A large number of observations can be found in the proceedings of the IAU Symposium on 'Galactic and Intergalactic Magnetic Fields (Beck et al. 1990) including detailed measurements of magnetic fields in different SNR's like the Crab Nebula with $B \simeq 0.3\,mG$ (Kennel & Coroniti 1984) or Kepler's SNR with $B \simeq 70\,\mu G$ (Matsui et al. 1984). A number of these remants exhibit a radial field but the magnetic field in the Cygnus Loop is tangential and van der Laan (1962) deduces a compression by a factor of 4 from the synchrotron emission (cf. also next section).

On the galactic scales this physical effect can be used over a wide range of radio frequences to determine the magnetic field structures of external galaxies showing also a large number of galaxies with rather ordered field structures. A few results are summarized in Table 4 (adopted from Krause 1990) to illustrate possible field configurations of our Galaxy. According to present galactic dynamo theories (cf. next section) the dynamo can generate most easily either an axisymmetric spiral structure (ASS) in its fundamental mode ($m = 0$) or a bisymmetric spiral structure (BSS) through the first excited mode ($m = 1$).

1.5.3 Remarks on galactic magnetic fields

Summarizing the observational facts a large scale ($L \sim 1\,\mathrm{kpc}$) ordered field permeates the Galaxy and a clear tendency for alignment along the spiral arms can be inferred from the data. In the vicinity of the solar system the magnetic field has a strength of $B_0 \simeq 1.6\,\mu\mathrm{G}$. However, a small scale ($1\,\mathrm{pc} \lesssim L \lesssim 100\,\mathrm{pc}$) irregular (turbulent) component is also visible having a typical average field strength of about $B_{\mathrm{gal}} \simeq 4\,\mu\mathrm{G}$. Adopting this value the corresponding energy density results in

$$U_{\mathrm{mag}} = \frac{B_{\mathrm{gal}}^2}{8\pi} \simeq 0.4\,\mathrm{eV\,cm^{-3}}. \tag{41}$$

However, as seen in the OH-data much larger field strengths occur at certain places within the ISM because the magnetic field can be amplified by several events.

Magnetic fields are compressed in shock waves. If the index 1 denotes the upstream quantities and 2 the downstream values the condition of flux freezing (49) yields

$$\frac{B_1}{\rho_1} = \frac{B_2}{\rho_2} \tag{42}$$

and in the case of a strong shock with an adiabatic index of $\gamma = 5/3$ the downstream field is

$$B_2 = 4\,B_1 \quad \text{but} \quad P_2 \gg \frac{B_2^2}{8\pi}, \tag{43}$$

i.e. the downstream thermal pressure P_2 usually exceeds the magnetic pressure.

Magnetic fields are also amplified by contraction and if we assume flux freezing (cf. Eq. 49) mass and magnetic flux conservation in a spherical symmetric contraction from initial values R_i, ρ_i and B_i to final values R, ρ and B yields

$$M \propto R_i^3 \rho_i = R^3 \rho$$
$$\Phi \propto B_i R_i^2 = BR^2$$
(44)

which relates the magnetic field strength B to the density ρ

$$\frac{B}{B_i} = \left(\frac{\rho}{\rho_i}\right)^\kappa$$
(45)

where in the spherical case $\kappa = 2/3$. However, this exponent κ depends strongly on the adopted geometry. If rotation inhibits a spherical collaps the magnetic field increases more rapidly, i.e. $2/3 \lesssim \kappa \lesssim 1$ (e.g. Mestel 1977). In the case where the matter can slip along the field lines κ decreases and from a sequence of equilibrium configurations Mouschovias (1976) deduces

$$B \propto \rho^{1/2}$$
(46)

which is also in good agreement with the observations of magnetic fields inside dense clouds. According to Troland & Heiles (1986) magnetic field strengths obtained for OH Zeeman splitting are $9\,\mu G$ for Cas A, $11\,\mu G$ for W22, $38\,\mu G$ for Ori B and $120\,\mu G$ for Ori A. In the cases of OH-maser emission located in regions of densities around 10^6cm^{-3} Lo et al. (1975) find for six masers magnetic fields between $2.5\,\text{mG}$ and $9.0\,\text{mG}$. Thus, magnetic fields of the order of mG are typical for very dense cores of interstellar clouds where Eq. (45) gives the right magnetic field amplification if $\kappa \simeq 1/2$.

The large scale evolution of magnetic fields (e.g. Parker 1979) can be described by an equation obtainable from Ohm's law in a conducting fluid

$$\mathbf{j} = \sigma \left(\mathbf{E} + \frac{1}{c} \mathbf{u} \times \mathbf{B} \right)$$
(47)

and Maxwell's equations yielding in the MHD-approximation (invariant under Galilei transformations)

$$\frac{\partial \mathbf{B}}{\partial t} = \nabla \times (\mathbf{u} \times \mathbf{B}) + \frac{c^2}{4\pi\sigma} \nabla^2 \mathbf{B}$$
(48)

where σ denotes the conductivity. Depending on the value of σ the magnetic field is either 'frozen' in the flow ($\sigma \to \infty$) or is characterized by a diffusion equation in the case without motions ($\mathbf{u} = 0$). The concept of frozen-in magnetic fields can be illustrated by defining the magnetic flux Φ through a closed material curve C

$$\frac{d\Phi}{dt} = - \oint_C \frac{c}{\sigma} \mathbf{j} \cdot d\mathbf{l}.$$
(49)

which remains constant in a perfectly conducting fluid, i.e. $\sigma \to \infty$. In this case we can also combine the equation of mass conservation with (48) and get

$$\frac{D}{Dt}\left(\frac{\mathbf{B}}{\rho}\right) = \frac{\mathbf{B}}{\rho} \cdot \nabla \mathbf{u} \tag{50}$$

showing that $\mathbf{B}/\rho$ statisfies the same equation for a moving line element. For example, in a spherical compressive flow $\mathbf{u} = (\alpha r, 0, 0)$ with $\alpha < 0$, in polar radial coordinates any line element decreases linearly in r while the density of a fluid element at position $r(t)$ increases as r^{-3}. Hence, the magnetic field $\mathbf{B}$ flowing a fluid element will increase as r^{-2} in accordance with the results obtained by Eq. (45).

Equation (48) has interesting consequences on the large scale evolution of plasmas since the magnetic field can link spatially distinct regions exerting forces on them which lead to a transport of angular momentum by Alfvén waves. Such a process is called magnetic braking (Ebert et al. 1960, Gillis et al. 1974, 1979, Mouschovias & Palaleogou 1979, 1980) and is able to remove the angular momentum from a contracting interstellar cloud. Without the loss of angular momentum the centrifugal forces inhibit the formation of dense cores and this so-called angular momentum problem (Mestel 1965) can be solved by magnetic fields. However, at some stage of star formation magnetic fields cannot remain frozen in the fluid because Eqs. (45) and (46) lead to an amplification of the magnetic field which is not observed in stars. On the other hand the theory has to explain why so little flux treading the original cloud has survived the process of star formation. Ambipolar diffusion (Mestel & Spitzer 1956) due to a drift between ions and neutrals lowers the initial flux through the cloud. The exact value of the critical particle density n_c when the field decouples from the matter is still controversial (Nakano 1984, Mouschovias et al. 1985) and ranges between $10^4 \mathrm{cm}^{-3}$ and $10^{11} \mathrm{cm}^{-3}$. Note that analytical solutions in simple symmetries of (48) show oscillations which can provide a natural explanation for retrograde rotating objects if the field can decouple at the time of such occurrences.

Figures 8a,b shows the results of full 3D-MHD calculations (Dorfi 1989) where the evolution of the total specific angular momentum J/M_{cl} of a cloud with mass M_{cl} and radius R_{cl} is plotted as a function of the dimensionless external Alfvén crossing time $R_{\mathrm{cl}}/v_{\mathrm{A,ext}}$. The three curves correspond to different inclinations between the axis of rotation and the magnetic field of $3\,\mu\mathrm{G}$, i.e. $\theta = 0°$ shown by the full curve, $\theta = 45°$ by the dashed one and $\theta = 90°$ by the dash-dotted one. They demonstrate that the typical braking time scales may differ only by a factor of 3. Note that already after a free-fall time $t_{\mathrm{ff}} \propto (G\rho_{\mathrm{cl}})^{-1/2}$ the initial homogeneous cloud develops very non-homologous structures under the joint action of gravity, rotation and magnetic fields. Such structures are typical for the observed interstellar clouds. An inclined magnetic field has an additional precessional torque (cf. Fig. 8b) which tends to align the angular momentum vector along the magnetic field direction and depending on the time when the field decouples from the matter almost any final angular momentum direction can be achieved. Fig. 8b depicts the evolution of the three components (J_x, J_y, J_z)

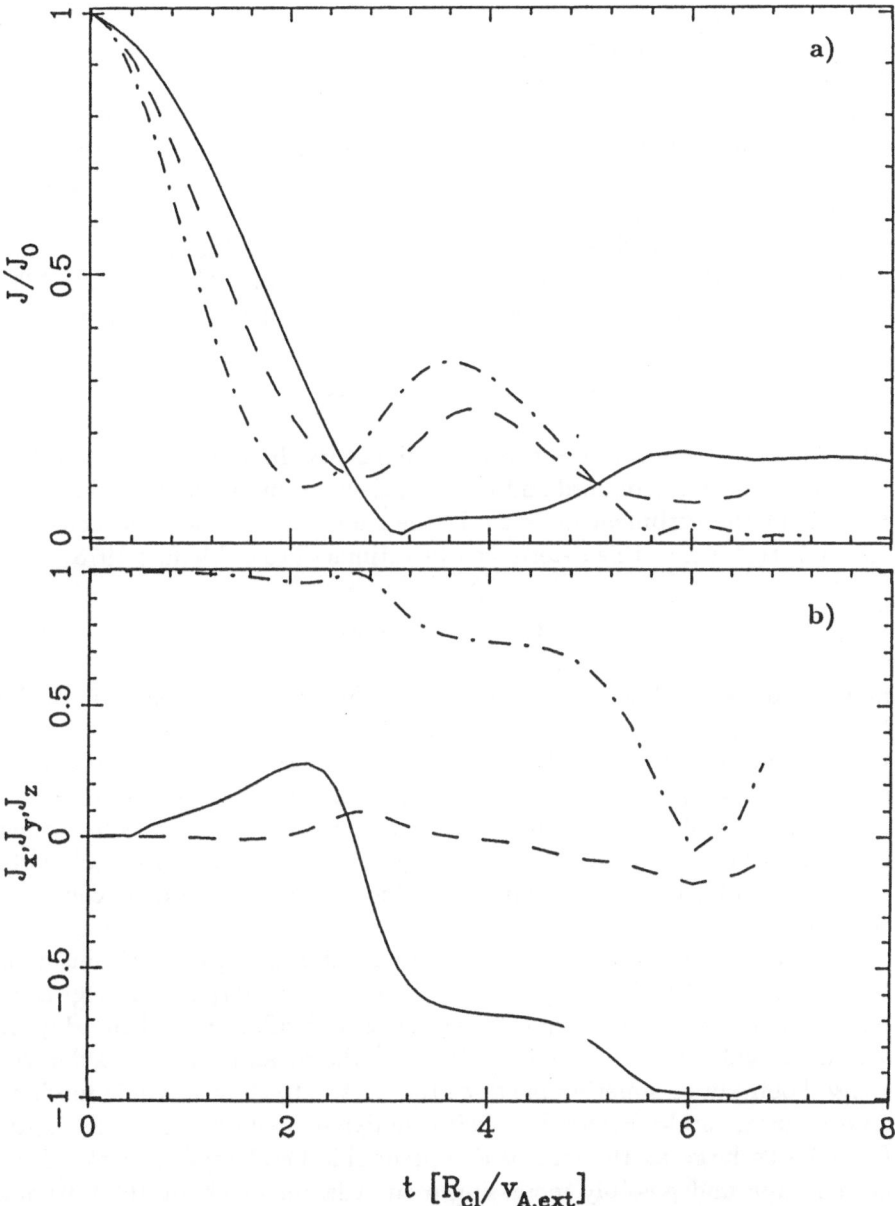

Fig. 8. a) The temporal evolution of the total specific angular momentum J in units of J/J_0 plotted against the Alfvén crossing time. The three curves correspond to different angles between the rotation axis and the initial magnetic field. b) the evolution of the three vector components (J_x, J_y, J_z) in the case of $\theta = 45°$

(full, dashed, dash-dotted) of the angular momentum vector in the inclined case of $\theta = 45°$. I have elucidated this example in more detail because the inclusion of magnetic fields introduces a number of new effects making important differences compared to a theory neglecting magnetic fields. This also results in a much better agreement with observations. A more thorough discussion of the interaction of magnetic fields with interstellar clouds can be found e.g. in Mestel (1990 and references therein).

In the case of small scale turbulent motions Eq. (48) has to be modified according to the so-called mean-field electrodynamics (Krause & Rädler 1980) which is obtained by averaging over the fluctuating component

$$\frac{\partial \mathbf{B}}{\partial t} = \nabla \times (\mathbf{u} \times \mathbf{B}) + \nabla \times (\alpha \mathbf{B}) + \eta \nabla^2 \mathbf{B} \qquad (51)$$

where η includes now also a turbulent diffusivity. In addition the so-called α-term has been introduced and the coefficients α and η 'hide' the information about the turbulent motions. In the simple case of homogeneous and isotropic turbulence this α-term can be estimated (e.g. Moffatt 1978)

$$\alpha = \frac{1}{3} \mathbf{u} \cdot (\nabla \times \mathbf{u}) \qquad (52)$$

and $\alpha \neq 0$ means that the flow has a finite helicity. This α-effect is at the heart of all modern dynamo theory. If we think of the mean field the α-term will generate a current parallel or antiparallel to $\mathbf{B}$ which can result in an enhancement of the mean field depending on the preferred sense of rotation. At this point I also want to mention that since flow lines can be regarded as directed lines a connection between dynamo theory and theory of knots has been established (e.g. Chap. 6 of Zeldovich et al. 1983 and references therein).

An estimate on the α-effect in the case of galaxies is given by Vainshtein & Ruzmainkin (1972) taking $\alpha \simeq l^2 \omega / H$ where $l \simeq 100\,\mathrm{pc} \simeq H$ (e.g. Burton & Liszt 1981) is the typical scale of an individual turbulent eddy in a confined galactic disk and $\omega \simeq 10^{-15}\mathrm{s}^{-1}$ the mean galactic angular velocity. Fountain or superbubble flows (cf. Sect. 1.6) show cyclonic motions on even larger scales generating their own dynamo effect. The scale length l can be as large as the size of a superbubble breaking out of the disk, i.e. $l \gtrsim \mathrm{kpc}$ and possibly increasing α by a factor of about 10. Without going into the details of dynamo theory I sketch a few basic features important for the magnetic field generation on a galactic scale. The observed regular properties of the galactic magnetic field (cf. Sect. 1.5.1) as well as of extragalactic fields (cf. Table 4) suggest dynamo action with the main modes $m = 0$ (axisymmetric, ASS) or $m = 1$ (bisymmetric, BSS) excited (Ruzmaikin et al. 1988). They also show that larger deviations from solid body rotation will decrease the growth rates of non-axisymmetric modes. However, up to now none of the models for galactic dynamos takes into account the spiral structure, tidal disturbances of the disk or the disk-halo connection (cf. Sect. 1.6.1). The basis of current dynamo theories is a purely kinematical point of view, i.e. the forces of the magnetic field on the flowing matter are neglected. Beside the large scale fields the small scale fields play an important role for the ISM, e.g. for the propagation of cosmic rays and their confinement or for star formation, support and stability of magnetized clouds.

1.6 Further remarks

In recent years the evidence has grown that the understanding of the ISM is also closely connected to the halo dynamics. Hence, this section is devoted to processes going beyond the galactic plane. A simple model for rising hot gas to a galactic corona at kiloparsec distances considers the repeated SN-explosions in young stellar groups (Chevalier & Gardner 1974). Such so-called superbubbles are discussed in more detail in the following section. If this rising gas is confined to the Galaxy the material will circulate and can fall back as cooler clumps to the galactic plane in the case a thermal instability operates in the halo. Such a scenario has been called a galactic fountain (Shapiro & Field 1976). However, if the gas is able to leave the Galaxy driven e.g. by cosmic rays and/or dissipation of Alfvén waves we have to deal with an important sink for the galactic ISM and several aspects of these galactic winds are discussed in Sect. 1.6.2.

1.6.1 Supershells

Most stars are born in clusters and become visible in the optical range as so-called OB-associations (e.g. Gies 1987). Since more massive stars evolve more rapidly than low mass stars the OB-association still exists when the first supernova (SN) explodes. Therefore SNe are not distributed randomly over our entire galaxy but a certain degree of clustering must be taken into account. Such arguments are favoured after giant HI-shells (also called 'worms' or 'chimneys') have been observed in the Milky Way (Heiles 1979, 1984). The typical dimensions run from $L \sim 100$ pc up to $L \sim 1$ kpc, the kinetic energy content is in the range of 10^{50} up to 10^{53} erg and the kinematical ages can be estimated between 10^7 and 10^8 years. Brinks & Shane (1984) have detected 140 holes found in the 21 cm emission of M31 having typical radii between $L \sim 125$ and $L \sim 300$ pc. Similar structures are known in the SMC and M101. Many of these holes contain OB-associations and about 20% have bright HII-rims due to the shorter life time of the brighter O-stars responsible for the ionizing radiation.

The theoretical explanation is based on the work of Weaver et al. (1977) where a model of repeated SN-explosions from OB-associations is developed by analogy with the theory of wind driven bubbles. Stellar evolution calculations and theory of Type II SNe predict that all stars with masses $M \gtrsim 7\,M_\odot$ will explode and release a large amount of energy in the region of such an OB-association. Using stellar evolution calculations the typical life time for explosive energy input is given by the evolution time scale of a $7\,M_\odot$-star, i.e. $\tau_{OB} \simeq 5\,10^7\,\mathrm{yr}$. If the energy released per explosion is denoted by E_{SN} the mean time averaged power can be written as

$$P_{OB} \simeq \frac{N_* E_{SN}}{\tau_{OB}} = 6.3\,10^{35}\,N_* E_{51}\ \mathrm{erg\ s^{-1}} \tag{53}$$

where N_* is the number of stars yielding a SN-explosion.

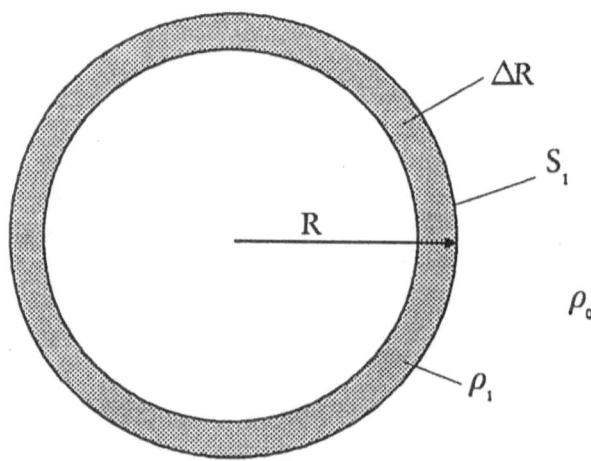

Fig. 9. Geometry of the thin shell approximation

In the case of a constant external density ρ_0 the expansion of such a wind driven bubble can been calculated according to the thin shell approximation, i.e. $\Delta R \ll R$ for all times. However, if a strong shock is assumed with $\rho_1 = 4\,\rho_0$ and if the shell mass equals the swept-up mass, the thickness of the shell can be estimated

$$M = \frac{4\pi}{3} R^3 \rho_0 = 4\pi R^2 \Delta R \rho_1 \quad \text{or} \quad \Delta R = \frac{R}{12}. \tag{54}$$

In this case the shell dynamics follows from Newton's law

$$\frac{d}{dt}\left(\Delta M \dot{R}\right) = F, \qquad \Delta M = \frac{4\pi}{3} R^3 \rho_0, \tag{55}$$

where F denotes the force acting on the shell. Introducing the pressure P the force is $F = 4\pi R^2 P$ and therefore

$$\frac{d}{dt}\left(\frac{1}{3}R^3\rho_0\dot{R}\right) = PR^2.$$ (56)

However, to solve this equation an expression for the interior pressure is needed. Under the assumption of energy conservation and an adiabatic index of $\gamma = 5/3$ the pressure is calculated from Eq. (53)

$$P = (\gamma - 1)E$$

$$P = \frac{2}{3}\frac{N_*E_{\rm SN}t}{\tau_{\rm OB}}\left(\frac{4\pi}{3}R^3\right)^{-1} = \frac{2}{3}P_{\rm OB}t\left(\frac{4\pi}{3}R^3\right)^{-1}$$ (57)

and Eq. (55) results in

$$\frac{d}{dt}\left(\frac{1}{3}R^3\rho_0\dot{R}\right) = \frac{P_{\rm OB}t}{2\pi R}.$$ (58)

This differential equation can be solved making the ansatz $R_{\rm SB}(t) = At^\eta$ getting

$$\eta = \frac{3}{5} \quad \text{and} \quad A = \left(\frac{25P_{\rm OB}}{14\pi\rho_0}\right)^{1/5}.$$ (59)

Inserting these expressions a supershell expands according to

$$R_{\rm SB}(t) = \left(\frac{25P_{\rm OB}}{14\pi\rho_0}\right)^{1/5}t^{3/5}, \quad t_7 = \frac{t}{10^7\text{yr}}$$ (60)

or in physical problem oriented units

$$R_{\rm SB}(t) = 132\left(\frac{N_*E_{51}}{n_0}\right)^{1/5}t_7^{3/5} \text{ pc}.$$ (61)

According to the last equation the typical radius of such a superbubble $R_{\rm SB}$ can be estimated if $N_*E_{51} = 25$ and $n_0 = 1\,\text{cm}^{-3}$ is inserted

$$R_{\rm SB}(t) = 251\ t_7^{3/5}\ \text{pc}$$ (62)

and the mass $M_{\rm SB}$ contained in this superbubble is given by the expression

$$M_{\rm SB}(t) = \frac{4\pi}{3}R_{\rm SB}^3\rho_0 = 9.9\,10^5\ t_7^{9/5}\ n_0\ M_\odot.$$ (63)

The kinetic energy in this shell can be determined and leads to

$$E_{\rm kin} = \frac{1}{2}M_{\rm SB}\dot{R}_{\rm SB}^2 = \frac{1}{2}\frac{4\pi}{3}\rho_0 R_{\rm SB}^3\dot{R}_{\rm SB}^2$$
$$= \frac{3}{7}P_{\rm OB}t,$$ (64)

i.e. 43% of the energy injected by the repeated explosion of SNe goes into kinetic energy of the bubble.

However, this simple theory does not consider important modifications for the late evolution of superbubbles. First, due to the large dimensions (cf. Eq. 62) is is very likely that the bubble will expand bejond the typical galactic scale height and will 'open' the galaxy towards a galactic halo. In this case the galactic gravitational potential $g(z)$ has to be included which results in deviations from spherical symmetry. The variations in the galactic acceleration cannot be neglected since also the external density, i.e. $n_0(z)$ decreases with the distance z from the galactic plane. Secondly, a large numer of interstellar clouds is overrun by the expansion shock wave and consequently, the density inside the superbubble increases through the evaporation of the compressed and stripped interstellar clouds. Thirdly, radiative cooling becomes important which can lead to a fragmentation of the expanding shell. As pointed out by McCray (1987) radiative cooling sets in after

$$t_{\text{cool,SB}} \simeq 4\,10^6 \zeta^{-1.5} \left(N_* E_{51}\right)^{0.3} n_0^{-0.7} \text{ yr} \tag{65}$$

because the interior starts radiating the energy away. ζ denotes the metalicity in solar system units. Around the time of $t_{\text{cool,SB}}$ the expansion is slowed down and as stated earlier the HII-emission dies out when the last O-star explodes in a SN.

From the above considerations it is clear that the radius of these superbubbles increases with galactocentric distance because the mean external density n_0 decreases and some fragments of these giant shell can be named 'worms' or 'chimneys' seen in HI. The structure of the ISM in a galaxy can be dominated by these large bubbles and it becomes plausible to define also a 2-dimensional analogon to the 3-dimensional porosity parameter Q of Eq. (16) by

$$Q_{\text{SB}} = \frac{R_{\text{SB}}}{N_* A_{\text{disk}}} A_{\text{SB}} \tau_{\text{SB}} \simeq 1 \tag{66}$$

where the typical area of a superbubble and their life time is given by

$$A_{\text{SB}} = \pi R_{\text{SB}}^2, \qquad \tau_{\text{SB}} = R_{\text{SB}} \bar{v}_{\text{ISM}}. \tag{67}$$

Note that more advanced models of the ISM also have to consider the galactocentric distance with respect to the type of SNe exploding at certain positions. The central region of a galaxy is dominated by old Type I SNe wheras in the 'molecular ring' region Type II SNe occur more frequently. At this point it should be clear that a galactic modelling of the ISM can depend strongly on the location within the galaxy.

Beside this different dimensionality (Eq. 66) the formation of clumps or clouds in this cooling shells will lead to an almost ballistical infall onto the galactic plane and the model of Bregman (1980) suggests a typical height of cloud formation at some 5 kpc. The infall velocities can reach up to $100\,\text{km s}^{-1}$ consistent with 21cm observations. The circulation of gas in supershells or galactic fountains can clearly influence the structure of the hot gas and correspondingly the filling factor f_{HIM} because of large mass and energy fluxes occurring in these superbubbles. Since the energy to drive such flows comes from spatially and temporally correlated SN-explosions the galactic fountains are dynamic phenomena connected to the combined remnants from many SNe. These superbubbles expand rapidly and they soon break out of the galactic disk (cf. Eq. 62) resulting in an upward flow to the halo. The consequences of this disk-halo interaction on the structure of the ISM as well as the modifications on the halo are discussed by Norman & Ikeuchi (1989).

The magnetic field will be convected along with this flow generating a complex field topology in the fountain but having also a magnetic field parallel to the flow, i.e. vertical to the disk. Reconnection can further rearrange the magnetic flux if this reconnection proceeds faster than the typical flow times through the fountain and if an amplification of the magnetic field is possible (Kahn 1991). After a typical cycle time of about $3\,10^7$ years the gas and the amplified magnetic field are falling down onto the galactic plane and dissipate some of their kinetic energy in shock waves. Hence, fountain-like flows can provide an important contribution to the galactic magnetic field and can also influence the operation of the galactic dynamo because of the forced magnetic field circulation resulting for example in a different value of α-term in Eqs. (51,52) and because of the change in the appropriate boundary conditions of a galactic dynamo.

1.6.2 Galactic winds

From the properties of cosmic rays as discussed in Sect. 1.4 it is clear that a number of high energy particles can leave the Galaxy as inferred from the energy dependent confinement time. The question of a galactic halo has been adressed by many authors (e.g. Burke 1968, Johnson & Axford 1971, Mathews & Baker 1971, Chevalier & Oegerle 1979) but all these models do not include the influence of high energy particles. Axford (1981b) suggests that the escape of cosmic rays can give rise to an expanding galactic corona and Ipavich (1975) has constructed a model of a galactic wind driven by cosmic rays and based on a one-dimensional spherical geometry. Given the estimates presented in Sect. 1.4.1 the energy loss of our Galaxy by cosmic rays can reach up to $\bar{L}_{gal,CR} \simeq 10^{41} \mathrm{erg\,s^{-1}}$ but the cosmic rays have a scale height larger than the thermal gas. Note that these high energy particles are unable to cool thermally in contrast to the gas. The large volume available to store energetic particles in the galactic halo has been employed by Biermann & Davis (1958) to explain the observed isotropy of cosmic rays. Such conclusions relating the cosmic ray data to properties of the halo are closely connected to the particle transport in the halo and correspondingly to the energy-dependence of the cosmic ray diffusion coefficient. The dependence of this cosmic ray escape length λ_e from the Galaxy seems to indicate a $\lambda_e \propto R^{-0.6}$ where R denotes the particles rigidity (becoming equivalent to the momentum at high energies). This power law can be extrapolated to rigidities up to about 20 MeV/c/nuc based on measurements of the HEAO-3 spacecraft (Engelman et al. 1985).

Since the work of Parker (1966) it has become evident that the system containing gas, cosmic rays and magnetic fields is unstable whenever a magnetic field line becomes elevated so that the gas slides down towards the galactic plane and the lower density regions get inflated by the cosmic ray pressure. We can consider this instability as a basic starting point for the generation of a galactic halo. On the other hand the combined effect of clustered SNe makes it very likely that hot gas is lifted to a galactic corona at kiloparsec distances as shown in the previous section. According to the survey of Gies (1987) about 70% of the O-stars occur in groups within a typical radius of about 10 pc. These SNe clustered in OB-assoziations can break through the extended neutral disk (Lockman et al. 1986) into the halo defined by regions above and below the galactic plane at $|z| \geq 1$ kpc. So taking both effects (cosmic rays and clustered SNe) together it seems plausible to discuss the existence of a galactic wind in more detail.

Following the recent work of Breitschwerdt et al. (1991) it is useful to adopt a simple flux tube geometry where the area cross section $A(z)$ scales like

$$A(z) = A_0 \left(1 + \left(\frac{z}{z_0} \right)^2 \right) \tag{68}$$

and $z_0 = 15$ kpc. Taking $z \ll z_0$ implies $A(z) \simeq$ const., whereas large distances $z \gg z_0$ yield a radial dependence of $A(z) \propto z^2$. This flux tube geometry allows a more realistic modelling of a galactic wind since in the vicinity of the galactic plane an almost linear structure is achieved whereas at larger distances the adiabatic losses are included due to $A(z) \propto z^2$. Together with a given gravitational galactic potential the set of steady state equations, i.e. the time-independent version of Eqs. (116)-(122) modified with respect to the geometry of $A(z)$ can be solved for an outflow passing a critical point (X-type singularity). In considering the halo dynamics it is essential to include selfexcited magnetic fluctuations δB (e.g. Lerche 1967, Wentzel 1968, Kulsrud & Pearce 1969) and these waves can couple strongly to the gas of the halo which is ionized by OB-stars, globular clusters and extragalactic sources (see e.g. Bregman & Harrington 1986). Therefore an additional wave pressure $P_w = \langle (\delta B)^2 \rangle / 8\pi$ acts on the ionized thermal gas. As long as the gas is unable to cool the ion-neutral friction (Kulsrud & Pearce 1969) cannot damp the waves. In the case of damping the cosmic rays stream through the gas without dragging the thermal plasma with them, i.e $\bar{\kappa} \to \infty$. Note that the regions below $|z| \leq 1$ kpc are dominated by the diffuse propagation of cosmic rays rather than by streaming along the field lines (e.g. Breitschwerdt el al. 1991). These regions ($|z| \leq 1$ kpc) are highly disturbed by SN-explosions and also subject to a z-dependent mass loading from SNRs caused by the z-dependence of O-stars becoming a SN. Observations suggest a typical scale height of SN of $H_{SN} \sim 55\,pc$ (Bregman 1980). Note that these models of Breitschwerdt et al. (1991) neglect the diffusion of cosmic rays ($\bar{\kappa} = 0$) and start at a reference level of $z = 1\,kpc$ where the boundary conditions are specified. However, according to the present cosmic ray propagation models (Ginzburg & Ptuskin 1985) a diffusive transport of high energy particles is likely up to distance of about $10\,kpc$ with typical mean diffusion coefficients of about $\bar{\kappa} \gtrsim 10^{29}\,cm^2s^{-1}$.

Summarizing the results of Breitschwerdt et al. (1991) it is possible to drive a supersonic mass loss from our Galaxy generated by the combined effect of cosmic rays and thermal pressure. The extragalactic pressure has to exceed $nT \sim 36\,cm^{-3}K$ to prevent the formation of an outflow. Integrating their typical values of the mass loss in each flux tube yields an upper limit for the total galactic mass loss rate of about $\dot{M}_{gal} \simeq 1\,M_{\odot}yr^{-1}$ for our galaxy. Assuming a life time of $t_{gal} \simeq 10^{10}$ years we get $\dot{M}_{gal}t_{gal} \simeq 10^{10}M_{\odot} \simeq M_{ISM}$, a value which is comparable to the numbers of Sect. 1.3 where the results from the diffuse γ-ray emission are discussed (before Table 3). It is also clear that the gas leaving the Milky Way is hot and affects the HIM as well as the overall chemical evolution of a galaxy since the rising gas comes mainly from SNe having a different chemical composition compared to the mean ISM. Chevalier & Oegerle (1979) argue on the other hand that the presence of clouds in the halo indicates that the energy input from SNe is lost due to radiation rather than used to power a galactic wind. However, up to now no clear observational indications can be collected to prove the existence of such galactic winds. Nevertheless, the obtained mass loss rates summed over the life time of a galaxy provide an important mass loss to the intergalactic medium. In the future a detailed calculation of the synchrotron emission from such cosmic ray driven halos could remove this uncertainty.

2 Supernova Remnants

As emphasized at several points in the previous section we have seen that SN-explosions are essential for controlling the overall properties of the ISM and since SN-explosions are the most violent events in our Galaxy the energy transferred to the ISM is also responsible for a large variety of individual phenomena. Therefore a more thorough theory of the evolution of a SNR has to be given in order to understand the dynamics of the ISM. On the other hand it should be clear that at this point we cannot be interested in modelling a particular event in great detail but have to consider the general features of SNR evolution. According to Woltjer (1972) the evolution of a SNR can basically be described by four phases depending on different physical processes relevant at each stage. The following sections are based on these distinctions but I have to mention that the transition between these phases is accompanied with several rather complex dynamical phenomena propagating through the remnant (at least in spherical symmetry).

From the observational facts SN-explosions are divided into two major classes depending whether hydrogen lines are absent (Type I) or present (Type II) in their spectra. The theoretical explanations for the distinct explosion mechanisms can reveal these observational differences by either employing a carbon deflagration of a white dwarf for Type I or a core collapse of a massive star with its hydrogen envelope for Type II (see e.g. the reviews by Woosley 1986, Woosley & Weaver 1986). However, in the present context I have to neglect these different classes of SNe and for our purposes the explosion energy will be set to the canonical number of $E_{\text{SN}} = 10^{51}$ erg bearing in mind that some SNe can have energies less than this value.

2.1 Free Expansion Phase

The strong shock wave generated inside the star moves outwards and we expect some differences in the early stages of the explosion (e.g. in the light curve) depending on the stellar structure and on a possible stellar wind from the SN-progenitor . The stellar atmosphere can be approximated by a power law $\rho \propto r^{-n}$ where n is 8 to 9 (e.g. Arnett 1987, 1988 in a model for SN1987A). The result of a SN-explosion is a free expansion with a simple velocity field given in Eq. (70) and after the shock wave has run through the atmosphere of the exploding star a characteristic flow pattern develops which can be seen schematically in Fig. 10 and as a result of numerical simulations in Fig. 11. A contact discontinuity CD separates the ejected material from the surrounding matter. A strong shock front S_1 compresses the ISM and the so-called reverse shock S_2 decelerates the ejecta. Note that the reverse shock is located in the SN-material and runs backwards in the expanding frame of the contact discontinuity.

During this phase the external medium has no influence on the violent expansion of the shock wave because the pressure and the velocity of the ambient material are neglibile. In the case of spherically symmetric flows (considered throughout these sections) the density structure surrounding the SN $\rho \propto r^{-a}$ is generally taken either to be constant, i.e. $a = 0$ or to be generated by a stellar wind of constant velocity and mass loss, i.e. $a = 2$. Since most of the SN-energy (except the emission of neutrinos) is kinetic energy we can relate the ejection velocity v_{ej} to the SN-energy E_{SN}

$$E_{\text{SN}} = \frac{1}{2} M_{\text{ej}} v_{\text{ej}}^2, \quad \text{or} \quad v_{\text{ej}} = \left(\frac{2 E_{\text{SN}}}{M_{\text{ej}}} \right)^{1/2} \tag{69}$$

and the shock radius scales as

$$R_s(t) = v_{\text{ej}} t. \tag{70}$$

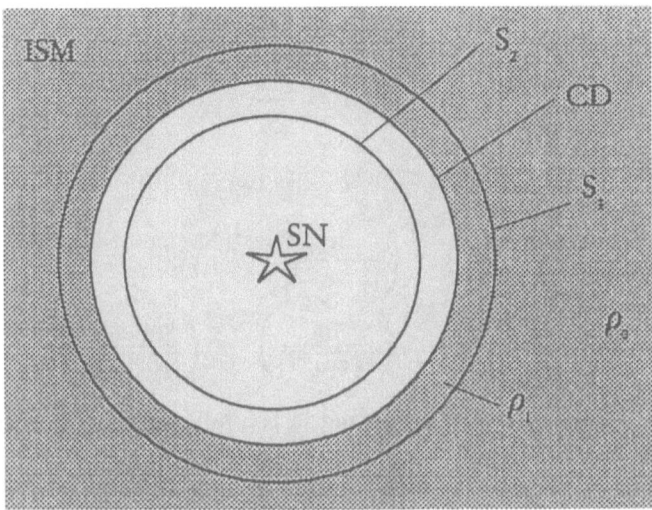

Fig. 10. Schematic flow pattern of a SN exploding into a homogeneous medium

Since we are mainly interested in the large scale evolution of a SNR and its influence on the ISM I will not discuss the possibility of clumped ejecta (e.g. Hamilton 1985) nor various self-similar solutions for young SNRs (e.g. Chevalier 1982, Nadyozhin 1985). The interaction of a stellar wind with the ambient medium creates a complicated flow structure containing the wind bubble, shocked wind material and shocked ambient material (e.g. Weaver et al. 1977). Assuming the SN-progenitor to be a red supergiant with a wind velocity of $10\,\mathrm{km\,s^{-1}}$ and a mass loss rate of $\dot{M} = 10^{-6}M_\odot\,\mathrm{yr}^{-1}$ a cavity of about 1 pc is formed in about 10^5 years containing $0.1\,M_\odot$. Taking a typical density of $1\,\mathrm{cm}^{-3}$ the swept-up mass of the ambient medium is also about $0.1\,M_\odot$ in a bubble with a radius of 1 pc. In this case the disturbed medium is rather small in mass and radius compared to the subsequent evolution of a SNR. Therefore I will ignore this phase of the interaction of the SN-ejecta with the stellar wind and will consider for the rest of this chapter only explosions occurring in a medium of a mean constant density ρ_0. Modifications caused by an inhomogeneous medium (embedded interstellar clouds) are discussed later (cf. Sect. 2.4.3) in the context of cloud evaporation or production of high energy γ-rays due to collisions of cosmic rays with the thermal plasma.

The accumulated mass of the external homogeneous medium of density ρ_0 compressed between the contact discontinuity and the forward shock equals the ejected mass M_{ej} at a certain time and distance. This condition defines the sweep-up radius R_{sw} as well as the sweep-up time t_{sw}, i.e.

$$R_{sw} = \left(\frac{3M_{ej}}{4\pi\rho_0}\right)^{1/3}$$

$$t_{sw} = \frac{R_{sw}}{v_{ej}} \tag{71}$$

or in more appropriate physical units ($\rho_0 = \mu m_H n_0$ with $\mu = 0.61$)

$$R_{sw} = 2.6 \left(\frac{M_{ej}}{[M_\odot]}\right)^{1/3} \left(\frac{n_0}{[1\,\mathrm{cm}^{-3}]}\right)^{-1/3} \mathrm{pc}$$

$$t_{sw} = 210 \left(\frac{M_{ej}}{[M_\odot]}\right)^{5/6} \left(\frac{n_0}{[1\,\mathrm{cm}^{-3}]}\right)^{-1/3} \left(\frac{E_{SN}}{[10^{51}\mathrm{erg}]}\right)^{-1/2} \mathrm{years}. \tag{72}$$

The results on SNR presented in the following sections are obtained by using a numerical method based on an adaptive grid which distributes the individual grid points at locations of strong gradients (Dorfi & Drury 1987). The equations of cosmic ray hydrodynamics (116)-(122) are discretized in a conservative and implicit form to ensure also numerically a conservation of global quantities like mass, momentum and energy and to avoid the restrictive timestep condition necessary for explicit discretizations. To illustrate the obtainable resolution individual grid points of the flow region around the contact discontinuity are plotted in Fig. 11 showing the forward shock running in the ISM as well as the reverse shock located in the ejecta decelerating them. Fig. 11d depicts the relative grid spacing $\Delta r/r$ where the three discontinuities are clearly resolved by a continuous decrease of the grid size. Note that this numerical treatment is independent of the physical processes involved in the transition region. More details on the numerical method applicable to SNRs are given in Dorfi (1990 and references therein).

The simple nature of such a freely expanding flow, i.e. $u = r/t$ can also be inferred from Fig. 12 where the evolution of a SNR is displayed up to a time of 3250 years. At the end of this phase the reverse shock propagates inwards and heats up the interior. The initial stellar profile is advected outwards and is clearly seen in the density structure (Fig. 12a) where the density decreases like $\rho \propto t^{-3} f(u)$ and the function $f(u)$ depends on the initial conditions and the explosion mechanism. However, the last time step $t = 1.02\,10^{11}$s shown in Fig. 12 indicates the end of the expansion phase because the velocity (Fig. 12b) becomes negative around $1.2\,10^{19}$cm. The gas pressure in Fig. 12c exhibits the growing separation between the two shock waves because the reverse shock moves inwards. Since these models include also the acceleration of cosmic rays in shock waves Fig. 12d depicts the particle pressure P_c obtained for a constant mean cosmic ray diffusion coefficient of $\bar{\kappa} = 10^{27}\mathrm{cm}^2\,\mathrm{s}^{-1}$ and a cosmic ray adiabatic index of $\gamma_c = 4/3$ and an injection parameter of $\eta_{inj} = 10^{-3}$ (see Sect. 2.5 for a detailed discussion of these quantities). During the expansion phase the effects of cosmic rays are negligible due to the large thermal pressure and due to the

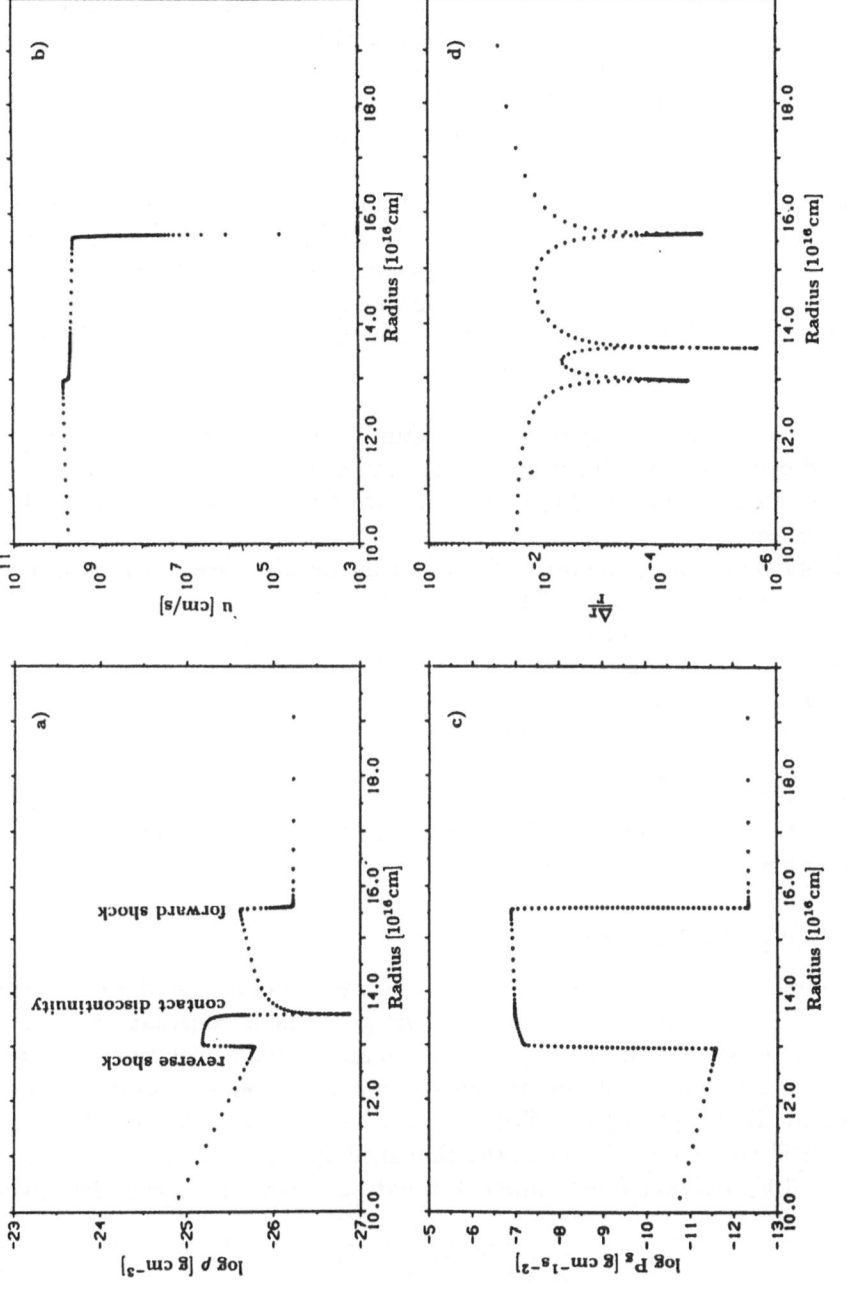

Fig. 11. The flow structure of a SN-explosion in a medium of constant density $n_0 = 0.3\,\mathrm{cm}^{-3}$ after $t = 1\,\mathrm{year}$.

small volume occupied by these high energy particles although the cosmic ray pressure can be several orders of magnitude above the value in the ISM. The large value of $\bar{\kappa}$ leads to a flat pressure structure in the interior of the SNR as long as the adiabatic expansion can be compensated by the diffusive flux. A typical expansion velocity of $u \simeq 10^8 \mathrm{km\,s^{-1}}$ and the diffusion coefficient of $\bar{\kappa}$ define a length scale of $\bar{\kappa}/u \simeq 10^{18}\mathrm{cm}$ where the adiabatic losses can be balanced by diffusion. However, smaller values of $\bar{\kappa}$ result in large gradients of cosmic rays in the interior of SNR during this expansion phase. Note that cosmic rays can influence the SNR evolution at late stages as demonstrated in the following sections.

Around the time of t_{sw} the flow structure changes due to the large amount of interstellar material that has been pushed away. The reverse shock starts travelling inwards and runs through the ejected material heating up the interior to very high temperatures. This transition can be accompanied by several running sound waves until a rather flat pressure distribution is achieved and we have now a pressure driven expansion described in the next section.

Note that a large number of young galactic SNR are observed during this free expansion phase. The well studied Cas A remnant shows freely expanding gas at a velociy of about $4000\,\mathrm{km\,s^{-1}}$ (e.g. Kamper & van den Bergh 1976) but also several knots with larger velocites are found within the remnant and their velocities scale also linearly with the distance from the center of the expansion indicating that they have not been decelerated significantly. Based on radio observations Bell (1977) deduces an upper limit of about $1700\,\mathrm{km\,s^{-1}}$ to the mean radial velocity but this velocity seems to belong to the gas shocked and slowed down by the reverse shock therefore having a smaller expansion rate.

2.2 Sedov-Taylor Phase

The further evolution is determined by the amount of interstellar gas swept up during the expansion phase. The energy remains constant and hence radiative losses are unimportant leading to an adiabatic evolutionary stage. An exact self-similar solution for such a pressure driven explosion has been given by Taylor (1950) and Sedov (1959) based on a rather cumbersome analytical treatment. However, the thin shell approximation introduced in Sect. 1.6.1 provides a much simpler estimate and Newton's second law gives

$$\frac{d}{dt}\left(\frac{1}{3}R^3\rho_0\dot{R}\right) = PR^2. \tag{73}$$

Together with the fact that during this phase the expansion of the remnant proceeds adiabatically. We can relate the gas pressure to the explosion energy

$$P = \frac{2}{3}E_{\mathrm{SN}}\bigg/\frac{4\pi}{3}R^3, \qquad \gamma = \frac{5}{3}. \tag{74}$$

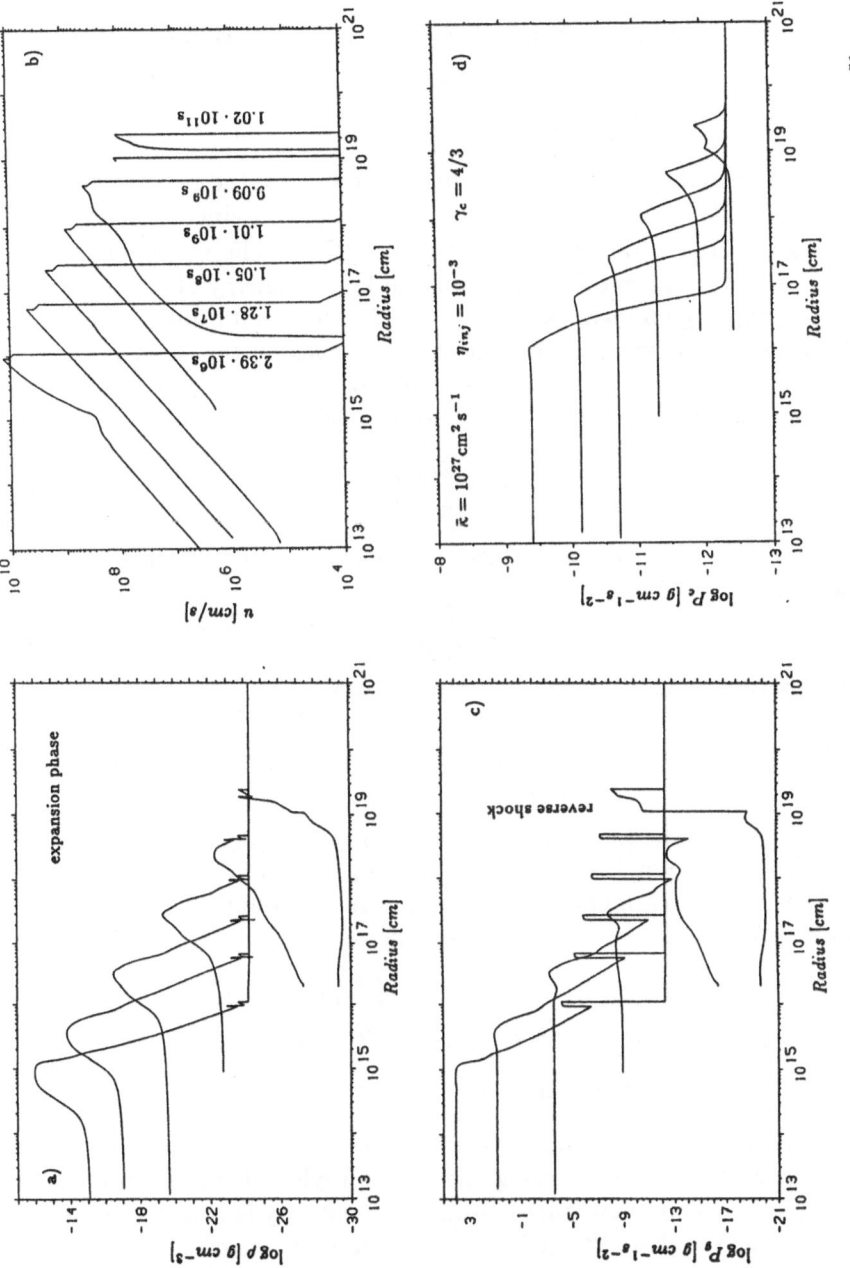

Fig. 12. Temporal evolution of a SNR during the free expansion phase. The SN-energy is set to $E_{SN} = 10^{51}$ erg, the ejected mass to $M_{ej} = 5\,M_\odot$ and the external density to $n_0 = 0.3\,\mathrm{cm}^{-3}$

Making again a power law ansatz of $R = At^\eta$ for the shock radius R_s we get

$$R_s(t) = \left(\frac{25}{4\pi}\frac{E_{SN}}{\rho_0}\right)^{1/5} t^{2/5}, \qquad \eta = \frac{2}{5}. \tag{75}$$

Note that the numerical value $25/4\pi$ gives 1.99 whereas the exact self-similar treatment yields 2.02. The shock velocity is easily obtained from $\dot{R}_s = v_s$ and from Eq. (75) it follows immediatly $v_s \propto R_s^{-3/2}$. The thermal energy is calculated to be $E_{th} = 0.72\,E_{SN}$ while the kinetic energy stays at $E_{kin} = 0.28\,E_{SN}$ adopting an ideal gas with $\gamma = 5/3$. The usual Rankine-Hugoniot conditions of shock waves expressing the basic conservation laws across the shock front (e.g. Landau & Lifschitz 1976) fix the postshock temperature. In a strong shock the downstream pressure is given by

$$P_2 = \frac{2\rho_1 v_s^2}{\gamma + 1} \tag{76}$$

and taking $\rho_2 = 4\rho_1$ and an ideal gas $P = \mathcal{R}\rho T/\mu$ we obtain the postshock temperature

$$T_s = \frac{3\mu}{16\mathcal{R}} v_s^2. \tag{77}$$

The adiabatic expansion results in a rapid decrease of this temperature with time, i.e.

$$T_s = \frac{3\mu}{100\mathcal{R}} \left(\frac{2.02 E_{SN}}{\rho_0}\right)^{2/5} t^{-6/5} \tag{78}$$

and also with radius, i.e. $T_s \propto R_s^{-3}$. In appropriate physical units $E_{51} = E_{SN}/[10^{51}\text{erg}]$, $n_0 = n/[1\,\text{cm}^{-3}]$ and $\rho_0 = \mu m_H n_0$ with $\mu = 0.61$ for a fully ionized gas and a Helium abundance of $n_{He}/n_H = 0.1$ and $t_4 = t/[10^4\text{years}]$ we can rewrite the aforementioned expressions

$$R_s(t) = 14.8\, E_{51}^{1/5} n_0^{-1/5} t_4^{2/5}\ \text{pc}$$
$$v_s(t) = 580\, E_{51}^{1/5} n_0^{-1/5} t_4^{-3/5}\ \text{km s}^{-1}. \tag{79}$$
$$T_s(t) = 4.6\,10^6\, E_{51}^{2/5} n_0^{-2/5} t_4^{-6/5}\ \text{K}$$

The acceleration of cosmic rays results in the development of a thick shell of energetic particles during the Sedov-Taylor phase as depicted in Fig. 13d. The region of enhanced cosmic ray intensity is confined to about $0.75 R_s$ containing about 60% of the SNR volume for times $t \gtrsim 10\, t_{sw}$. The maximum cosmic ray pressure of about $12\, P_{c,ext}$ is reached for the adopted value of $\bar{\kappa} = 10^{25} \text{cm}^2\,\text{s}^{-1}$ at an age of $2\,10^4$ years. The expansion proceeds and the gas cools adiabatically according to Eq. (78). At a certain time t_c the temperature will decreases below a critical value where radiative losses become important.

Tycho's SNR is a well observed example of a Type I SN-explosion in the Sedov-Taylor phase. The explosion happened in the year 1572 and hence we

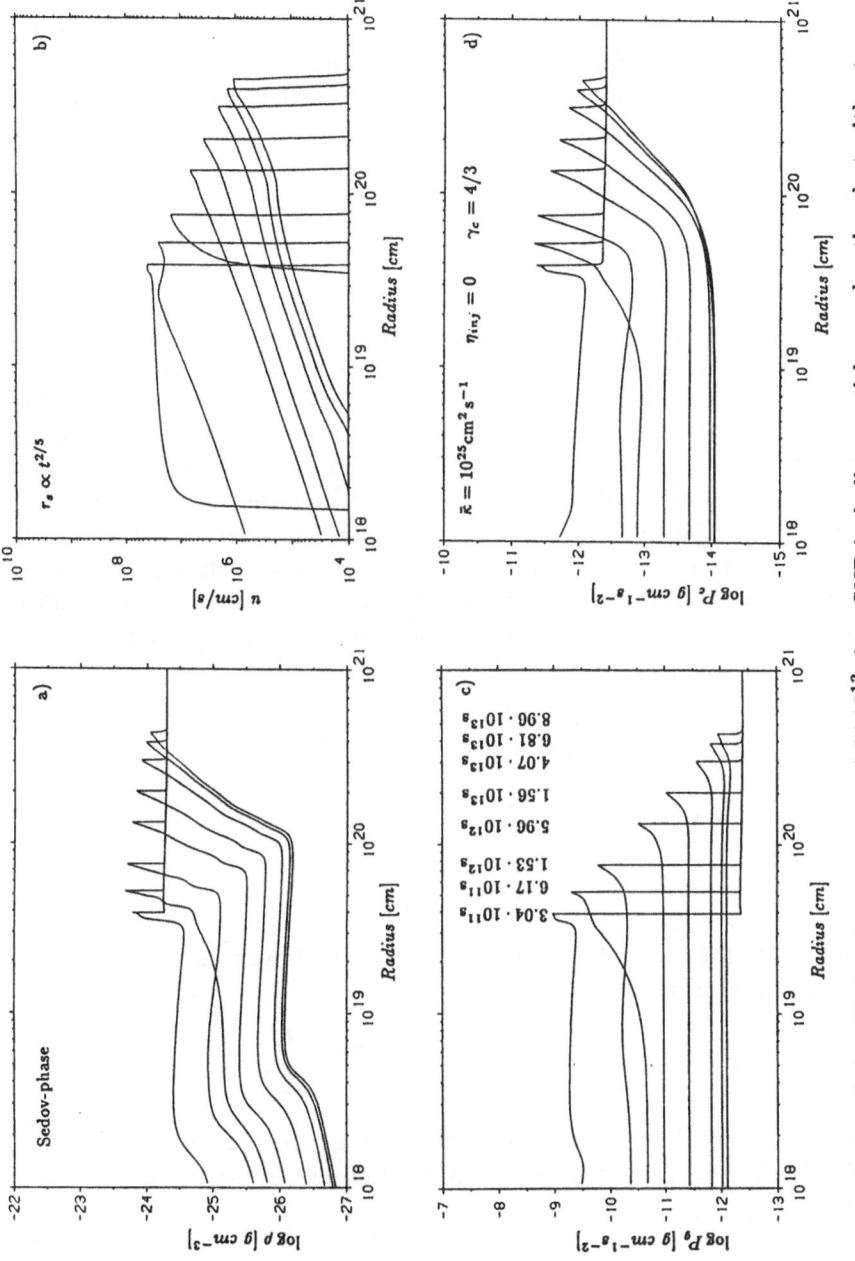

Fig. 13. The Sedov-Taylor phase up to $t = 8.96\,10^{13}$ s in a SNR including particle acceleration but without radiative cooling

have now an age of $t = 420$ years. The proper motions of optical filaments have been studied by Kamper & van den Bergh (1978) and they identify a shock wave increasing with time like t^η where $\eta = 0.38 \pm 0.01$. This value is in close agreement to the theoretical value of $\eta = 0.4$ for an adiabatic blast wave. As discussed in Sect. 2.7 most of X-rays emitted during this stage are produced in the immediate postshock region.

2.3 Cooling Phase

Around the age of 10^4 years the temperature (79) is about 10^6K and the most abundant ionized elements start to capture free electrons and their subsequent excitation allows the energy to be effectively radiated away. This effect is visible in the large increase of the cooling function $\Lambda(T)$ around 10^6K plotted in Fig. 2. Accordingly the thermal pressure in the postshock region decreases and a thin cool shell forms where the shocked gas comes almost at rest (shown in Fig. 14). At later times so-called secondary shocks (Falle 1975) are generated as a consequence of rapid cooling in the downstream region (cf. also Sect. 2.7 for effects on the production of γ-quanta).

The corresponding cooling time can be estimated analytically from the energy equation (Field 1965)

$$\frac{d}{dt}\left(\frac{3}{2}kT\right) = -n\Lambda(T) \tag{80}$$

and by introducing the cooling time t_c we get

$$t_c \simeq \frac{3kT}{2n\Lambda(T)}. \tag{81}$$

To calcualate the time when a thin shell forms we have to set the evolution time t_{dyn} equal to this cooling time, i.e.

$$t_c \simeq t_{\text{dyn}}, \quad \frac{3kT(t_c)}{2n\Lambda(T_c)} = \frac{R_s}{v_s} = \frac{5}{2}t_c \tag{82}$$

by use of Eq. (75). If we assume that the compression corresponds to a strong shock $\gamma = 5/3$ and adopting simple Kahn's cooling law Eq. (11),

$$n = 4n_0, \quad \Lambda(T) = C\,T^{-1/2} \tag{83}$$

Eq. (82) can be explicitly solved under the condition that the Sedov-Taylor phase is valid up to this point

$$\frac{kT(t_c)}{4n_0CT^{-1/2}t_c} \simeq 1 \tag{84}$$

yielding the time, radius and velocity when a thin shell forms

$$t_{\text{shell}} \simeq 2.13\,10^4\,E_{51}^{3/14} n_0^{-4/7} \text{ yr}$$

$$R_{\text{shell}} \simeq 20\,E_{51}^{2/7} n_0^{-3/7} \text{ pc} \qquad (85)$$

$$v_{\text{shell}} \simeq 368\,E_{51}^{1/14} n_0^{1/7} \text{ km s}^{-1}.$$

More accurate estimates on the cooling time and the formation of a dense shell are due to Cox (1972) or Kahn (1976) yielding slightly different exponents.

In the example of Fig. 14 the external density is set to $n_0 = 5\,\text{cm}^{-3}$, a value typical for Cas A leading to a significantly shorter sweep-up time of $t_{\text{sw}} = 1.47\,10^{10}$s at a sweep-up radius of $R_{\text{sw}} = 6.58\,10^{18}$ cm. The corresponding cooling time can be calculated from Eq. (85) yielding $t_{\text{shell}} \simeq 2.68\,10^{11}$s. Note that in the evaluation of the cooling time t_c (81) cosmic ray effects can be taken into account by increasing n, i.e. $n = 7n_0$ for $\gamma_c = 4/3$ which results in a smaller t_c (see also Sect. 2.4.2). The radial dependence of the density, velocity, cosmic ray energy density and temperature is depicted in Fig. 14 for a radius interval between 10^{19}cm and 10^{20}cm.

The curve labeled $t = 2.05\,10^{11}$s exhibits a compression ratio of $r = 4$ and so neither radiative cooling nor cosmic rays modify the shock structure at this time where the energies are divided up as $E_{\text{th}} = 0.69\,E_{\text{SN}}$, $E_{\text{kin}} = 0.28\,E_{\text{SN}}$, $E_{\text{cr}} = 0.02\,E_{\text{SN}}$ and $E_{\text{cool}} = 0.99\,E_{\text{SN}}$. The effects of cooling become more pronounced in the case of higher ambient densities since the radiated power depends on the density squared (cf. Eq. 10). The compression ratio increases to $r = 7.9$ (Fig. 14a) where the cooling sets in more effectively but we can still separate the shock front from the cooling zone in the downstream region. If we insert the value of $n_0 = 5\,\text{cm}^{-3}$ into Eq. (85) we estimate the radius $R_{\text{shell}} \simeq 3.1\,10^{19}$cm at which a dense shell forms. The good agreement of this value with the numerical calculations can be inferred from Fig. 14. The temperature structure clearly exhibits the shock with a downstream temperature of $6\,10^5$K decreasing to $2.2\,10^5$K in the cooling region (Fig. 14d). We get inward velocities for radii $R < 10^{19}$cm (Fig. 14b). The cosmic ray precursor in the velocity can be seen in Fig. 14b growing in time. At $t = 8.01\,10^{11}$s a very dense shell is built up and the density increases up to $r = 66.3$ (Fig. 14a) with a postshock temperature of $5\,10^5$K and a minimum of $8\,10^3$K in the cooling zone. I want to emphasize that the velocity jump in the postshock region is followed by a local minimum which both develop in time to a further secondary shock. This fact can be deduced from the cosmic ray structure (Fig. 14c) showing not only a single sharp peak like the one at earlier times. The time evolution of the cosmic ray acceleration depicts the growing precursor which is best detectable in velocity and cosmic ray energy. However, up to that age cosmic rays have only a small influence on the density and temperature structure in the upstream region.

Note that at the time of t_{shell} the interior of the remnant is still hot and cools adiabatically, i.e. for the remnants volume V

94

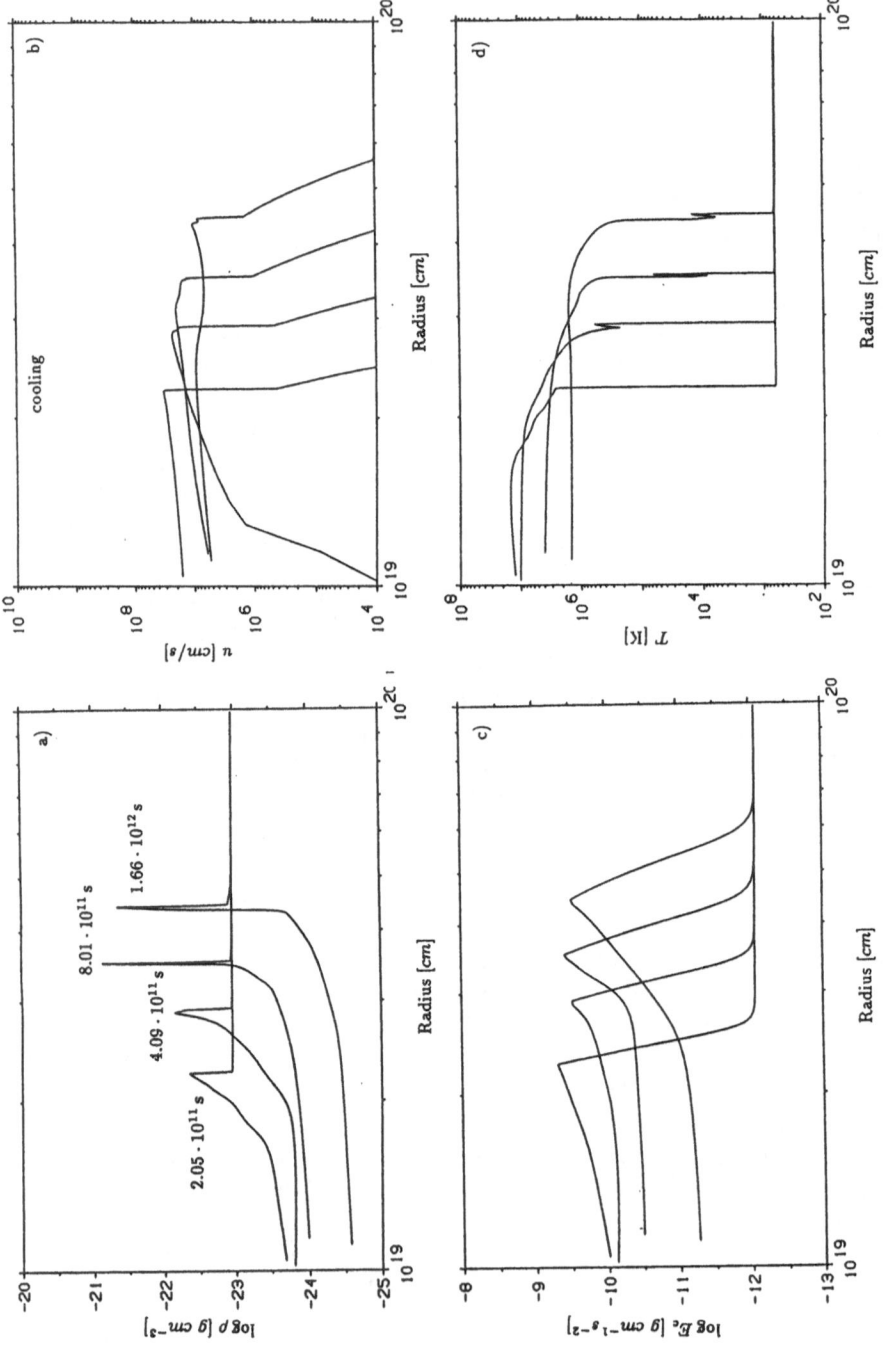

Fig. 14. Formation of a thin shell in the post shock region for a SN exploding into a medium with $n_0 = 5\,\mathrm{cm}^{-3}$

$$PV^\gamma = const., \quad \gamma = 5/3 \quad \text{and} \quad V \propto R_s^3. \tag{86}$$

At this stage the remnant evolves by accumulating the swept-up material in the shell and no significant internal motions are expected. The ram pressure is then equal the internal pressure, i.e.

$$\rho_0 v_s^2 \simeq P \tag{87}$$

and the SNR is entering the so-called snowplow phase where

$$v_s^2 R_s^5 = const. \tag{88}$$

The power law ansatz for the shock radius $R_s \propto t^\eta$ leads to

$$\eta = \frac{2}{7}, \quad R_s \propto t^{2/7}, \quad v_s \propto t^{-5/7} \tag{89}$$

and we can take $R_s \simeq R_{\text{shell}}\, t^{2/7}$ for $t \simeq t_{\text{shell}}$ and calculate the shock velocity of $v_s \simeq v_{\text{shell}} t^{-5/7}$. This motion will continue until the velocity decreases to the mean velocity in the ISM $\bar{v}_{\text{ISM}}$.

2.4 Final SNR Evolution

2.4.1 Final SNR Radius and Cloud Motions

Employing the last estimate (89) the final SNR radius can be obtained by setting the shock velocity v_s to the mean ISM velocity $\bar{v}_{\text{ISM}} \simeq v_{\text{A,ext}} \simeq 20\,\text{km s}^{-1}$ where the Alfvén velocity (27) is calculated from $n_0 = 0.3\,\text{cm}^{-3}$ and $B_0 = 5\,\mu\text{G}$.

$$\tau_{\text{SNR}} = t(v_s = \bar{v}_{\text{ISM}}), \quad \frac{\tau_{\text{SNR}}}{t_{\text{shell}}} = \left(\frac{\bar{v}_{\text{ISM}}}{v_{\text{shell}}}\right)^{-7/5}. \tag{90}$$

At this time the SNR is dispersed into the ISM and breaks up into several fragments which can move even further in space by conserving their momentum. Typical values are then

$$\begin{aligned} \tau_{\text{SNR}} &\simeq 1.26\,10^6\, E_{51}^{11/35} n_0^{-13/35}\ \text{yr} \\ R_{\text{SNR}} &\simeq 64\, E_{51}^{11/35} n_0^{-13/35}\ \text{pc} \end{aligned} \tag{91}$$

Up to this stage the analytical description of a SNR exploding into a homogeneous medium is summarized in Fig. 15.

Spitzer (1978) has given an estimate how much kinetic energy of the SN can be converted into cloud motions when the clouds are pieces of the SNR shell breaking apart. Such a situation is very likely because the hotter and more tenuous interior pushes the dense cooling shell which leads to a Rayleigh-Taylor unstable configuration (e.g. Lamb 1945). Since the late phases are characterized by a conservation of the outward momentum we

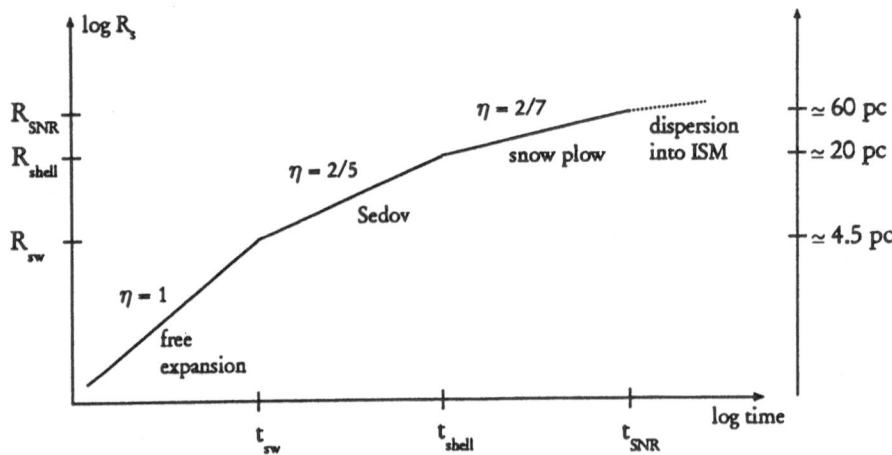

Fig. 15. Schematic picture of a SNR evolution up to the time of dispersion in the ISM. The values at the right vertical scale are calculated for a SN-energy $E_{SN} = 10^{51}$erg, an external density $n_0 = 1\,\mathrm{cm}^{-3}$ and an ejected mass of $M_{ej} = 5\,M_\odot$

take the mass M_t and velocity v_t of the shell at the time this condition is applicable

$$M_t v_t = M v. \tag{92}$$

At this age $t \gg t_{sw}$ the swept-up mass is very large compared to the ejected mass and hence we write $M = 4\pi/3\,\rho_0 R_s^3$ or

$$M_t v_t = \frac{4\pi}{3}\rho_0 R_s^3 \frac{dR_s}{dt} \tag{93}$$

which can easily be integrated over time

$$t = \frac{\pi\rho_0 R_s^4}{3M_t v_t} = \frac{M R_s}{4M_t v_t}. \tag{94}$$

Denoting by M_c the mass of the clouds moving with a velocity v_c the kinetic energy of them is given in units of the explosion energy E_{SN} by

$$\epsilon_c = \frac{M_c v_c^2}{2E_{SN}} \tag{95}$$

where ϵ_c is the remaining energy fraction and

$$\epsilon_c = \frac{M_c v_c^2}{M_t v_t^2}\frac{M_t v_t^2}{2E_{SN}}. \tag{96}$$

Using Eq. (92) and $1/2\,M_t v_t^2 \simeq 1/4\,E_{SN}$ for $\gamma = 5/3$ we finally find

$$\epsilon_c \simeq \frac{v_c}{4v_t} = 0.023 \qquad (97)$$

according to the following estimates. As pointed out in Sect. 1.3.2 the rms velocity of clouds is about $v_c \simeq 8\,\mathrm{km\,s^{-1}}$ (McKee & Ostriker 1977) and we fix the velocity v_t by a typical temperature of 10^5 K appropriate for the momentum conserving phase and by Eqs. (79) which gives $v_t \simeq 85.5\,\mathrm{km\,s^{-1}}$. Hence, at the end of the final SNR phases only about $0.023\,E_{SN}$ are needed to explain the typical rms velocities of interstellar clouds. The mass M_c can be calculated form (95) yielding

$$M_c = \frac{2E_{SN}}{0.023 v_c^2} \simeq 6.8\,10^7\,\frac{E_{SN}}{[10^{51}\mathrm{erg}]}\,M_\odot. \qquad (98)$$

2.4.2 Cosmic Ray Effects

As shown in detail in the next sections cosmic ray effects are important at the late stages of SNR evolution. For this purpose we can redo the previous estimates assuming that the SNR is now dominated by high energy particles, i.e. $\gamma = 4/3$ and a corresponding compression ratio in strong shock of $r = 7$. Since the cooling time t_c (81) is sensitive to the density squared we expect the shell formation to happen at earlier times and smaller radii. Equations. (85) with $n = 7n_0$ yield then

$$t_{shell} \simeq 1.74\,10^4\,E_{51}^{3/14} n_0^{-4/7}\ \mathrm{yr}$$
$$R_{shell} \simeq 18.5\,E_{51}^{2/7} n_0^{-3/7}\ \mathrm{pc} \qquad (99)$$
$$v_{shell} \simeq 415\,E_{51}^{1/14} n_0^{1/7}\ \mathrm{km\,s^{-1}}.$$

The numerical factors change by 0.82, 0.92 and 1.13 compared to the thermal case with $\gamma = 5/3$.

The final SNR radius is also affected by cosmic rays since the snowplow phase (Sect. 2.3) has a softer expansion law than Eqs. (86)-(89), i.e.

$$PV^{4/3} = const., \quad V \propto R_s^3, \quad \rho_0 v_s^2 \simeq P \propto R_s^{-4}$$
$$\eta = \frac{1}{3}, \quad R_s \propto t^{1/3}, \quad v_s \propto t^{-2/3} \qquad (100)$$

opposite to $\eta = 2/7$ in the case of $\gamma = 5/3$. The same procedure as in Sect. 2.4.1 yields instead of Eq. (91)

$$\tau_{SNR} \simeq 1.64\,10^6\,E_{51}^{9/28} n_0^{-5/14}\ \mathrm{yr}$$
$$R_{SNR} \simeq 84.2\,E_{51}^{3/14} n_0^{-5/21}\ \mathrm{pc}. \qquad (101)$$

These numerical factors of (101) can be regarded as upper limits since we do not expect $\gamma = 4/3$ during the entire late evolutionary phases of a SNR.

2.4.3 Remarks on SNRs in inhomogeneous media

As stated in the previous sections the ISM is far from being a homogeneous medium although the SNR models presented so far explode in constant surroundings. Most of the mass is contained in dense clouds (cf. Tables 1,2) occupying only small volumes whereas most of the volume is filled with hot but very tenuous gas (HIM). Hence, I will make some remarks on the evolution of a SNR in a cloudy medium but refer e.g. to McKee (1988) for a more profound discussion of this topic.

An expanding SNR shock overtakes interstellar clouds of various sizes and drives a shock wave into them. Numerical simulations without embedded magnetic fields (Woodward 1976, Krebs & Hillebrandt 1983) as well as ones including a magnetic field (Nittman 1981, Oettl et al. 1985) clearly show the development of Rayleigh-Taylor and Kelvin-Helmholtz instabilities although some of the results suffer from poor resolution of the grid used and/or numerical diffusion at the cloud-intercloud interface. A strong SNR shock with velocity v_s hits a spherical cloud of density ρ_{cl} and radius R_{cl}. Neglecting radiative effects the ram pressures must be comparable (McKee & Cowie 1975), i.e. $\rho_{cl} v_{s,cl}^2 \simeq \rho_0 v_s^2$ where $v_{s,cl}$ is the shock speed in the cloud; thus

$$v_{s,cl} \simeq \left(\frac{\rho_0}{\rho_{cl}}\right)^{1/2} v_s. \tag{102}$$

The physical effects involved in this cloud crushing can be illustrated by defining three relevant time scales, namely the cloud crushing time t_{cc}, the intercloud crossing time t_{ic} and the age t of the remnant

$$t_{cc} = \frac{R_{cl}}{v_{s,cl}} = \frac{R_{cl}}{v_s}\left(\frac{\rho_0}{\rho_{cl}}\right)^{1/2}, \quad t_{ic} = \frac{2R_{cl}}{v_s}, \quad t = \frac{2}{5}\frac{R_s}{v_s} \tag{103}$$

having assumed a Sedov-Taylor expansion law (79). Since the density ratio between the clouds and the intercloud medium is usually large we expect $t_{cc} > t_{ic}$. Furtheron small, medium or large clouds can be distinguished (McKee 1988) depending on these time scales where small clouds ($t > t_{cc}$) are compressed by the enclosing shock wave but large clouds ($t_{ic} > t$) are subject to smaller cloud shocks and the SNR shock weakens during the cloud passing time. The net acceleration of a cloud is made up by the ram pressure exerting a drag on the cloud as well as by the shock accelerating the gas. However, the ultimate stage of a cloud hit by a SNR shock is not well established since this non-linear interaction requires accurate multidimensional numerical simulations with sufficient resolution. Of course such simulations are essential to determine a shock induced collapse as well as subsequent star formation.

The mentioned instabilities on the surface of these shocked clouds provide a natural way of transferring mass from the clouds to the intercloud medium and this mass-loading can dramatically alter the dynamics of a

SNR. It is obvious that the cooling time (85) can give only an upper limit in such a case because a mass-loaded remnant cools earlier since the energy radiated away depends on the mean density squared. A similar effect comes from the evaporation of clouds embedded in a hot medium due to thermal conduction (McKee & Ostriker 1977). Without going too much in details it should be clear that a remnant with mass loading evolves different from those without but it is almost impossible to disentangle this effect observationally from SNRs exploding into a medium stratified by the progenitors stellar wind. However, more refined observational techniques in the γ-range may provide better resolved observations if enough cosmic rays are accelerated at the SNR shock. These particles collide with the dense clouds and therefore produce neutral pions which decay and become visible as local regions of enhanced γ-ray emission (cf. Sect. 2.7).

2.5 Particle Acceleration in SNR's

As pointed out in the previous sections cosmic rays lead to several effects during the SNR evolution but the solutions presented so far are based on the simplifying assumption that γ_c and $\bar{\kappa}$ are taken to be constant. Hence this section is devoted to numerical simulations where the time-dependence of the acceleration process in SNR's is discussed in more detail. Such calculations are necessary to determine the amount of SN-energy E_{SN} converted into energetic particles and hence how much energy is still available for producing the HIM.

2.5.1 Cosmic Ray Hydrodynamcis

This sections deals with the basic equations describing cosmic rays as a fluid, i.e. *cosmic ray hydrodynamics*. Starting from the cosmic ray transport equation (Skilling 1975) for the particle distribution function $f(r, p, t)$

$$\frac{\partial f}{\partial t} + \mathbf{u} \cdot \nabla f = \nabla \cdot (\kappa \nabla f) + \frac{1}{3} \nabla \cdot \mathbf{u} \, p \frac{\partial f}{\partial p} \tag{104}$$

we can take moments ending up with the particle pressure P_c and the cosmic ray energy density E_c defined through

$$P_c = \frac{4\pi}{3} \int_0^\infty p^3 v(p) f(p) dp$$

$$E_c = 4\pi \int_0^\infty p^2 T(p) f(p) dp \tag{105}$$

where $v(p)$ is the particle velocity and $T(p)$ the particle kinetic energy. By taking a moment of the transport equation (104) we obtain a hydrodynamical equation for the cosmic rays

$$\frac{\partial E_c}{\partial t} + \nabla\cdot(E_c \mathbf{u}) - P_c \nabla\cdot\mathbf{u} = \nabla\cdot(\bar{\kappa}\nabla E_c). \qquad (106)$$

In doing so we have introduced an effective mean diffusion coefficient $\bar{\kappa}$ for the cosmic rays by

$$\bar{\kappa} = \frac{\int_0^\infty \kappa(r,p)p^2 T(p)\frac{\partial f}{\partial p}\,dp}{\int_0^\infty p^2 T(p)\frac{\partial f}{\partial p}\,dp}. \qquad (107)$$

It is clear that the internal energy density is related to the pressure through an adiabatic exponent $4/3 \le \gamma_c \le 5/3$ of the cosmic rays

$$P_c = (\gamma_c - 1)E_c. \qquad (108)$$

Note that the cosmic ray quantities $\bar{\kappa}$ and γ_c have now to be specified during the SNR evolution.

Another problem comes from the injection of energetic particles in shock waves as seen also in the bow shock of the earth (e.g. Lee 1983). There is no quantitative theory available for this complicated process and the injection is usually treated by introducing an injection flux across the shock front

$$F_c^{\text{inj}} = \eta_{\text{inj}}\frac{\rho_1 u_1(u_1^2 - u_2^2)}{2} \qquad (109)$$

where η_{inj} denotes the fraction of the kinetic energy flux dissipated and put to high energy particles. From solar wind observations the typical values of η_{inj} are in the range from 10^{-4} up to 10^{-2}.

As a first step we can take the mean effective diffusion coefficient $\bar{\kappa}$ of cosmic rays as well as the adiabatic coefficient γ_c to be constant and some results of such SNR calculations for different values of $\bar{\kappa}$, γ_c and η_{inj} are shown in Table 5. The time is given in units of $t_{\text{sw}} = 3.75\,10^{10}$s.

Table 5. The amount of energy transferred into the cosmic ray energy E_{cr} at different times in units of $0.01\,E_{\text{SN}}$.

$\bar{\kappa}[\text{cm}^2\,\text{s}^{-1}] =$	10^{23}	10^{25}	10^{25}	10^{27}	10^{27}	10^{27}	10^{27}	10^{27}
$\gamma_c =$	$4/3$	$4/3$	$4/3$	$4/3$	$4/3$	$3/2$	$3/2$	$3/2$
$\eta_{\text{inj}} =$	0	0	10^{-3}	0	10^{-3}	0	10^{-3}	10^{-2}
$t_{\text{sw}} = 1$	0.3	0.4	0.2	0.2	0.2	0.2	0.2	0.3
10	0.6	0.5	0.7	< 0.1	< 0.1	< 0.1	0.3	0.9
30	1.1	1.2	1.3	< 0.1	< 0.1	0.1	0.1	1.3
100	1.2	3.1	3.4	< 0.1	0.1	0.1	0.3	1.6
300	2.8	7.7	7.6	0.3	0.4	0.3	0.4	2.1
1000	7.3	18.8	18.8	1.5	1.7	1.1	1.7	3.2
3000	18.4	32.0	32.0	3.7	5.1	1.9	3.4	6.7

However, as pointed out by Achterberg et al. (1984) and Heavens (1984a) the values of $\bar{\kappa}$ and γ_c cannot be assigned in a simple way to certain numbers but they are time-dependent functions as long as particle acceleration proceeds in a SNR. Due to variations in the contributions from non-relativistic and relativistic cosmic rays this time-dependence has to be considered if the amount of SN-energy converted to energetic particles is extracted from SNR evolutionary simulations. Therefore the next section is devoted to the time-dependence of shock acceleration.

2.5.2 Time-dependent Effects in Cosmic Ray Shocks

In the case of effective particle acceleration we expect that the energetic particles can totally smooth the shock structure. Therefore only an adiabatic compression is left for the gas as shown in Fig. 7 (Sect. 1.4.3). Such shock transitions occur also in SNRs if particle acceleration becomes very effective. In addition to this smoothing of the shock we expect a time-dependence of quantities related to cosmic rays like the mean cosmic ray diffusion coefficient $\bar{\kappa}$ and the adiabatic index γ_c. Since the time-dependence of $\bar{\kappa}$ and γ_c during the SNR evolution can exactly be calculated only from the full cosmic ray transport equation (104) we can introduce an approximate theory (Drury et al. 1989) to determine $\bar{\kappa}$ and γ_c and still use the simpler moment equation (106). These estimates are based on the so-called test particle picture in plane geometry where the reactions of the cosmic rays on the thermal flow are neglected. An important result obtained in the test particle picture is the cut-off momentum $p_{\max}(t)$ of accelerated particles

$$\frac{dp_{\max}}{dt} = \frac{p_{\max}}{\tau_{\mathrm{acc}}} \qquad (110)$$

where the acceleration time τ_{acc} in a shock wave with the upstream values (index 1) and downstream values (index 2) is taken from Drury (1983)

$$\tau_{\mathrm{acc}} = \frac{3}{u_1 - u_2} \left(\frac{\kappa_1}{u_1} + \frac{\kappa_2}{u_2} \right). \qquad (111)$$

Another estimate is employed to relate the momentum-dependent diffusion coefficient $\kappa(p)$ to the mean diffusion coefficient $\bar{\kappa}$ of Eq. (107) needed in cosmic ray hydrodynamics. We adopt a fully developed turbulence and can therefore apply the Bohm-limit for a particle with mass m and charge Ze, i.e.

$$\kappa(p) = \frac{1}{3} \frac{mc^3}{ZeB} \frac{(p/mc)^2}{\sqrt{1 + (p/mc)^2}}. \qquad (112)$$

If the mean diffusion coefficient $\bar{\kappa}$ is dominated by the highest particle momenta and if the spatial dependence in Eq. (107) can be ignored the following ansatz is quite plausible

$$\bar{\kappa} = \delta_k \kappa(p_{\max}), \qquad \delta_k = \frac{1}{\max(3, \ln(p_{\max}/mc)}, \tag{113}$$

since δ_k can be estimated reasonable well (Drury et al. 1989).

The next step includes the determination of γ_c which should clearly be bound between $4/3 \leq \gamma_c \leq 5/3$. Here the energetic particles are divided into a non-relativistic and a relativistic part, where the non-relativistic particle energy $E_{c,n}$ is given by the amount of injected particles at the shock wave and the advected energy into the shock

$$E_{c,n} = 4\pi \int_{p_{inj}}^{mc} \frac{p^4}{2m} f_{inj}(p) dp + (3\gamma_{c,0} - 4) E_{c,0}. \tag{114}$$

$E_{c,0}$ and $\gamma_{c,0}$ denote the upstream values advected into the shock. The last expression includes the injected particle spectrum $f_{inj}(p)$ ranging from the injection momentum p_{inj} up to momentum mc which is used to define the upper limit of the integral. Note that this assumption relies also on $p_{\max} > mc$. The rest of the local cosmic ray energy must be contained in relativistic particles and we get

$$\begin{aligned}
\gamma_c &= 1 + \frac{P_c}{E_c} = 1 + \frac{(E_c - E_{c,n})/3 + 2E_{c,n}/3}{E_c} \\
&= \frac{4E_c + E_{c,n}}{3E_c}, \qquad \text{if} \qquad E_{c,n} \leq E_c
\end{aligned} \tag{115}$$

and set $\gamma_c = 5/3$ in the case of $E_{c,n} > E_c$. Starting without accelerated particles γ_c is equal to $5/3$ but decreases when the particle pressure is built up and levels at $4/3$ if $E_{c,n} \ll E_c$. The analysis of Blandford (1980) shows that a simple power-law with index $4 < q < 5$ without cut-offs yields the remarkable result of $\gamma_c = q/3$ and hence we may conclude that the approximation (115) to the full particle distribution function catches the essential physics.

In Fig. 16 a typical solution of the above system of equations is presented including the aforementioned time-dependence of γ_c and $\bar{\kappa}$, no cooling, no dissipation of Alfvén waves, i.e. $\alpha_H = 0$ and injection of particles at the shock with $\eta_{inj} = 10^{-4}$. As mentioned earlier a transition happens from gas dominated to cosmic ray dominated shocks before the Sedov-phase begins. Hence, this process also occurs in spherical calculations if the cosmic ray pressure exceeds a certain level which is almost independent of the detailed knowledge of the particle distribution function. Figure 16 plots the different energy contributions in units of the initial explosion energy E_{SN} between 10^{10}s and $8\,10^{13}$s. At about $3\,10^{10}$s we see a rapid increase of the cosmic ray energy E_{cr} and at $t = 1.3\,10^{11}$s the reverse shock changes the kinetic, thermal and cosmic ray energies. Note that most of the supernova energy is transferred to high energy particles and after about 10^{12}s the total cosmic ray energy E_{cr} stays almost at a level of about $0.7\,E_{SN}$. SNR models

103

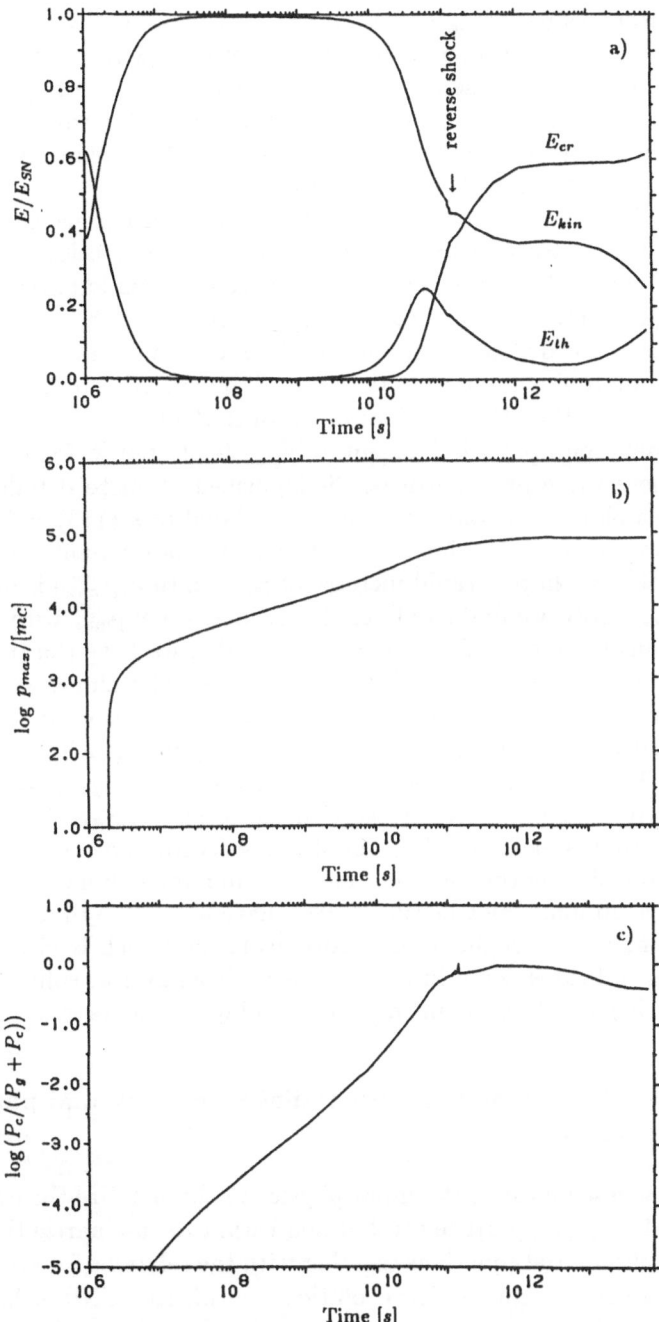

Fig. 16. The temporal evolution of the energies a) in units of $E_{\rm SN}$ between 10^6s and 10^{14}s, b) maximum particle momentum $p_{\rm max}$, c) ratio $P_c/(P_c + P_g)$ at the shock front

with constant values of γ_c and $\bar{\kappa}$ do not give such high cosmic ray effects (cf. Table 5) although the injection of energetic particles at the shock front has been included in some models. From this particular example we clearly see the importance of a time-dependent and careful treatment of γ_c and $\bar{\kappa}$ which is necessary to simulate more realistic SNR's. This result exhibits a very efficient conversion of the supernova explosion energy into high energy particles. However, these high cosmic ray energy densities accompanied with low plasma temperatures have to be compared to observational constraints, in particular to the X-ray luminosities of young remnants (cf. Sect. 2.6). Adopting such an amount of E_{cr} we may conclude that the estimate (32,33) on the power needed to maintain the observed cosmic ray energy density can easily be fulfilled and that the bulk of the galactic cosmic rays can be produced by SNR's (cf. Blandford & Ostriker 1980).

The maximum particle momentum p_{max} is plotted in Fig. 16b in units of mc where three phases can be distinguished. A more detailed description of this plot is necessary since p_{max} is related to $\bar{\kappa}$ (113) which controls the diffusive losses to the interior of the expanding remnant. At the beginning $t \simeq 10^6$s we have a rapid increase of p_{max} until $\kappa(p_{max})$ is of the order of $v_s R_s$. Secondly, we find an almost linear increase of p_{max} with time while the remnant remains in free expansion. Thirdly, we have the transition to the Sedov-phase after about 10^{11}s where p_{max} remains almost constant. For these computations we have assumed a magnetic field of $B = 5\,\mu$G and have scaled $B \propto \rho^{1/2}$ which is equivalent to keeping the Alfvén velocity v_A constant. The maximum particle energy corresponds to $cp_{max} = 9.3\,10^{13}$eV.

The ratio of the particle pressure to the total pressure $P_c/(P_c + P_g)$ at the shock front is plotted in Fig. 16c showing a variation over several orders of magnitude during the SNR lifetime. Self-similar solutions for SNR's with cosmic rays obtained by Chevalier (1983) have to assume that this quantity is kept constant throughout the entire evolution which is clearly not the case. This indicates some limitations of such analytical solutions if non-linear particle acceleration in shock waves plays an important role.

2.5.3 Evolution of SNR's with radiative cooling and particle acceleration

Having discussed most of the input physics to characterize the evolution of SNR's with ongoing particle acceleration I want to summarize the simplest system of differential equations together with the relations for $\bar{\kappa}(t)$ and $\gamma_c(t)$ and show several examples. The evolution of a spherical SNR is described by the usual two-fluid equations of one-dimensional *cosmic ray hydrodynamics*. The gas density is denoted by ρ, the velocity by $\mathbf{u}$, the gas pressure by P_g. We have to specify the two energy equations including also radiative equilibrium cooling $\mathcal{L}(\rho, E_g)$ due to lines and bremsstrahlung (cf. Fig. 2, but $\mathcal{L}(\rho, E_g)$ is now written as a function of E_g and not T), heating caused by dissipation of Alfvén waves (cf. Sect. 1.6.2) as well as the injection of particles at the shock front R_s. This effect is parameterized by the term $R_{c,\mathrm{inj}}$ where a small fraction η_{inj} of the incoming kinetic energy is carried off by the cosmic rays (e.g. Drury, 1983).

$$\frac{\partial \rho}{\partial t} + \nabla \cdot (\rho \mathbf{u}) = 0 \tag{116}$$

$$\rho \left[\frac{\partial \mathbf{u}}{\partial t} + (\mathbf{u} \cdot \nabla) \mathbf{u} \right] + \nabla (P_g + P_c) = 0 \tag{117}$$

$$\frac{\partial E_g}{\partial t} + \nabla \cdot (E_g \mathbf{u}) - P_g \nabla \cdot \mathbf{u} - \alpha_H v_A |\nabla E_c| +$$
$$+ \mathcal{L}(\rho, E_g) + R_{c,\mathrm{inj}} \delta(r - R_s) = 0 \tag{118}$$

$$\frac{\partial E_c}{\partial t} + \nabla \cdot (E_c \mathbf{u}) - P_c \nabla \cdot \mathbf{u} + \alpha_H v_A |\nabla E_c| -$$
$$- \nabla \cdot (\bar{\kappa}(t) \nabla E_c) - R_{c,\mathrm{inj}} \delta(r - R_s) = 0. \tag{119}$$

The system is closed by the equations of state for the thermal plasma

$$P_g = (\gamma_g - 1) E_g, \qquad \gamma_g = \frac{5}{3} \tag{120}$$

and for the cosmic rays where $\gamma_c(t)$ is defined by (115)

$$P_c = (\gamma_c(t) - 1) E_c. \tag{121}$$

Equations (118) and (119) contain the injection term $R_{c,\mathrm{inj}}$ which denotes the fraction of kinetic energy dissipated per second into cosmic rays and is therefore given by

$$R_{c,\mathrm{inj}} = \eta_{\mathrm{inj}} \nabla \cdot \left(\frac{\rho u^2}{2} \mathbf{u} \right). \tag{122}$$

This energy injection rate added to the cosmic ray energy density E_c at the location of shock fronts R_s has to be subtracted from the energy equation of the background gas (118) to guarantee the conservation of the total energy. Integration of (122) over an infinitely small volume around the shock front leads to the flux of energetic particles injected at the shock front given in Eq. (109). Typical values of η_{inj} are in the range of $[0, 10^{-2}]$ (cf. Drury et al. 1989 for some discussion of different injection terms).

The effect of dissipation of Alfvén waves is parameterized by α_H and we take $\alpha_H = 1$ in the case of Alfvénic heating. The dependence of the solutions on α_H is explored by Markiewicz et al. (1990) in the case of simplified SNR models. Hydromagnetic waves are generated by a resonant instability (Lerche 1967, Wentzel 1968, Kulsrud & Pearce 1969) if the energetic particles show a sufficiently strong anisotropy in the frame of the background plasma. This can happen through a streaming caused by a cosmic ray gradient along the magnetic field. Under many circumstances these magnetic fluctuations may grow and become comparable to the average magnetic field

(McKenzie & Völk 1982) and so no general theory exists beyond quasi-linear approximations. However, assuming strong damping of the waves the dissipation of the fluctuations leads to a heating of the thermal plasma which can be modelled by an extra term proportional to $v_A \nabla P_c$ acting in the precursor region of the forward shock (Völk et al. 1984).

Some typical solutions of the system (116)-(122) are shown in the following figures (17,18,19). The SN explodes into a homogeneous medium and modifications of the external medium due to a possible stellar wind from a SN-progenitor are not taken into account. The supernova explosion energy is set to $E_{SN} = 10^{51}$erg, the ejected mass to $M_e = 5\,M_\odot$. The external density of $n_0 = 0.3\,\mathrm{cm}^{-3}$ and the temperature of $T_0 = 8000\,\mathrm{K}$ correspond to the usual WIM parameter (cf. Table 2).

The time evolution of the different energy contributions is displayed on a linear scale up to $7\,10^5$yr in Fig. 17. Note that the radial structure in the vicinity of the shock waves is plotted in Figs. 18 and 19. Unfortunately, the final amount of cosmic rays accelerated in the expanding SNR depends on the injection of particles at the shock waves and hence I intend to discuss two extreme cases, one with almost no cosmic rays and another one where a large fraction of the SN-energy is put into energetic particles showing the possible range of different but homogeneous SNR's.

The case depicted in Fig. 17a is characterized by a low conversion efficiency from the explosion energy E_{SN} to the total cosmic ray energy E_{cr} and E_{cr} stays below $0.04\,E_{SN}$ during the entire evolution. Cooling effects become important after $4.24\,10^4$ years according to (85). I emphasize the importance of radiative cooling where most of the SNR thermal energy content is already radiated away after about $5\,10^4$ years and E_{cool} denotes the remaining SNR energy. Fig. 17b exhibits the temporal dependence of the energies in the case of a more efficient cosmic ray production. This model includes also heating of Alfvén waves in the precursor region of the forward shock which yields an increase of the thermal energy E_{th} although radiative cooling decreases the total energy of the remnant. The injection of particles at the shock wave is given by the fraction $\eta_{inj} = 10^{-3}$ of the incoming kinetic energy carried off by cosmic rays. For the larger values of E_{cr} (cf. Fig. 17b) cooling becomes important at earlier times because the thermal gas remains cooler during the Sedov-phase. We can estimate the onset of cooling at $3.4\,10^4$ years but E_{cool} decreases more slowly in time compared to Fig. 17a.

The shock structure of the low cosmic ray case (Fig. 17a) is explored in the shock frame on a linear scale between $0.8R_s$ and $1.3R_s$ in Fig. 18. As a consequence of cooling the density (Fig. 18a) is enhanced by a factor of 30 at $t = 4.01\,10^{11}$s associated with $E_{cool} = 0.41E_{SN}$ and a postshock velocity of $80\,\mathrm{km\,s}^{-1}$ (Fig. 18b). The cosmic ray pressure suppresses higher compression ratios because the total postshock pressure cannot drop below the level of P_c. Since a number of particles is accelerated at the shock

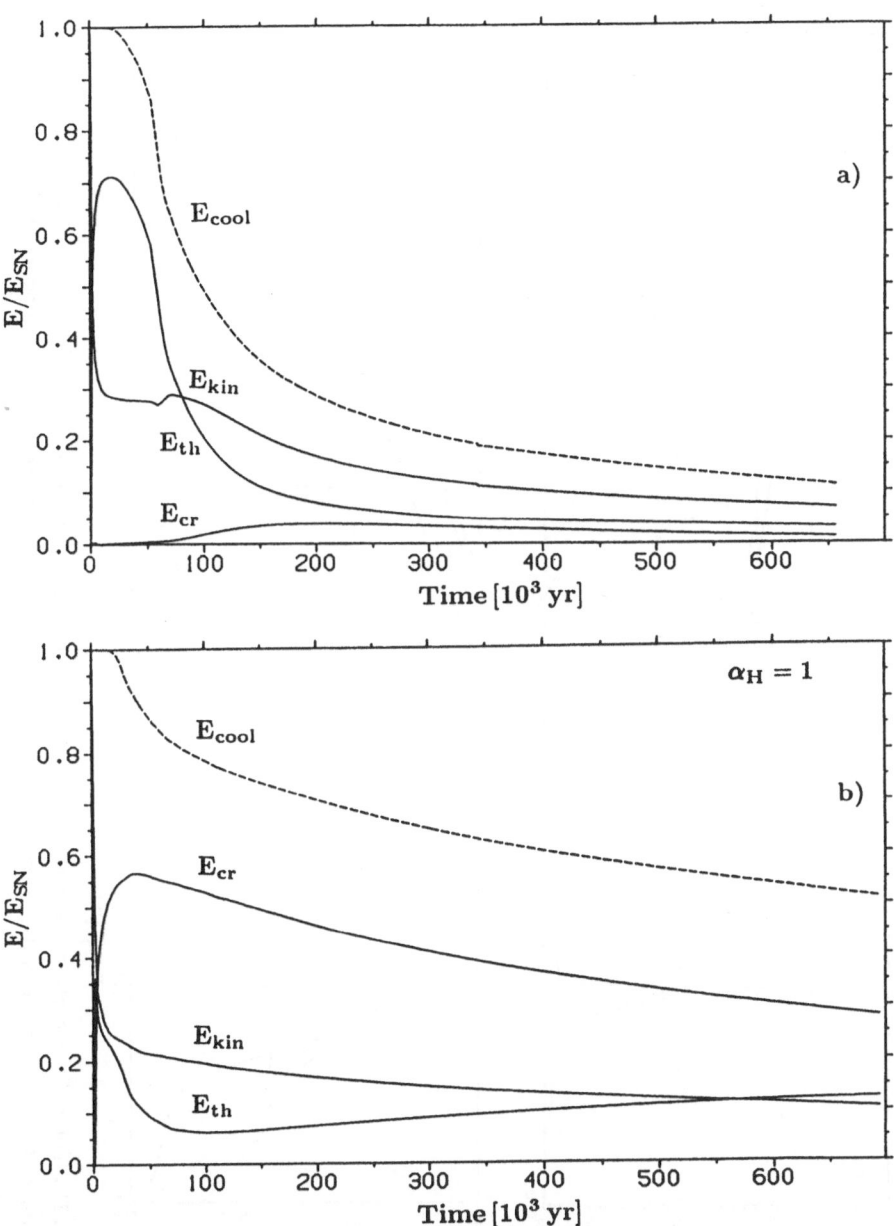

Fig. 17. The temporal evolution of the total energies as a function of time in units of the initial SN explosion energy E_{SN} in the case a) of low cosmic ray production and b) for a higher conversion of E_{SN} into E_{cr}. Case b) includes Alfvénic heating and injection of particles

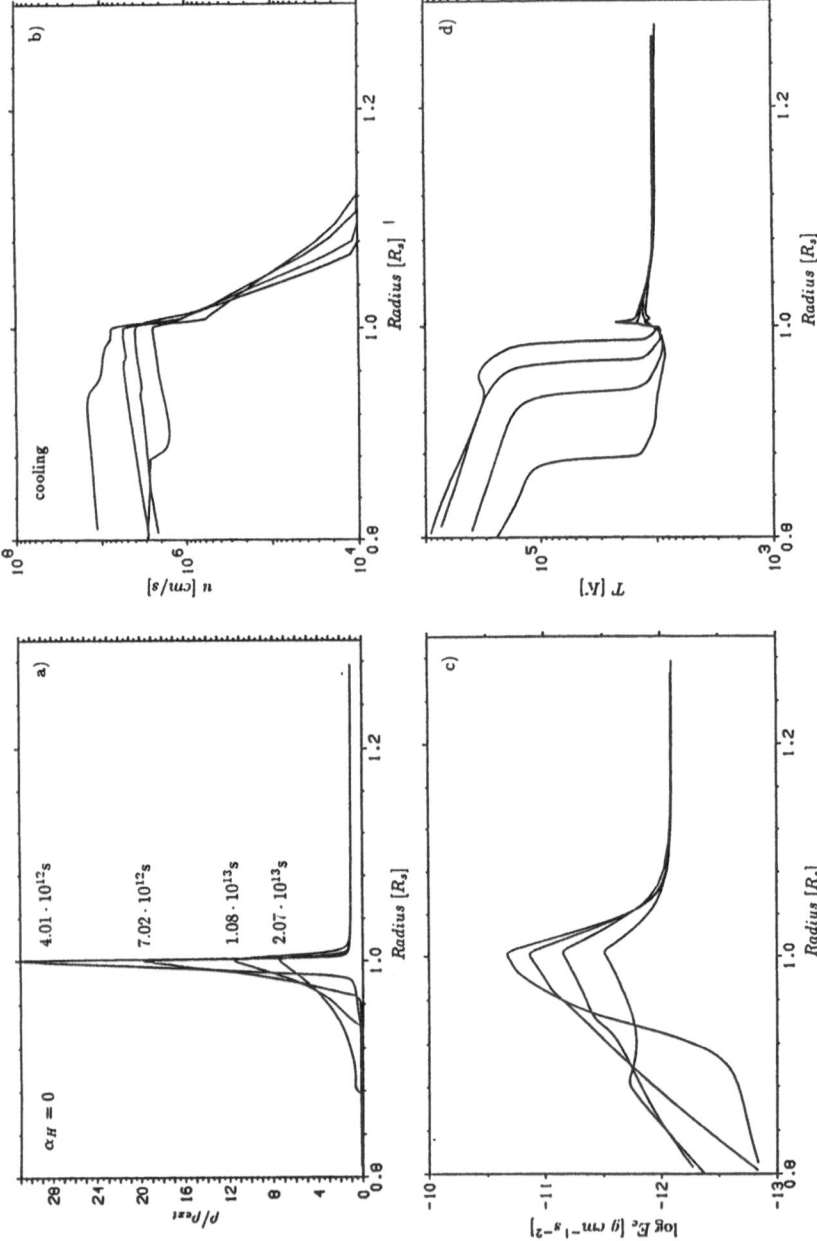

Fig. 18. The radial structure of the physical variables in the shock frame in units of the shock radius R_s. a) density in units of the external density, b) velocity with cosmic ray precursor and thermal subshock, c) cosmic ray energy density, d) temperature

this is an effective mechnanism to avoid catastrophic cooling (Falle 1975, 1981). The small temperature peak demonstrates the importance of cooling and during the depicted time interval the temperature varies less than a factor of two across the shock. The radial increase of the temperature in the postshock region is caused by hot gas which is left over from gas shocked at earlier times. The low densities prohibit a more rapid cooling since the density drops by about a factor of 20 compared to the external density. This density drop represents the contact discontinuity between the ejected stellar material and the shocked ISM (see also Fig. 12 in Sect. 2.1) yielding a slower cooling rate by a factor of 20^2. However, I emphasize that a cooling front is moving inwards seen in this shock frame. At $t = 4.01\,10^{11}$s the cooling front is located at $0.99R_s$ decreasing to $0.97R_s$ at $t = 7.02\,10^{12}$s, to $0.94R_s$ at $t = 1.08\,10^{13}$s and to $0.87R_s$ at an age of $t = 2.07\,10^{13}$s. The variations in the cold downstream region are due to cosmic ray gradients and at the time of $t = 2.07\,10^{13}$s the cosmic rays show a second peak around $0.88R_s = 1.54\,10^{20}$cm produced by a secondary shock which rises the temperature of the tenuous interior from 10^4K to about 10^5K.

Radiative cooling alters also the radial structure in the cooling region behind the forward shock which leads to the development of so-called secondary shocks (Falle 1975, 1981). Including the acceleration of energetic particles we have an additional pressure which limits catastrophic cooling even more efficient when a large number of cosmic rays is produced. In this case also a smooth expansion of the forward shock is expected (cf. Fig. 7, Sect. 1.4.3). In Fig. 19 we present the density a), the velocity b), the cosmic ray energy c) and the gas temperature d) in the shock frame for the model depicted also in Fig. 17b. Note that the cosmic ray pressure inside the remnant is about two orders of magnitude above the external value. The density in units of the external density of $n_0 = 0.3\,\mathrm{cm}^{-3}$ reveals the increase by cooling but also a cosmic ray precursor in front of the shock. The temperature (Fig. 19d) exhibits the almost isothermal shock wave and the cooling front moving inwards relative to the shock frame can be easily traced. Note also the density enhancement due to cooling and the occurrence of complicated structures in the downstream region of the shock wave which yield also fluctuations in the collisional production of neutral pions and the associated γ-ray fluxes (cf. Sect. 2.7).

Although large compression ratios are observed in the downstream regions all computations carried out so far show that the compression ratio r in the radiative shock wave remains below $r < 70$ limited by the downstream cosmic ray pressure avoiding the situation of catastrophic cooling. The postshock velocity is still of the order of $100\,\mathrm{km\,s}^{-1}$ showing a small cosmic ray precursor growing in time (cf. Fig. 19b). The last figures explore that radiative cooling as well as Alfvénic heating are important to determine the amount of E_{SN} which can be converted into cosmic rays. The effects of radiative cooling alter the overall energetics of a SNR but between 10% and

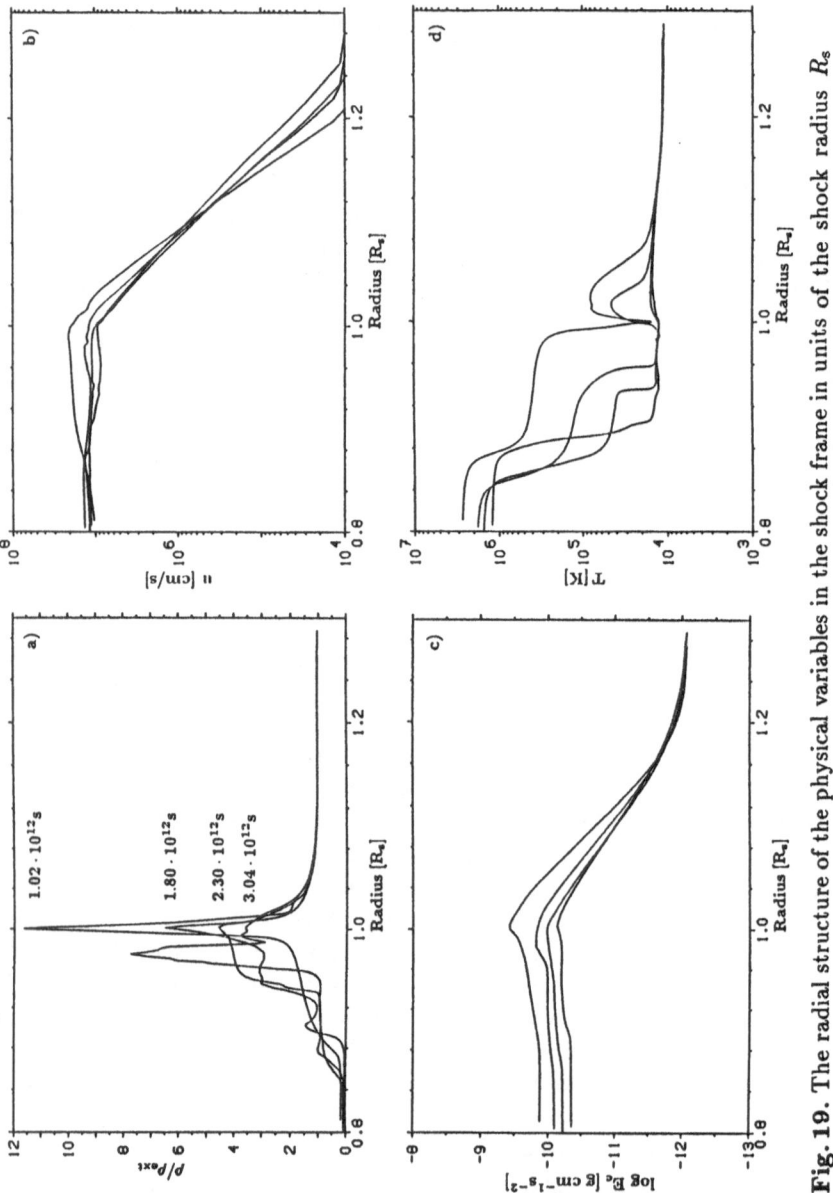

Fig. 19. The radial structure of the physical variables in the shock frame in units of the shock radius R_s between 10^{12} s and $3 \cdot 10^{12}$ s showing the development of downstream features due to radiative cooling. a) density, b) velocity, c) cosmic ray energy density and d) gas temperature

30% of the initial explosion energy E_{SN} is converted to high energy particles and can explain the power needed to maintain the observed cosmic ray energy density. The production of cosmic rays leads to lower temperatures and consequently radiative cooling is already important at earlier times during the SNR evolution. Although radiative cooling is relevant for the thermal structure of the remnant the motion of the shock wave is also influenced by the cosmic ray pressure (cf. Figs. 18,19). Several small scale structures (secondary shocks) are created in the downstream region and pushed by the pressure of the high energy particles, causing fluctuations of the downstream cosmic ray pressure. In the cases of large amounts of cosmic rays the energy can be returned to the thermal gas and cooling acts indirectly on the cosmic rays (cf. Fig. 17b). Hence, the thermal energy content of the remnant can remain almost constant during the cooling phase because heating due to dissipation of Alfvén waves transfers energy back to the thermal plasma from the cosmic ray energy gained in the Sedov-phase. I also want to stress that radiative cooling and acceleration of cosmic rays in shock waves need accurate numerical methods to resolve the physical length scales in order to obtain correct results.

2.6 X-ray Emission from SNR's

Many SNR's are prominent X-ray sources which reveal several properties of the evolution of the remnant itself as well as of the interstellar environment since a thermal plasma at temperatures around 10^6K emits in the X-ray range. During the entire evolution SNR's contain hot gas radiating at X-ray temperatures. Recently, a compilation of the X-ray surface brightness of 47 galactic SNR's has been published based on the observations of the *Einstein Observatory* (Seward 1990). Since the plasma is heated by several shock waves other non-thermal processes occurring at the shock transition can alter the amount of explosion energy transferred into the radiating gas. In the case of the Tycho remnant Heavens (1984b) has discussed the contraints on the efficiency of generating cosmic rays in shock waves based on such arguments. Moreover, the cosmic ray precursor leads to an adiabatic heating of the incoming plasma in the precursor region and this effect has been invoked to explain some observed X-ray halos around SNR's (Morfill et al. 1984).

However, the purpose ot this section is not to describe a particular remnant in detail but to show a number of global consequences of the acceleration process on the overall X-ray emission. In general differences caused by particle acceleration are expected only after $t \gtrsim t_{sw}$ looking at the time scales of the acceleration process (cf. Table 5, Eqs. 111). Therefore older remnants will exhibit differences depending on particle acceleration and for these older remnants we expect a relevant influence of the structure of the ambient medium on the X-ray appearence. Moreover, if cooling becomes important instabilities will develop changing more and more the spherical shape of the remnant which has been assumed in these calculations. Hence, to study the effects of particle acceleration on the emission of X-rays possibly leading to observable effects we have to follow the evolution of a SNR throughout the cooling phase (Sect. 2.3) up to the time where the expansion velocity becomes comparable to the typical velocities encountered in the surrounding interstellar medium. As inferred from the Figures 17,18,19 (Sect. 2.5.3) cosmic rays limit the pressure drop in the cooling postshock region. In numerical simulations of SNR's with cooling but without particle acceleration (Falle 1975, 1981) a magnetic field has been included which is dynamically unimportant until the thermal gas cools thereby compressing the magnetic field. As a consequence the magnetic pressure inhibits a further collapse of the cooling region. In a study of SNR's in M33 Blair et al. (1981) have found evidence for a limited compression ratio which can be attributed to such magnetic fields but also to the pressure of the high energy particles accelerated at the shock waves (Chevalier 1983, Dorfi 1991). Up to now no direct observational evidences for shock acceleration of particles other than electrons have been detected in SNR's. The production of relativistic electrons is seen through synchrotron emission in a broad frequency range. Since we are mainly interested in the overall changes of the X-ray emission due to particle acceleration we can assume the same chemical abundances across the whole remnant neglecting the differences in the cooling curves depending on the chemical composition (e.g. Böhringer & Hensler 1989). As mentioned before a detailed modelling of a certain remnant should take into account such chemical changes as well as time-dependent ionization processes but for exploring general features of the influence of particle acceleration on the X-ray luminosites we can ignore these differences.

If the acceleration is effective the energy dissipated in the shock waves is converted into energetic particles and hence the thermal gas remains cooler resulting in a softer X-ray radiation. This distinct thermal history of the shocked gas also leads to a different X-ray history which can be traced in the spatial X-ray structure of a remnant. Again, the two extreme cases concerning the amount of production of cosmic rays of Fig. 17 are discussed in more detail. The first model (case a) describes the evolution of a SNR where most of the kinetic energy is transferred to the thermal energy of the plasma and the total cosmic ray energy stays below $0.07E_{\mathrm{SN}}$. The second model (case b) corresponds to an extreme case where up to $0.6\,E_{\mathrm{SN}}$ of the explosion energy is used to accelerate cosmic rays. For both cases the radial dependence of the physical variables has been shown in the previous section. The associated γ-ray fluxes due to the decay of neutral pions continuously produced are depicted in the next section.

The case of larger amounts of cosmic rays (Fig. 20) shows a similar X-ray spectrum compared to the case of low cosmic ray production (not shown here) for evolutionary times less than the sweep-up time. The dif-

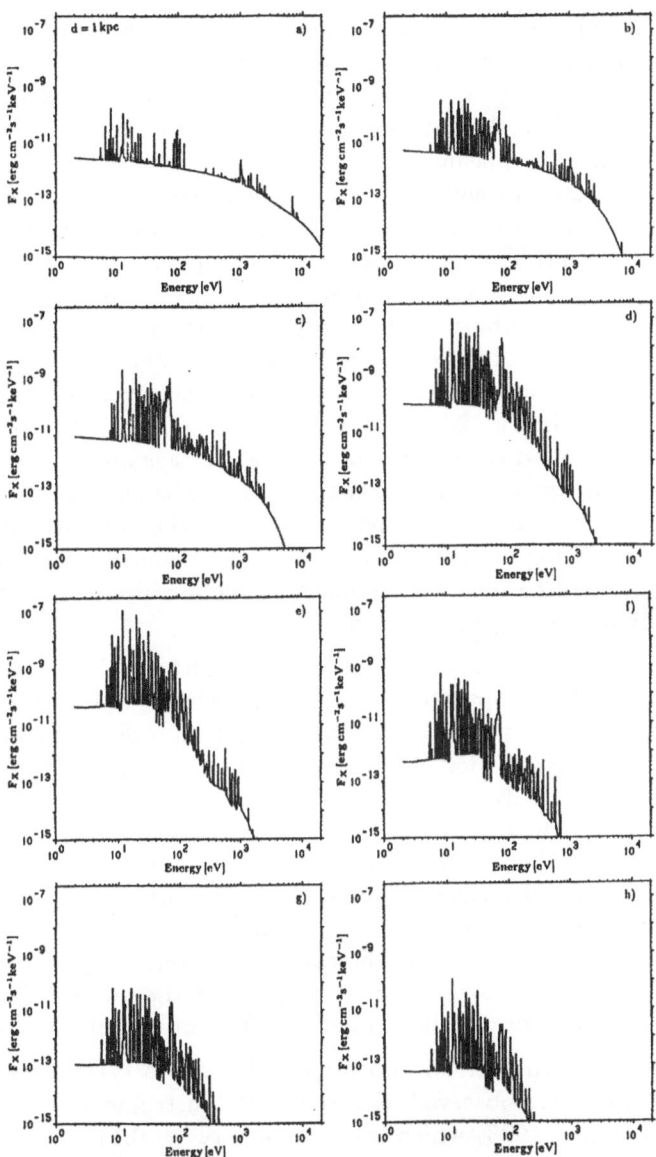

Fig. 20. Theoretical X-ray spectra in the energy range between 2 eV and 20 keV at different radii R_s of the forward shock in the case of high cosmic ray production. The values are normalized to a SNR distance of $d = 1$ kpc and the figures a)-h) correspomd to shock radii $R_s = 2.0$ pc, 4.9, 8.9, 16.5, 19.5, 29.9, 39.9 and 49.9

ferent radial distribution of hot gas can already be seen at $R_s = 4.9\,\mathrm{pc}$.
Since a large fraction of the explosion energy is converted into high energy
particles the remnant stays cooler as can be inferred from the X-ray spec-
tra of Fig. 20b,c,d. Note that already at $t = 4.26\,10^3\,\mathrm{yr}$ and $R_s = 8.9\,\mathrm{pc}$ a
notable fraction of $10^6\,\mathrm{K}$-plasma exists in this case and becomes even more
prominent in Fig. 20c through the X-ray lines around $70\,\mathrm{eV}$. In contrast
to the low cosmic ray case no important high temperature plasma is produced in this remnant. The feature at $35 - 50\,\mathrm{eV}$ points at a thermal gas
of $10^5\,\mathrm{K}$. However, in the interior we find a very tenuous hot plasma which
cannot significantly contribute to the X-ray emission. In addition radiative
losses are important already at earlier times (cf. Fig. 17) and the maximum
in the X-ray flux is located at $t = 3.24\,10^4\,\mathrm{yr}$ with an expansion radius of
$R_s = 19.5\,\mathrm{pc}$ plotted in Fig. 20e. Note that this maximum is less by a factor of about 5 compared to the case of low cosmic rays and occurs about
two times earlier in the evolution of the remnant. The subsequent phases
exhibit again rather similar X-ray spectra although the absolute flux level
decreases as the remnant expands. In general the main difference caused
by the acceleration of cosmic rays becomes visible as a spectral feature in
X-rays typically in the range above $1\,\mathrm{keV}$.

The late phases of $R_s = 39.9\,\mathrm{pc}$ and $R_s = 49.9\,\mathrm{pc}$ of Figs. 20g,h are
characterized again by comparable X-ray spectra relative to the case of low
cosmic rays but the emissivity is lower by at least an order of magnitude.
The cosmic ray energy equation (119) used in this model of a SNR contains
also the dissipation of Alfvénic fluctuations in regions of large cosmic ray
gradients which leads to a heating of the thermal plasma in the precursor
region of the forward shock. We can state that only warm gas is produced
not able to radiate significantly at X-ray wavelengths although this Alfvénic
heating enhances the total thermal energy of the remnant. Some secondary
shocks triggered by gradients of cosmic rays in the downstream region are
more frequent than in the low cosmic ray cases but they do not heat the
thermal gas enough to increase the X-ray flux in the interior of the remnant.

Fig. 21 shows the theoretical appearance of the two typical remnants
(cf. Fig. 17) if they were observed with the PSPC instrument on board of
ROSAT (e.g. Trümper 1983) which is sensitive between $0.1\,\mathrm{keV}$ and $2\,\mathrm{keV}$
and has an effective detector area of about $220\,\mathrm{cm}^2$ at $1\,\mathrm{keV}$. The observed
luminosities are scaled to a source distance of $d = 1\,\mathrm{kpc}$ and I have assumed
a low column density of hydrogen of $n_H = 10^{20}\mathrm{cm}^{-2}$ along the line of sight.
Note that the dip in the calculated ROSAT-spectra around $0.5\,\mathrm{keV}$ is due
to an absorption at the entrance window of the PSPC. All curves are labeled with the shock radius R_s. As stated before the observable difference
only becomes visible for shock radii $R_s \gtrsim R_{sw}$. Hence, comparing Fig. 21a
vs. Fig. 21b the theoretical ROSAT-spectra at $R_s = 2\,\mathrm{pc}$ are identical but
exhibit an increasing diversity as the remnant expands. At $R_s = 16.7\,\mathrm{pc}$
(cf. Fig. 21a) we still see a significant contribution from hot gas at energies

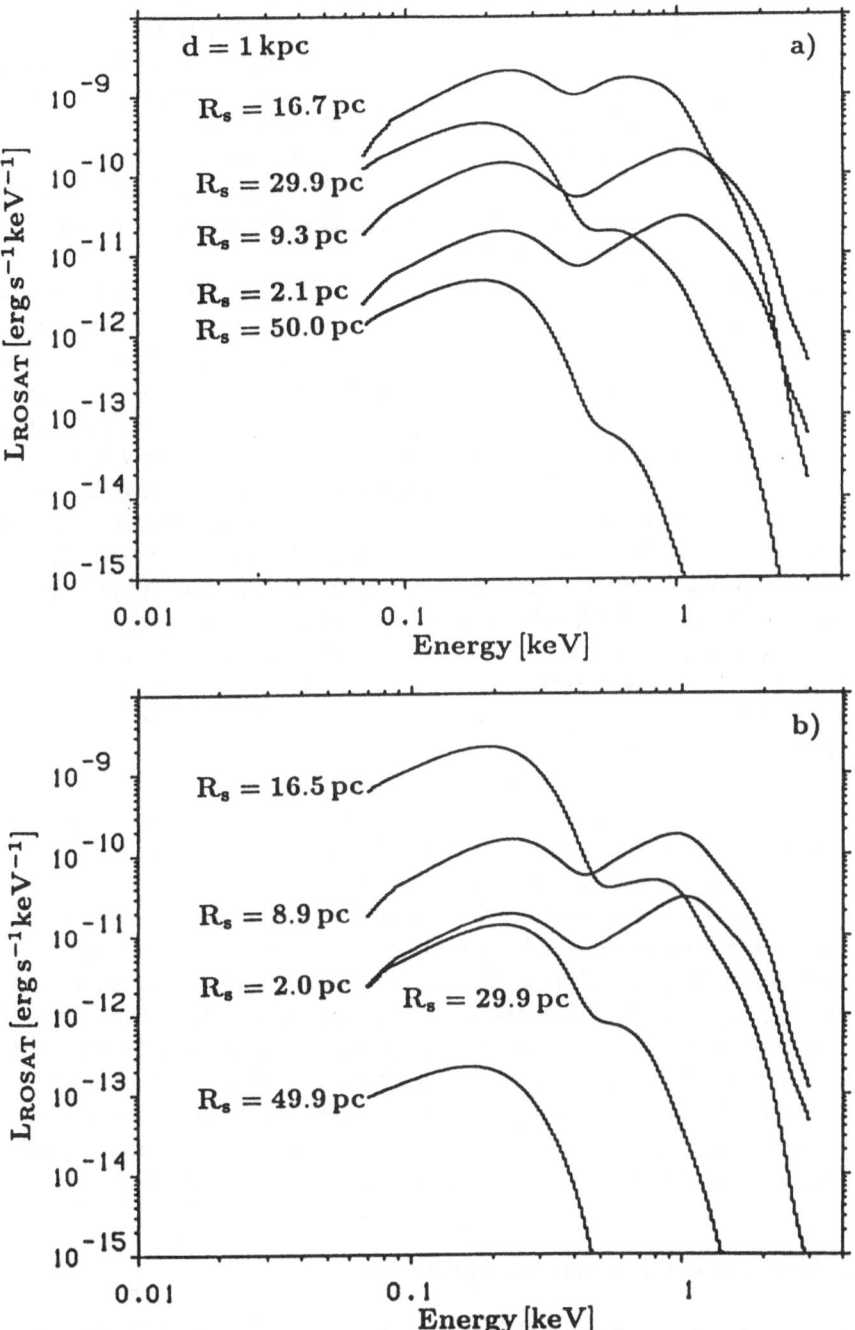

Fig. 21. The theoretical spectral appearance of two remnants (observed by the PSPC instrument of ROSAT) for different radii R_s of the forward shock. a) denotes a case of low cosmic ray pressure and in case b) the total cosmic ray energy can reach up to $0.6E_{SN}$

above 1 keV in the case of low cosmic rays. A remnant with an effective cosmic ray production does not exhibit this amount of hot gas at $R_s = 16.5$ pc (cf. 21b). Such a difference remains valid at all later expansion phases, e.g. $R_s = 29.9$ pc but the X-ray flux level can be shifted down by a factor up to 20 if cosmic rays are accelerated. Again, this can easily be seen from the lowest curves of Fig. 21a,b labeled by $R_s = 50$ pc and $R_s = 49.9$ pc where the latter model has a theoretical ROSAT-luminosity of less than 10^{-15} erg s^{-1}keV^{-1} above 0.5 keV.

The evolution of the X-ray flux and the ROSAT-luminosity in the energy range between 0.1 keV and 2.4 keV is depicted as a function of time in Fig. 22, again normalized to a source distance of $d = 1$ kpc. The two curves correspond to the two remnants differing by the production of energetic particles. As explained in the previous sections cosmic rays alter the shock structure and the associated temperature structure only after $t \gtrsim t_{sw}$ or in this case $t \gtrsim 10^{11}$s. At this age the remnant of case b) (effective cosmic ray acceleration) starts already cooling, increasing the X-ray flux in the ROSAT sensitivity range. The maximum is reached at about $5\,10^{11}$s whereas the case with no significant cosmic rays (case a, upper curve at late times) exhibits its maximum around $1.2\,10^{12}$s which is a factor of 4 more luminous than case b). After that event both X-ray luminosities decrease almost parallel as seen in Fig. 22a, i.e. $t \gtrsim 10^{12}$s and the relative difference is given by about a factor of 20. Qualitatively the same behaviour can be observed in ROSAT-luminosity of Fig. 22b but the folding of the theoretical X-ray spectra with the spectral sensitivity of the ROSAT PSPC leads to some differences in the time interval between 10^{11}s and 10^{12}s due to the very different radial structure of gas temperatures (cf. Figs. 18,19).

This difference in the efficiency of particle acceleration can be traced indirectly in the X-rays originating from the thermal plasma. However, the diversity of SNR's and of the surrounding interstellar medium necessitates a detailed modelling of each individual remnant before definite conclusions on the efficiency of particle acceleration can be drawn from the observations of X-ray fluxes. Since the X-ray fluxes depend on the particle acceleration these effects also alter the X-ray statistics of SNRs as well as the contribution of the galactic X-ray background because if a thermal plasma is heated up to X-ray emitting temperatures by shock waves cosmic rays are accelerated at the same astrophysical site changing thereby the shock structure.

2.7 Gamma-ray Emission from SNR's

Most of the γ-rays above 100 MeV are due to the decay of neutral pions created by collisions of cosmic ray protons in the energy range between 1 MeV and 30 MeV with the thermal plasma (e.g. Stecker 1973, Stephens & Badhwar 1981). Contributions by bremsstrahlung or inverse Compton scattering of cosmic electrons are important for the production of γ-rays at

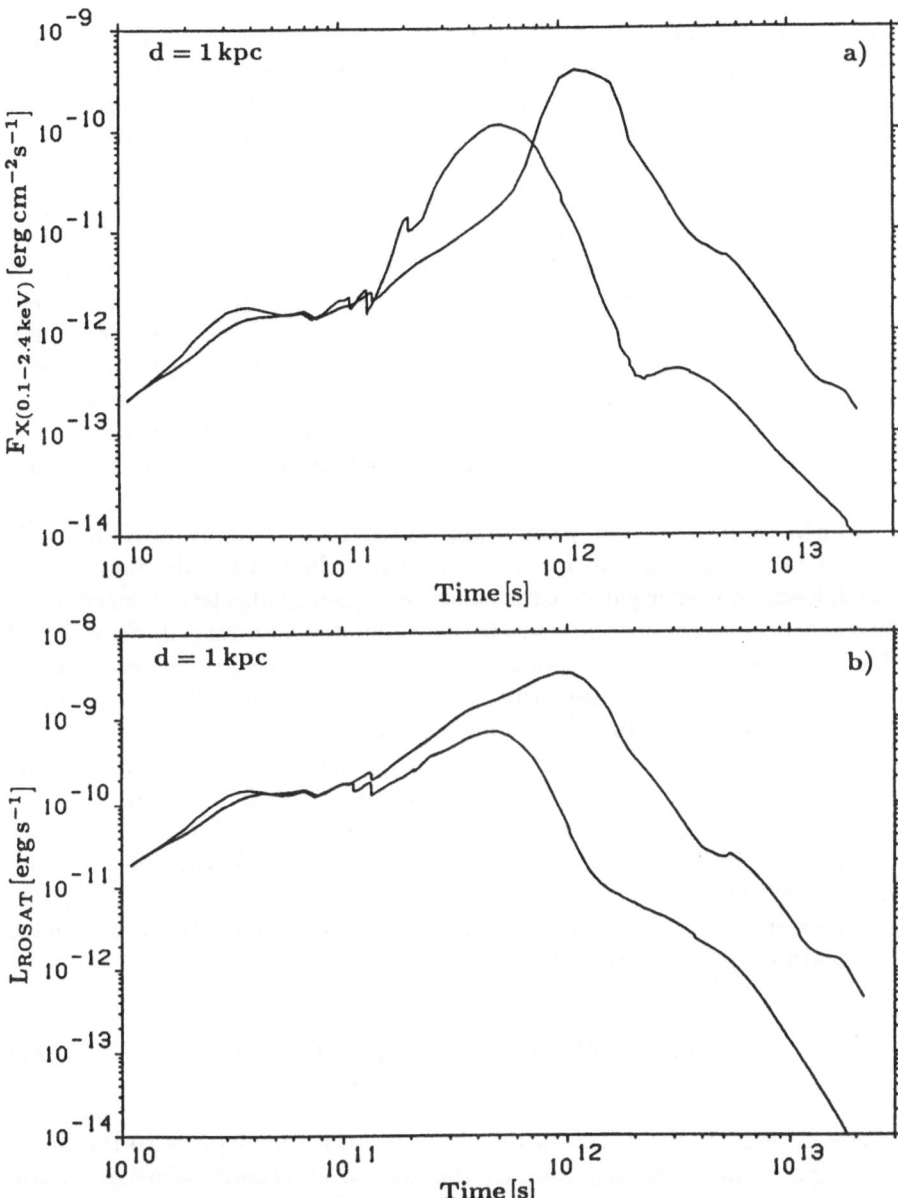

Fig. 22. The temporal evolution of the X-ray flux a) in the energy range between 0.1 keV and 2 keV and the corresponding ROSAT-flux b) normalized to source distance $d = 1$ kpc

energies lower than 100 MeV. These three processes produce different γ-ray spectra and so their relative contributions can be calculated if the physical conditions of the galactic environment are known. The observations of cosmic electrons are affected at lower energies by solar wind modulation (for bremsstrahlung). At higher energies the inverse Compton scattering lies outside the frequencies which can be observed by earth-bound radio telescopes. Therefore the spectrum of cosmic electrons is poorly known (e.g. Kniffen & Fichtel 1981) and consequently the contribution of the two latter processes to the gamma luminosity of our galaxy is not well determined (e.g. Sacher & Schönfelder 1983 and Sect. 1.3). The acceleration of high energy particles is therefore closely connected to the production of γ-ray photons above 100 MeV. Collisions of the cosmic ray protons with the thermal plasma create neutral pions decaying into observable γ-quanta. Hence, the observed γ-ray flux above 100 MeV depends on the cosmic ray energy density times the gas density.

When calculating the γ-ray luminosities from π^o-decay in SNR's it is in particular important to include radiative cooling of the thermal plasma which leads to higher gas densities and consequently also to enhanced γ-ray fluxes. Since we are working on the hydrodynamical level (c.f. Sect. 2.5.1) the particle distribution function is averaged out leaving no information on the resulting cosmic ray spectrum as well as on the neutral pion spectrum created by collisions. But in a simple estimate of the γ-luminosity of a SNR producing cosmic rays we can assume a particle spectrum similar to the observed one and calculate a yielding factor $q_\gamma = 1.4\,10^{-13}\,\mathrm{cm^3 erg^{-1} s^{-1}}$ for the production of γ-rays above 100 MeV by π^o-decay (Higdon & Lingenfelter 1975) and can convert the energy density of the cosmic rays into a γ-ray flux above 100 MeV.

To evaluate the total γ-ray flux from the expanding shock wave we have to integrate over the remnant

$$F_{\gamma,s}(> 100\,\mathrm{MeV}) = \frac{q_\gamma}{4\pi d^2} \int_0^{R_s} n E_c 4\pi r^2 \, dr \qquad (123)$$

where R_s denotes the radius of the shock front, d the distance to the SNR, n the gas number density and E_c the cosmic ray energy density. In the case of a cosmic ray precursor we calculate the integral γ-ray flux $F_{\gamma,p}$ up to the radius of the precursor R_p where the cosmic ray energy density E_c declines to the value of $1/e$ compared to the maximum located at the shock front. Hence, the two upper curves in each plot correspond to the γ-ray fluxes integrated up to the shock radius R_s or the radius of the precursor R_p. The lower curves give the amount of γ-ray flux expected from a bubble of radius R_s filled with the interstellar cosmic ray energy density $E_{c,0}$ (cf. Eq. 29) and thermal gas n_0 describing the ambient medium, i.e. $F_{\gamma,b}(t) = 4\pi/3\,n_0 E_{c,0} R_s^3(t)$. During the Sedov-Taylor phase (79) we therefore have $F_{\gamma,b}(t) \propto t^{6/5}$. Note that in Figs. (23) and (24) these γ-ray fluxes are normalized to a source distance of $d = 1\,\mathrm{kpc}$.

Fig. 23a is typical for the case where a small fraction of E_{SN} is converted to cosmic rays. The γ-ray flux $F_{\gamma,s}$ within the shock radius does not differ from the γ-ray flux within the precursor $F_{\gamma,p}$ because no significant modifications of the shock front are present up to the time when cooling becomes important (85) and a dense shell is formed. In this case at the time of $t = 4.05\,10^{11}\mathrm{s}$ a strong secondary shock is clearly visible triggered by radiative cooling which decreases the cosmic ray energy density as well as the gas density in the vicinity of the forward shock yielding also a lower γ-ray flux $F_{\gamma,s}$. At the time of $1.6\,10^5$years the γ-ray fluxes (Fig. 23a) reach their maximum of $F_{\gamma,s} = 3.3\,10^{-8}\mathrm{ph\,cm^{-2}s^{-1}}$ and $F_{\gamma,p} = 3.6\,10^{-8}\mathrm{ph\,cm^{-2}s^{-1}}$ decreasing towards $2\,10^{-8}\mathrm{ph\,cm^{-2}s^{-1}}$ at $6.5\,10^5$years and $R_s = 56.8\,\mathrm{pc}$.

In the case of higher cosmic ray pressures (cf. Fig. 23b) we get an enhancement of the γ-ray fluxes already at earlier times around $2\,10^4$years but small scale shock waves in the downstream region cause more visible fluctuations on the π^o-production because the cooler gas is pushed more easily by the cosmic ray pressure gradients. The larger difference of Fig. 23b between the $F_{\gamma,p}$- and $F_{\gamma,s}$-fluxes compared to Fig. 23a is due to the important contribution of creating neutral pions in the precursor region of the forward shock. This model includes Alfvénic heating in the precursor region (cf. Fig. 17b) and hence we observe a faster decrease of the γ-ray flux in time relative to Fig. 23a because cosmic ray energy is transferred to the thermal plasma. On the other hand the shock radius $R_s(t)$ is a smooth function of time for both models as can be derived from the lowest curves in Figs. 23a,b, denoting the background flux $F_{\gamma,b}$ of γ-rays. From this fact we can conclude that radiative cooling does not affect the motion of the shock front compared to simulations without cosmic rays (Falle 1975, 1981). The evolution of the γ-ray flux is calculated up to the age of 10^6 years when the SNR has reached a radius of $73.4\,\mathrm{pc}$ and the shock velocity has dropped to $30\,\mathrm{km\,s^{-1}}$ differing by less than a factor of two compared to the final SNR radius (91). The γ-ray flux decreases at that time to about $F_{\gamma,s} = 4\,10^{-8}\mathrm{ph\,cm^{-2}s^{-1}}$.

In Figure 24 the distribution of the energies in units of E_{SN} as well as the γ-ray fluxes $F_{\gamma,s}$ and $F_{\gamma,p}$ above $100\,\mathrm{MeV}$ are shown for a SN exploding into a homogeneous medium with an ambient density of $n_0 = 5\,\mathrm{cm^{-3}}$. The sweep-up time (72) is given in this case by $t_{sw} = 1.47\,10^{10}\mathrm{s}$ corresponding to 466 years. Radiative cooling effects become important already after 8490 years (85). In this case we can distinguish three reverse shocks occurring at 790, 1050 and 2640 years until the transition from the free expansion to the pressure driven Sedov-Taylor phase is accomplished. At the later time of $12840\,\mathrm{yr}$ a more pronounced secondary shock wave is running inwards and reducing the γ-ray fluxes $F_{\gamma,s}$ and $F_{\gamma,p}$. Between this event and 25400 years the compression ratio increases from $r = 7.9$ to its maximum value of $r = 66.3$. Further in time several smaller secondary shocks lead to

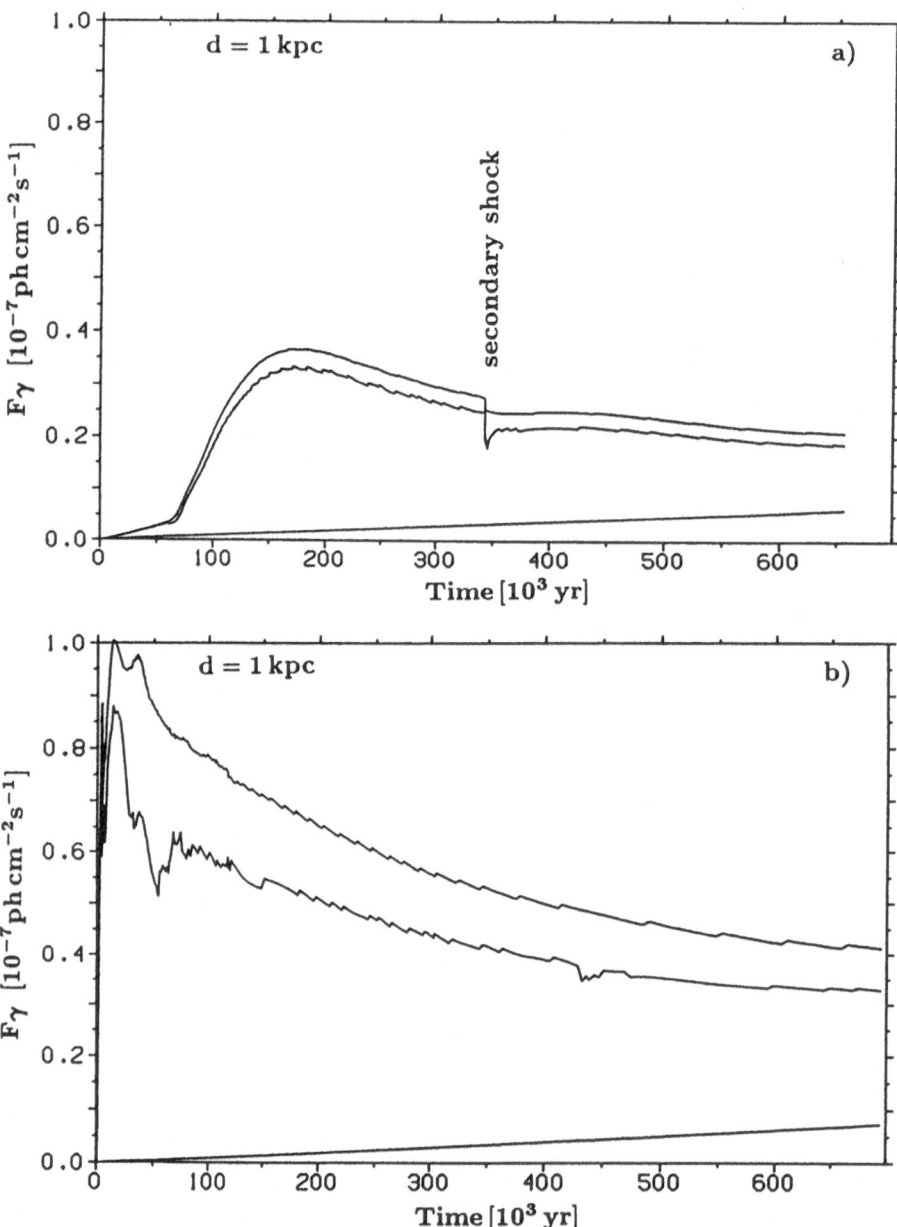

Fig. 23. The time evolution of the γ-ray fluxes in units of $10^{-7}\,\mathrm{ph\,cm^{-2}s^{-1}}$ up to $7\,10^5\,\mathrm{yr}$ in the case a) of $E_{\mathrm{cr}} \lesssim 0.07 E_{\mathrm{SN}}$ and case b) where $E_{\mathrm{cr}} \lesssim 0.6 E_{\mathrm{SN}}$. The fluctuations are caused by small scale shocks developing in the cooling region of the forward shock. All values are normalized to a source distance of $d = 1\,\mathrm{kpc}$

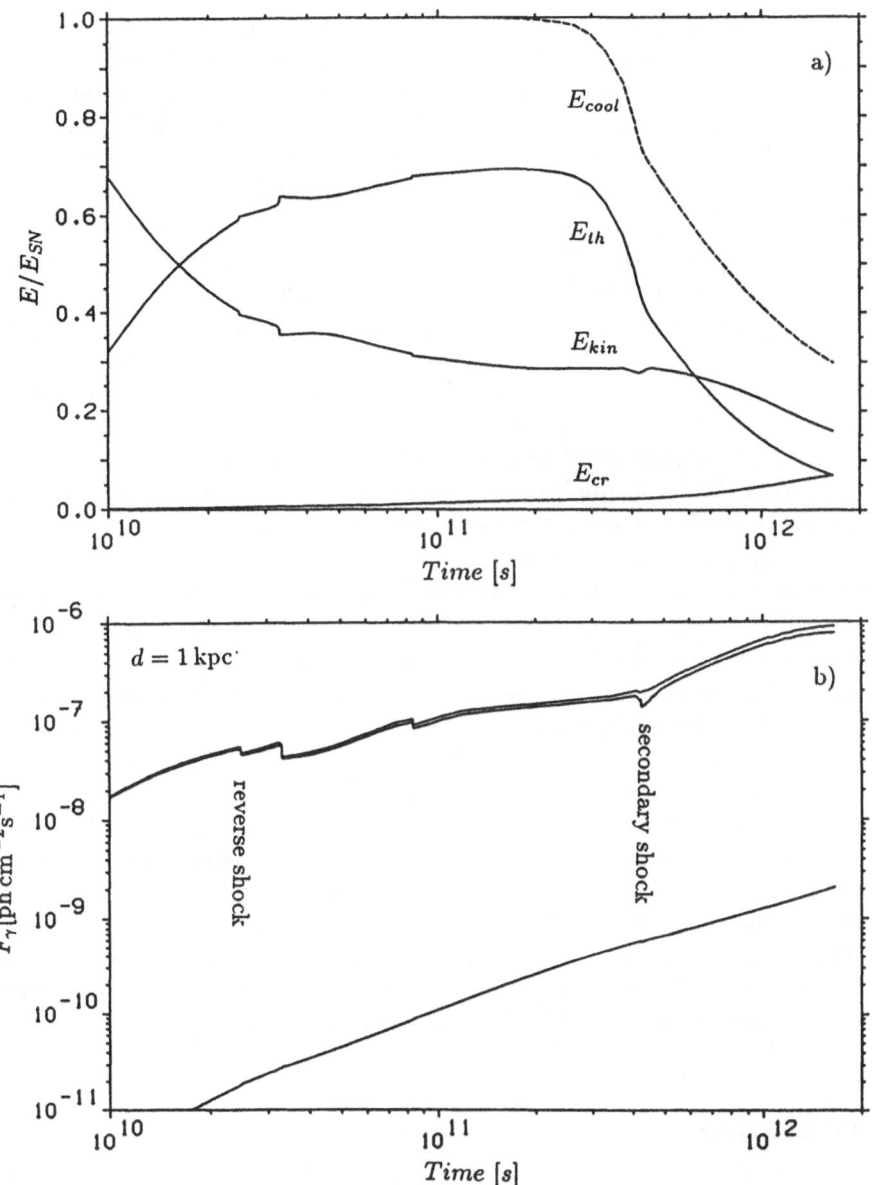

Fig. 24. The temporal evolution of the total energies a) in units of $E_{\rm SN} = 10^{51}$ erg and the γ-ray flux b) above 100 MeV in the case of an external density of $n_0 = 5\,{\rm cm}^{-3}$ typical for Cas A $(d = 1\,{\rm kpc})$

variations in the γ-ray fluxes at level of a few percents. The higher ambient density is also responsible for increased γ-ray fluxes which level at about 10^{-6}ph cm^{-2}s^{-1} for ages later than $5\,10^4$ years although the total cosmic ray energy remains at about $E_{cr} \leq 0.07\,E_{SN}$ during the entire evolution. The ongoing particle acceleration smoothens the shock front and the increasing precursor spreads the two curves $F_{\gamma,s}$ and $F_{\gamma,p}$. The last calculated model at $5.26\,10^5$ years exhibits a shock radius of 14.4 pc indicating again that the motion of the forward shock is not affected by radiative cooling if the acceleration of particles in shock waves is taken into account.

The γ-ray fluxes from π^0-decay yield 10^{-10} ph cm^{-2}s^{-1} for a typical remnant with $n_0 = 0.3$ cm^{-3} with a small amount of cosmic ray energy at a time of 300 years and increase almost like R_s^3 due to geometric effects. Remnants with higher fractions of cosmic rays or remnants evolving into a medium of higher ambient density produce γ-ray fluxes of up to about 10^{-6} ph cm^{-2}s^{-1} (normalized to a SNR at a distance of 1 kpc). In the latter case the γ-ray fluxes are almost constant between 10^4 and 10^6 years.

Summarizing the γ-ray results we have seen that radiative cooling and Alfvénic heating are essential to determine the γ-ray luminosities originating from neutral pion decay in evolving SNR's. I emphasize the density enhancements due to cooling and the occurrence of complicated structures in the downstream region of the shock wave which produce fluctuations in the collisional production of neutral pions and the associated γ-ray fluxes (Figs. 23,24). Since most SN are exploding into an inhomogeneous medium cloud evaporation and crusching as well as non-radial motions can produce similar fluctuations of the γ-ray flux. In a cloudy medium such γ-ray variations will occur even in the case of a constant cosmic ray energy density (cf. McKee 1988 for a recent review on SNR evolution into an inhomogeneous medium). Ambient densities of a $n_0 \simeq 5$ cm^{-3} can lead to γ-ray fluxes of up to 10^{-6}ph cm^{-2}s^{-1} (normalized to a source distance of $d = 1$ kpc) even in the cases where the total cosmic ray energy E_{cr} remains at a few percent of the initial SN-explosion energy E_{SN}.

Conclusions

Since this is the written version of the lectures given in Graz I have to admit that it is almost impossible to give a comprehensive review about the interstellar medium and supernova remnants. However, I hope that some of the unsolved problems concerning a theory of the interstellar medium have become clear during this course and I am looking forward to the solutions which might emerge from this audience of young scientists in the next years.

Acknowledgements

I like to thank my colleagues at the institute in Vienna Drs. M. Feuchtinger and A. Hajek who helped me with some figures. I am greatful to Dr. H. Böhringer (MPE, Garching) who provided me with the cooling data. Many thanks go to the scientific as well as the local organizing committee who made this 4th EADN school a very pleasant and stimulating event.

References

Achterberg, A., Blandford, R.D., Periwal, V. (1984): Astron. Astrophys. **132** 97

Arnett, W.D. (1987): Astrophys. J. **319** 136

Arnett, W.D. (1988): Astrophys. J. **331** 377

Axford, W.I. (1981a): Proc. Int. School and Workshop on Plasma Astrophysics, Varenna, ESA SP-161, 425

Axford, W.I. (1981b): Proc. 17th Int. Cosmic Ray Conf., Paris, **2** 299

Axford, W.I., Leer, E., Skadron, G. (1977): Proc. 15th Int. Cosmic Ray Conf., Plovdiv, **11** 132

Axford, W.I., Leer, E., McKenzie, J.F. (1982): Astron. Astrophys. **111** 317

Barsuhn, J., Walmsley, C.M. (1977): Astron. Astrophys. **54** 345

Beck, R., Kronberg, P.P., Wielebinski, R. (Eds.) (1990): 'Galactic and Intergalactic Magnetic fields', IAU Symp. 140, Kluwer, Dordrecht

Bell, A.R. (1977): Mon. Not. Roy. Astron. Soc. **179** 573

Bell, A.R. (1978a): Mon. Not. Roy. Astron. Soc. **182** 147

Bell, A.R. (1978b): Mon. Not. Roy. Astron. Soc. **182** 443

Biermann, L., Davis, L.Jr. (1958): Zeitschrift f. Astrophys. **51** 19

Blair, W.P., Kirshner, R.P., Chevalier, R.A. (1981): Astrophys. J. **247** 879

Blandford, R.D. (1980): Astrophys. J. **238** 410

Blandford, R.D. (1988): in 'Supernova Remnants and the Interstellar Medium', eds. R.S. Roger and T.L. Landecker, Cambridge Univ. Press, Cambridge, p. 309

Blandford, R.D., Ostriker, J.P. (1978): Astrophys. J. **221** L29

Blandford, R.D., Ostriker, J.P. (1980): Astrophys. J. **237** 793

Bloemen, H. (1989): Ann. Rev. Astron. Astrophys. **27** 469

Böhringer, H., Hensler, G. (1989): Astron. Astrophys. **215** 147

Bowyer, S., Leinert, C. (Eds.) (1989): 'The Galactic and Extragalactic Background Radiation', IAU Symp. 139, Kluwer, Dordrecht

Bregman, J.N. (1980): Astrophys. J. **236** 577

Bregman, J.N., Harrington, J.P. (1986): Astrophys. J. **309** 833

Breitschwerdt, D., McKenzie, J.F., Völk, H.J. (1991): Astron. Astrophys. **245** 79

Brinks, E., Shane, W.W. (1984): Astrophys. J. Suppl. **55** 179

Burke, J.A. (1968): Mon. Not. Roy. Astron. Soc. **140** 241

Burton, W.B., Liszt, H.S. (1981): in 'Origin of Cosmic Rays', IAU Symp. No. 94, eds. G. Setti, G. Spada, A.W. Wolfendale, D. Reidel, Dordrecht, p. 231

Cesarsky, C.J. (1980): Ann. Rev. Astron. Astrophys. **18** 289

Cesarsky, C.J., Montmerle, T. (1983): Space Sci. Rev. **36** 173

Cesarsky, C.J., Völk, H.J. (1978): Astron. Astrophys. **70** 367

Chevalier, R.A. (1982): Astrophys. J. **258** 790

Chevalier, R.A. (1983): Astrophys. J. **272** 765

Chevalier, R.A., Gardner, J. (1974): Astrophys. J. **192** 457

Chevalier, R.A., Oegerle, W.R. (1979): Astrophys. J. **227** 398

Cox, D.P. (1972): Astrophys. J. **178** 159

Cox, D.P., Reynolds, R.J. (1987): Ann. Rev. Astron. Astrophys. **25** 303

Cox, D.P., Smith, B.W. (1974): Astrophys. J. **189** L105

Dalgarno, A. (1987): in 'Physical Processes in Interstellar Clouds', eds. G.E. Morfill and M. Scholer, D. Reidel, Dordrecht, p. 219

Dalgarno, A., McCray, R.A. (1972): Ann. Rev. Astron. Astrophys. **10** 375

de Jong, T., Dalgarno, A., Boland, W. (1980): Astron. Astrophys. **91** 68

Dorfi, E.A. (1989): Astron. Astrophys. **225** 507

Dorfi, E.A. (1990): Astron. Astrophys. **234** 419

Dorfi, E.A. (1991): Astron. Astrophys. **251** 597

Dorfi, E.A., Drury L.O'C. (1987): J. Comput. Phys. **69** 175

Drury, L.O'C. (1983): Rep. Prog. Phys. **46** 973

Drury, L.O'C., Völk, H.J. (1981): Astrophys. J. **248** 344

Drury, L.O'C., Markiewicz, W., Völk, H.J. (1989): Astron. Astrophys. **225** 179

Ebert, R., von Hoerner, S., Temesvary, S. (1960): in 'Die Entstehung von Sternen durch kondensation diffuser Materie', Springer, Berlin, p. 311

Engelmann, J.J., Goret, P., Juliusson, E., Koch-Miramond, L., Lund, N., Masse, P., Rasmussen, I.L., Soutoul, A. (1985): Astron. Astrophys. **148** 12

Falle, S.A.E.G. (1975): Mon. Not. Roy. Astron. Soc. **172** 55

Falle, S.A.E.G. (1981): Mon. Not. Roy. Astron. Soc. **195** 1011

Fermi, E. (1949): Phys. Rev. **75** 1169

Field, J.B. (1965): Astrophys. J. **142** 531

Field, J.B., Goldsmith, D.W., Habing, H.J. (1969): Astrophys. J. **155** L49

Garcia-Muñoz, M., Mason, G.M., Simpson, J.A. (1977): Astrophys. J. **217** 859

Gies, D.R. (1987): Astrophys. J. Suppl. Ser. **64** 545

Gillis, J., Mestel, L. Paris, R.B. (1974): Astrophys. Sp. Sci. **27** 167

Gillis, J., Mestel, L. Paris, R.B. (1979): Mon. Not. Roy. Astron. Soc. **187** 311

Ginzburg, V.L., Ptuskin, V.S. (1985): Sov. Sci. Rev. E. Astrophys. Space Phys. **4** 161

Ginzburg, V.L., Syrovatskii, S.I. (1965): Ann. Rev. Astron. Astrophys. **3** 297

Guelin, M., Langer, W.D., Wilson, R.W. (1982), Astron. Astrophys. **107** 107

Hamilton, A.J.S. (1985): Astrophys. J. **291** 523

Hamilton, P.A., Lyne, A.G. (1987): Mon. Not. Roy. Astron. Soc. **224** 1073

Heavens, A.F. (1984a): Mon. Not. Roy. Astron. Soc. **210** 813

Heavens, A.F. (1984b): Mon. Not. Roy. Astron. Soc. **211** 195

Heiles, C. (1979): Astrophys. J. **229** 533

Heiles, C. (1984): Astrophys. J. Suppl. **55** 585

Heiles, C. (1987): Astrophys. J. **315** 555

Heiles, C. (1991): in: 'The Interstellar Disk-Halo Connection in Galaxies', ed. H. Bloemen, IAU Symp. 144, Kluwer, Dordrecht, p. 433

Henderson, A.P., Jackson, P.D., Kerr, F.J. (1982): Astrophys. J. **263** 182

Higdon, J.C., Lingenfelter, R.E. (1975): Astrophys. J. **198** L17

Honda, M. (1979): Proc. 16th Int. Cosmic Ray Conf., Kyoto, **14** 159

125

Ikeuchi, S., Habe, A., Tanaka, Y.D. (1984): Mon. Not. Roy. Astron. Soc. **207** 909

Ipavich, F. (1975): Astrophys. J. **196** 107

Jenkins, E.B. (1978a): Astrophys. J. **219** 845

Jenkins, E.B. (1978b): Astrophys. J. **220** 107

Jenkins, E.B., Meloy, D.A. (1974): Astrophys. J. **193** L121

Johnson, H.E., Axford, W.I. (1971): Astrophys. J. **165** 381

Jokipii, J.R. (1966): Astrophys. J. **146** 480

Kahn, F.D. (1976): Astron. Astrophys. **50** 145

Kahn, F.D. (1991): in 'The Interstellar Disk-Halo Connection in Galaxies', ed. H. Bloemen, IAU Symp. 144, Kluwer, Dordrecht, p. 1

Kamper, K., van den Bergh, S. (1976): Astrophys. J. Suppl. Ser. **32** 351

Kamper, K., van den Bergh, S. (1978): Astrophys. J. **224** 851

Kennel, C.F., Coroniti, F.V. (1984): Astrophys. J. **283** 710

Kennel, C.F., Edmiston, M., Hada, T. (1985): J. Geophys. Res. **90** A1

Kniffen, D.A., Fichtel, C.E. (1981): Astrophys. J. **250** 389

Kniffen, D.A., Fichtel, C.E., Thompson, D.J. (1977): Astrophys. J. **215** 765

Krause, M. (1990): in 'Galactic and Intergalactic Magnetic fields', eds. R. Beck, P.P Kronberg and R. Wielebinski, IAU Symp. 140, Kluwer, Dordrecht, p. 187

Krause, F., Rädler, K.-H. (1980): 'Mean-field Magnetohydrdynamics and Dynamo Theory', Pergamon Press, Oxford

Kraushaar, W. (1970): in 'Gamma Ray Astrophysics Colloqium', ESRO SP-58, p. 15

Krebs, J. Hillebrandt, W. (1983): Astron. Astrophys. **128** 41

Krymsky, G.F. (1977): Dokl. Nauk. SSR **234** 1306, (Engl. trans.: Sov. Phys. Dokl. **23** 327)

Kuslrud, R.M., Pearce, W. (1969): Astrophys. J. **156** 445

Lamb, H. (1945): 'Hydrodynamics', Dover Reprint, New York

Landau, L.D., Lifschitz, E.M. (1976): 'Fluid Mechanics', Pergamon Press, New York

Lee, M.A. (1982): J. Geophys. Res. **87** 5063

Lee, M.A. (1983): J. Geophys. Res. **88** 6109

Lerche, I. (1967): Astrophys. J. **147** 689

Lo, K.Y., Walker, R.C., Burke, B.F., Moran, J.M., Johnston, K.J., Ewing, M.S. (1975): Astrophys. J. **202** 650

Lockman, F.J., Hobbs, L.M., Shull, J.M. (1986): Astrophys. J. **301** 380

Lyne, A.G. (1990): in 'Galactic and Intergalactic Magnetic fields', eds. R. Beck, P.P Kronberg and R. Wielebinski, IAU Symp. 140, Kluwer, Dordrecht, p. 41

Manchester, R.N. (1972): Astrophys. J. **172** 43

Manchester, R.N. (1974): Astrophys. J. **188** 637

Markiewicz, W.J., Drury, L.O'C, Völk, H.J (1990): Astron. Astrophys. **236** 487

Mathews, W.G., Baker, J.C. (1971): Astrophys. J. **170** 241

Mathis, J.S., Mezger, P.G., Panagia, N. (1983): Astron. Astrophys. **128** 212

Mathis, J.S. (1990): Ann. Rev. Astron. Astrophys. **28** 37

Matsui, Y., Long, S.K., Dickel, J.R., Greisen, E.R. (1984): Astrophys. J. **287** 295

Mayer-Hasselwander, H.A., Pfeffermann, E., Pinkau, K., Rothermel, H., Sommer, M. (1972): Astrophys. J. **175** L23

McCammon, D., Sanders, W.T. (1990): Ann. Rev. Astron. Astrophys. **28** 657

McCray, R. (1987): in 'Physical Processes in Interstellar Clouds', eds. G.E. Morfill and M. Scholer, D. Reidel, Dordrecht, p. 95

McKee, C.F. (1988): in 'Supernova Remnants and the Interstellar Medium', eds. R.S. Roger and T.L. Landecker, Cambridge Univ. Press, Cambridge, p. 205

McKee, C.F., Cowie, L.L. (1975): Astrophys. J. **195** 715

McKee, C.F., Ostriker, J.P. (1977): Astrophys. J. **247** 908

McKenzie, J.F., Völk, H.J. (1982): Astron. Astrophys. **116** 191

Mestel, L. (1965): Quart. J. Roy. Astron. Soc. **6** 161

Mestel, L. (1977): in 'Star formation', IAU Symp. 75, eds. T. de Jong and A. Maeder, D. Reidel, Dordrecht, p. 213

Mestel, L. (1990): in 'Galactic and Intergalactic Magnetic fields', eds. R. Beck, P.P. Kronberg, R. Wielebinski, IAU Symp. 140, Kluwer, Dordrecht, p. 259

Mestel, L., Spitzer, L.Jr. (1956): Mon. Not. Roy. Astron. Soc. **116** 503

Metzger, A.E., Anderson, E.C., Van Dilla, M.A., Arnold, J.R. (1964): Nature **204** 766

Meyer, J.P. (1981): Proc. 17th Int. Cosmic Ray Conf., Paris, **2** 265

Mezger,P.G. (1990): in 'The Galactic and Extragalactic Background Radiation', eds. S Bowyer and C. Leinert, Kluwer, Dordrecht, p. 63

Moffatt, H.K. (1978): 'Magnetic field generation in electrically conducting fluids', Cambridge Univ. Press, Cambridge

Morfill, G.E., Drury, L.O'C., Aschenbach, B. (1984): Nature **311** 358

Mouschovias, T.Ch. (1976): Astrophys. J. **207** 141

Mouschovias, T.Ch., Palaleogou, E.V. (1979): Astrophys. J. **228** 475

Mouschovias, T.Ch., Palaleogou, E.V. (1980): Astrophys. J. **237** 877

Mouschovias, T.Ch., Palaleogou, E.V., Fiedler, R.A. (1985): Astrophys. J. **291** 772

Mouschovias, T.Ch., Shu, F.H., Woodward, P.R. (1974): Astron. Astrophys. **33** 73

Nadyozhin, D.K. (1985): Astrophys. Space Sci. **112** 225

Nakano, T. (1984): Fundam. Cosmic Phys. **9** 139

Nittman, J. (1981): Mon. Not. Roy. Astron. Soc. **197** 699

Norman, C.A., Ikeuchi, S. (1989): Astrophys. J. **345** 372

Oettl, R., Hillebrandt, W., Müller, E. (1985): Astron. Astrophys. **151** 33

Parker, E.N. (1966): Astrophys. J. **145** 811

Parker, E.N. (1979): 'Cosmical Magnetic Fields', Clarendon Press, Oxford

Prishchep, V.L., Ptuskin, V.S. (1975): Astrophys Space Sci. **32** 265

Ptuskin, V.S. (1981): Astrophys. Space Sci. **76** 265

Purcell, E.M. (1979): Astrophys. J. **231** 404

Raymond, J.C., Smith, B.W. (1977): Astrophys. J. Suppl. **35** 419

Reynolds, R.J. (1990): in 'The Galactic and Extragalactic Background Radiation', eds. S Bowyer and C. Leinert, Kluwer, Dordrecht, p. 157

Ruzmaikin, A.A., Shukurov, A.M., Sokoloff, D.D. (1988): 'Magnetic Fields of Galaxies', Kluwer, Dordrecht

Sacher, W., Schönfelder, V. (1983): Space Sci. Rev. **36** 249

Schwartz, D., Gursky, H. (1974): in 'X-Ray Astronomy', eds. R. Giacconi and H. Gursky, D. Reidel, Dordrecht, p. 359

Sedov, L.I. (1959): 'Similarity and Dimensional Methods in Mechanics', Academic Press, New York

Seward, F.D. (1990): Astrophys. J. Suppl. Ser. **73** 781

Shapiro, P.R., Field, G.B. (1976): Astrophys. J. **205** 762

Simpson, J.A. (1983): Ann. Rev. Nucl. Part. Sci. **33** 323

Skilling, J. (1975): Mon. Not. Roy. Astron. Soc. **172** 557

Spitzer, L. Jr. (1978): 'Physical Processes in the Interstellar Medium', Wiley & Sons, New York

Spitzer, L. Jr. (1990): Ann. Rev. Astron. Astrophys. **28** 71

Stecker, F.W. (1973): Astrophys. J. **185** 499

Stephens, S.A. (1991): in 'The Interstellar Disk-Halo Connection in Galaxies, IAU Symp. 144, ed. H. Bloemen, Kluwer, Dordrecht, p. 323

Stephens, S.A., Badhwar, G.D. (1981): Astrophys. Space Sci. **76** 213

Taylor, G.I. (1950): Proc. R. Soc. London **A201** 159

Tielens, A.G.G.M., Allamandola, L.J. (Eds.) (1989): 'Interstellar Dust', IAU Symp. 135, Kluwer, Dordrecht

Troland, T.H., Heiles, C. (1986): Astrophys. J. **301** 339

Trombka, J.I., Metzger, A.E., Arnold, J.R., Matteson, J.L., Reedy, R.C., Peterson, L.E. (1973): Astrophys. J. **181** 737

Trümper, J. (1983): Adv. Space Res. **2** No. 4 241

Vainshtein, S.I., Ruzmainkin, A.A. (1972): Sov. Astron. **15** 714

van der Laan, H. (1962): Mon. Not. Roy. Astron. Soc. **124** 125

Vedrenne, G., Albernhe, F., Martin, I., Talon, R. (1971): Astron. Astrophys. **15** 50

Verschuur, G.L. (1969): Astrophys. J. **156** 861

Völk, H.J. (1987): Proc. 20th Int. Cosmic Ray Conf., Moscow, **7** 157

Völk, H.J., Biermann, P.L. (1988): Astrophys. J. **333** L65-68

Völk, H.J., Drury, L.O'C., McKenzie, J.F. (1984): Astron. Astrophys. **130** 19

Vrba, F.J., Strom, S.E., Strom, K.M. (1976): Astron. J. **81** 958

Weaver, R., McCray, R., Castor, J., Shapiro, P., Moore, R., (1977): Astrophys. J. **218** 377

Webber, W.R., Simpson, G.A., Cane, H.V. (1980): Astrophys. J. **236** 448

Wentzel, D.G. (1968): Astrophys. J. **152** 987

Weyman, R. (1967): Astrophys. J. **147** 887

Woltjer, L. (1972): Ann. Rev. Astron. Astrophys. **10** 129

Woodward, P.R. (1976): Astrophys. J. **207** 484

Woosley, S.E. (1986): in 'Nucleosynthesis and Chemical Evolution', 16th Advanced Cource, Swiss Society of Astrophysics and Astronomy, Saas Fee 1986

Woosley, S.E., Weaver, T.A. (1986): Ann. Rev. Astron. Astrophys. **24** 205

Zeldovich, Ya.B., Ruzmaikin, A.A., Sokoloff, D.D. (1983): 'Magnetic Fields in Astrophysics', Gordon and Breach, New York

HIGH ENERGY EMISSION FROM NORMAL STARS

Loukas Vlahos

Department of Physics, University of Thessaloniki,
54006 Thessaloniki, Greece

Abstract: Emission from high-energy electrons and ions in normal stars is a complex phenomenon since it is related to: (1) the energy release and particle acceleration processes, (2) the structure of the ambient magnetic field, and (3) the structure of the ambient atmosphere of these stars. Energy release, heating and acceleration are processes that are not well understood yet, so theoretical estimates of the high-energy emission is bound to use *"artificial"* velocity distribution functions for the energetic particles. We review the flow of energy in normal stars, the fundamental emission and absorption processes for spontaneous and collective emission and apply our results on selective observations. Our goal is to derive from the observations and the physics of radiation processes the properties of **high-energy particles**.

1 Introduction

Standard text books in astrophysics (e.g., Shu, 1982; Bowers and Deeming, 1984) discuss at length the *low-energy or atomic emission from normal stars*. An extensive literature exists on these topics (see Athay, 1972; Sobolev, 1963; Griem, 1964; Sobelman,1979 and references therein). In these lectures we analyse the emission of the high-energy electrons ($E > 100keV$). Ions are accelerated efficiently but they do not radiate as efficiently as the electrons since $m_i = 1873m_e$, where m_i is the mass of the ion and m_e is mass of the electron. When ions reach very high energies they excite collisionally nuclear lines. This very interesting topic is beyond the scope of these lectures (see Ramaty and Murphy, 1987).

An important factor that influences the characteristics of the high-energy emission (intensity, spectra, time evolution, polarisation, etc.), is the detailed shape of the distribution of the accelerated electrons (e.g. energy dependence, angular shape of the velocity distribution, etc). This information is related to the dynamics of the energy release process, the acceleration process and the transport of energetic particles from the acceleration volume to the radiation volume (these two volumes are not always co-spatial).

Energy release and acceleration of particles are processes that are closely related to the flow of energy in normal stars. We should follow first the flow of energy inside a normal star and show qualitatively that electrons can reach very high energies at the surface of the star.

1.1 Energy flow and particle acceleration

Assuming that a normal star is composed of a hydrogen burning core, a radiation zone and a convection zone we can reach several important conclusions: (1) the turbulent convection zone generates magnetic flux tubes (turbulent dynamo, see Stix, 1989), (2) Buoyancy elevates the flux tubes to the surface of the star (Parker, 1979), and (3) since one portion of the emerged magnetic field is still inside the convection zone, where the ratio of plasma pressure to magnetic pressure is much larger than one ($\beta = \frac{n_0 k_B T}{B^2/8\pi}$, where n_0 is the ambient density, B the magnetic field intensity, T the plasma temperature and k_B Boltzmann's constant), and the other is inside the tenuous atmosphere of the star where the opposite is true, the fluid motion inside the convection zone controls the motion and disturbs the magnetic field. Disturbances and interactions of flux tubes will transfer and dissipate magnetic energy in the stellar atmosphere. A fraction of the dissipated energy will be transferred to the high-energy particles. The available magnetic energy inside a volume V is $W_F = (\frac{B^2}{8\pi}) \times V$. Assuming that $B \approx 100G$ and $V = (10^{10}cm)^3$ the *available* energy is $10^{32}ergs$. Let us also assume that the plasma density in the low atmosphere of the star is $10^{10}cm^{-3}$ and only 0.1% of the ambient plasma is accelerated, then the average energy per electron will exceed $\approx 100keV$ if the acceleration efficiency is 0.1%.

The topology of the emerged magnetic field above a turbulent convection zone is a poorly understood subject. Complex magnetic topologies evolve by dissipating energy or diffusion of magnetic flux in space. It is important to understand (a) how magnetic fields dissipate energy and (b) how a part of this energy will be deposited to charged particles. Reviewing the state of the art on these problems will lead us away from our main subject but it is important to mention the trends in the current research on these topics.

We mentioned already that turbulent flows inside the convection zone disturb the magnetic field that has already emerged from the surface of the star. MHD waves excited from these disturbances will propagate along the magnetic field and dissipate part of their energy inside the atmosphere. Magnetic fields filter the spectrum of waves that exist in the photosphere and allow the propagation of selected frequencies that can resonate with the structures that exist. Propagation of these waves and dissipation in the chromosphere or corona is one way energy is constantly transmitted from the convection zone to the stellar atmosphere (Heyvaerts, 1989).

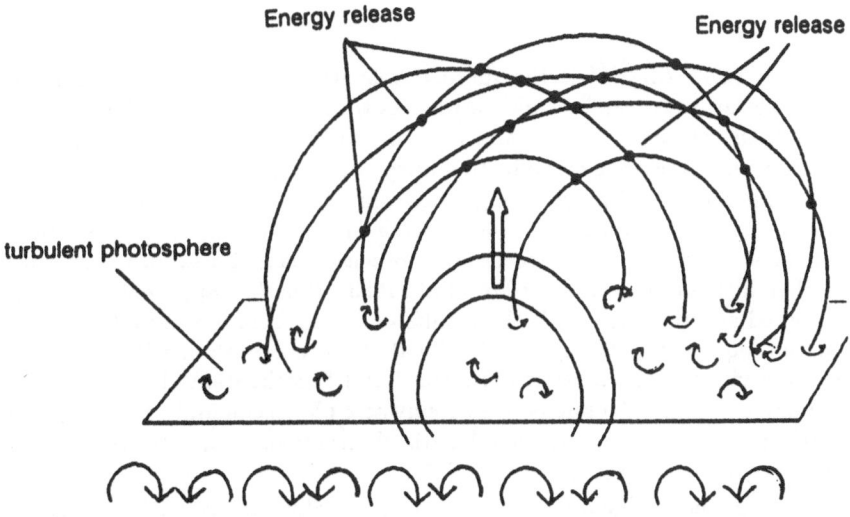

Fig. 1. Active region magnetic field and dynamic formation of neutral sheets

Magnetic fields are also split and filamented by the randomly moving coherent structures below the photosphere, new flux is constantly emerging and/or submerging (see review by Martin, 1990). Such small-scale structures are observed regularly in the Sun and probably are common in most normal stars. Emerged magnetic structures interact and form 'singular points' which are called **neutral sheets**. Formation of neutral sheets and sudden release of energy is also a poorly understood topic. In Fig.1, we demonstrate graphically the interaction of pre-existing magnetic field lines with newly emerged structures and show the formation of many independent neutral sheets. It is well known that a neutral sheet will heat locally the plasma, accelerate a small portion of electrons and form two opposite moving slow shocks (Priest, 1990). The presence of many neutral sheets simultaneously above the active region will form an environment where many local acceleration events act simultaneously on the particles (see Anastasiadis and Vlahos, 1991; Vlahos, 1989). We conclude that in normal stars, acceleration of high-energy particles is part of the evolution of the active region through the formation of thousands of reconnection sites.

The terms "heating" and "acceleration" need some discussion especially when applied to astrophysical objects and related to observations with poor spatial and temporal resolution. It is widely accepted that "heating" is called the process that will raise the mean energy of the plasma, i.e., increase the temperature of the Maxwellian distribution

$$f(v) = \frac{n_0}{2\pi^{3/2}V_j^3} exp(-\frac{v^2}{2V_j^2}) \tag{1.1}$$

where $V_j = (k_B T_j/m_j)^{1/2}$ is the thermal velocity, $j = e,i$ (for electrons and ions). Acceleration is usually called a process that will "energise" the electrons and ions with velocity $v >> V_j$ and form a non-Maxwellian tail e.g., a power law distribution. A serious complication arises when: (1) the heating is localised and the high-energy tail propagating away from the heated volume will appear in other parts of the stellar atmosphere as a "beam" of high-energy particles (Raoult et al. 1990), and (2) acceleration processes sometimes form super-hot Maxwellian distributions (see Vas'kov et al., 1983). It is then apparent that when the observed spectrum fits with the results obtained from a Maxwellian distribution this does not mean automatically that we are dealing with a heating process and vice versa. The terms "heating" and "acceleration" have been confused many times in discussions of astrophysical plasmas mainly from the lack of detail modelling and multi-wave length observations(e.g., soft X-ray, Hard X-ray, radio and γ-ray data).

In summary, energy generated in the core flows along the radiation zone, sets up turbulence in the convection zone, and generates magnetic fields in the form of flux tubes, buoyancy forces the magnetic field outside the convection zone and finally this partially submerged, partially emerged, field serves as a transmission line for the transfer of energy from turbulent flow to the magnetic field to the plasma. Transmitted energy will be dissipated in steady-state form to provide the "coronal heating", and/or in impulsive form releasing the energy observed during "flares". The lack of a convection zone will eliminate most of the high energy emission from normal stars.

1.2 Basic concepts

Radiation, once it is generated, will propagate inside the stellar atmosphere and the interstellar medium before reaching the Earth or a satellite inside the solar system. Transfer of electromagnetic radiation inside a plasma is a very complex phenomenon and requires detailed knowledge of the propagation medium. Radiation may be re-absorbed by the atmosphere of the star and never reach the observer or the medium may absorb selectively certain frequencies (absorption lines).

Emission is defined by the *monochromatic emission coefficient* j_ν, which is measured in energy dE (erg) per unit time dt (s^{-1}) per volume dV (cm^{-3}) per solid angle $d\Omega$ ($ster^{-1}$) and unit frequency $d\nu$ (Hz^{-1}), according to

$$dE = j_\nu d\nu dt d\Omega dV \tag{1.2}$$

The absorption coefficient is defined as $\alpha_\nu (cm^{-1})$. The *radiation transfer equation* is (Bowers and Deeming, 1984)

$$\frac{dI_\nu}{ds} = -\alpha_\nu I_\nu + j_\nu \tag{1.3},$$

where I_ν is the radiation intensity measured in $ergs^{-1} ster^{-1} Hz^{-1} cm^{-2}$. If we define the optical depth as

$$d\tau_\nu = \alpha_\nu ds \tag{1.4}$$

then eq. (1.4) takes a simpler form

$$\frac{dI_\nu}{d\tau_\nu} = -I_\nu + S_\nu, \tag{1.5}$$

where $S_\nu = \frac{j_\nu}{\alpha_\nu}$ is called the source function. The transfer equation is simple to solve

$$I_\nu(\tau_\nu) = I_\nu(0)e^{-\tau_\nu} + \int_0^{\tau_\nu} e^{-(\tau_\nu - \tau_\nu')} S_\nu d\tau_\nu' \tag{1.6}$$

We can simplify eq. (1.6) if the medium is in thermodynamic equilibrium since then the source function is independent of τ_ν and has the form

$$S_\nu = B_0(\omega, T) = \frac{2h\nu^3/c^2}{e^{\frac{h\nu}{k_B T}} - 1} \tag{1.7}$$

which is the well known Plank law. Combining eqs.(1.6) and (1.7) we have

$$I_\nu(\tau_\nu) = I_\nu(0)e^{-\tau_\nu} + S_\nu(1 - e^{-\tau_\nu}). \tag{1.8}$$

Several important problems related to various approximations on the transport of radiation, such as scattering, radiation diffusion, etc., can be found in standard textbooks on radiation (see e.g. Rybicki and Lightman, 1979).

Electrons accelerated or decelerated in vacuum will emit radiation. In the non relativistic limit the total energy radiated away is (in the dipole approximation)

$$W = \frac{2}{3c^3}\left(\frac{d^2d}{dt^2}\right)^2 \tag{1.9}$$

where $d = \sum q_i r_i$ is the dipole moment. The energy per unit frequency is (Rybicki and Lightman, 1979)

$$\frac{dW}{d\omega} = \frac{8\pi\omega^4}{3c^3}|d(\omega)|^2, \tag{1.10}$$

where

$$d(\omega) = \int_{-\infty}^{\infty} d(t)e^{i\omega t}dt.$$

Eq. (1.9) is used extensively for the estimate of the spontaneous emission of non-relativistic particles. Another important parameter used extensively in astrophysics is *the brightness temperature* T_b, for which the intensity (I_ν) from a given source is equal to the black body emission,

$$I_\nu = B_0(\omega, T_b). \tag{1.11}$$

In radio astronomy where the Rayleigh-Jeans law is appliccable ($h\nu << k_B T$), T_b is

$$T_b = 4.8 \times 10^{10}\left[\frac{I_\nu}{100mJy}\right]\left[\frac{0.48}{\alpha}\right]\left[\frac{5GHz}{\nu}\right]^2 K \tag{1.12}$$

where intensity (I_ν) is measured in milliJansky, the apparent radius of the star in milliarc seconds (mas), normalised with the apparent radius of the Sun at 10 pc (0.48 mas).

We can define two general classes of radiation: (1) spontaneous or incoherent emission, e.g., synchrotron and bremsstrahlung emission, and (2) collective or coherent radiation, e.g., plasma emission or electron cyclotron maser emission. One important characteristic that distinguishes spontaneous or incoherent emission from collective emission is the brightness temperature. When the brightness temperature is $T_b >> 10^{10} K$ the emitting source has probably achieved an order of coherence. Other observational factors which suggest coherence are high polarisation (100%) and narrow bandwidth ($\frac{\Delta\omega}{\omega} << 1$).

We will discuss first the spontaneous emission, which is most common in normal stars and then continue with the collective emission.

2 Spontaneous emission

2.1 Bremsstrahlung emission

Radiation due to the acceleration of a charge in the Coulomb field of another charge is called free-free emission or bremsstrahlung. It is obvious that if the scattering center is a like particle (electron-electron, proton-proton) the dipole moment ($\sum q_i r_i$) is proportional to the center of mass ($\sum m_i r_i$) which is a constant of the motion (e.g. $\sum_i q_i \frac{dv_i}{dt} = \frac{d^2}{dt^2}\sum_i \frac{q_i}{m_i}m_i r_i(t) = \frac{q_i}{m_i}\frac{d^2}{dt^2}\sum_i m_i r_i(t) =$

135

0, since $\sum_i m_i r_i = R \sum m_i$, where R is the radius vector of the center of mass) and the free-free emission in the dipole approximation is zero (see eq. (1.9)).

In Fig. 2 we show an electron with charge e moving rapidly in the field of an ion. The distance between the line of flight of the electron and the ion is called impact parameter (b).

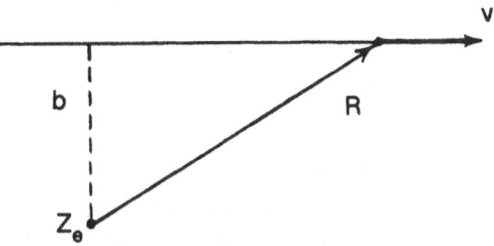

Fig. 2. Collision of a fast moving electron with an ion

The small angle scattering is the most frequent and we are going to use this approximation here. The second derivative of the dipole moment is $\frac{d^2 d}{dt^2} = -e\frac{dv}{dt}$, since the ion is assumed stationary. The Fourier transform of the dipole moment is

$$-\omega^2 d(\omega) = -\frac{e}{2\pi} \int_{-\infty}^{+\infty} \frac{dv}{dt} e^{i\omega t} dt \qquad (2.1)$$

The electron is in close interaction with the ion for a period $\tau = \frac{b}{v}$, which is called "collision time". We estimate the dipole moment when $\omega\tau << 1$, $d(\omega) \approx \frac{e}{2\pi\omega^2}\Delta v$ and ignore the contribution of the integral where $\omega\tau >> 1$. The emitted radiation is given by eq. (1.10) and when $\omega\tau << 1$ we have

$$\frac{dW}{d\omega} \approx \frac{2e^2}{3\pi c^3} |\Delta v|^2. \qquad (2.2)$$

We estimate Δv, assuming that the path is linear and the changes are normal to the path,

$$\Delta v = -\frac{eE_y}{m_e}\Delta t = \frac{Ze^2}{m_e}\int_{-\infty}^{\infty} \frac{bdt}{(v^2 t^2 + b^2)^{3/2}}. \qquad (2.3)$$

The total spectrum of radiation from a medium with ion density N_i and electron density N_e and for fixed velocity v for all electrons is estimated as follows. The flux of electrons incident to the ion is $N_e v$ and the element area around the ion is $2\pi bdb$, finally the total emission is

$$\frac{dW}{d\omega dV dt} = \frac{16e^6}{3c^3 m_e^2 v} N_e N_i Z^2 \int_{b_{min}}^{b_{max}} \frac{db}{b}. \qquad (2.4)$$

We denote with b_{max} the maximum distance beyond which the emission is not important and b_{min} is the closest approach for the electron.

The maximum distance is estimated from our assumption $\omega\tau \ll 1$ or $b_{max} \ll v/\omega$, the exact value is uncertain but since b_{max} appear inside the logarithm its exact value is not important. We will use $b_{max} = \frac{v}{\omega}$ here. The b_{min} can be found either classically or using quantum mechanics. In the classical limit the approach used above is not valid when the linear trajectory approximation does not apply, and this occurs when $\Delta v \approx v$ or $b_{min} \approx \frac{4Ze^2}{\pi m v^2}$. In general eq. (2.4) can be written

$$\eta_\omega = \frac{dW}{d\omega dV dt} = \frac{16e^6\pi}{3\sqrt{3}c^3m_e^2 v}N_e N_i Z^2 g_{ff}(v,\omega) \tag{2.5}$$

where g_{ff} is the Gaunt factor and η_ω is the emissivity ($erg cm^{-3}sec^{-1}Hz^{-1}$). The Gaunt factor depends on the velocity of the electron and the frequency of the emitted photon. An accurate estimate of the free-free emission needs a quantum treatment but a classical approximation gives the correct functional dependence for most of the physical parameters, quantum corrections will improve the accuracy of the Gaunt factor. Extensive tables and graphs exist in the literature (Karzas and Latter, 1961).

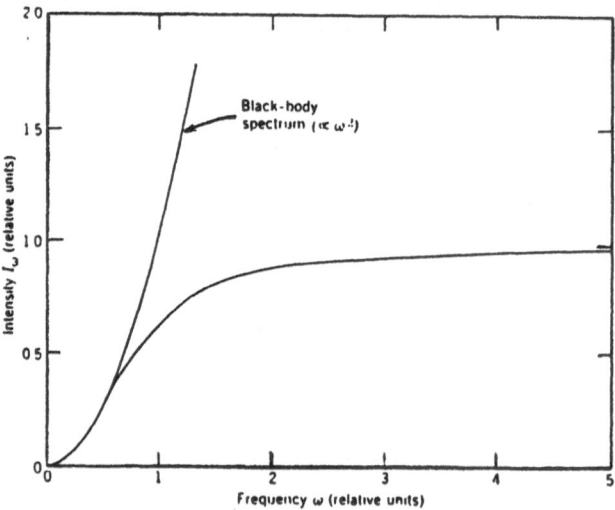

Fig. 3. The spectrum of the thermal bremsstrahlung emission

The emission coefficient from a distribution of particles $f(v)$ is given by the relation

$$j_\omega = \int_{v_{min}}^{\infty} \eta_\omega f(v)dv \tag{2.6}$$

where v_{min} is estimated from the relation $v_{min} = (2h\nu/m_e)^{1/2}$, since otherwise a photon with frequency ω cannot be created, and $f(v)$ is the normalised velocity distribution of the incident electrons. Using eq. (2.6) we can estimate the emission from a thermal distribution, when the distribution function is Maxwellian or from a non-thermal distribution using a power law or other types of non-thermal distributions. For a Maxwellian distribution (and since $d\omega = 2\pi d\nu$) we obtain

$$j_\nu = \frac{2^4 e^6}{3mc^3}(\frac{2\pi}{3km})^{1/2}T^{-1/2}Z^2 N_i N_e e^{\frac{-h\nu}{k_B T}} g_{ff} \qquad (2.7)$$

or

$$j_\nu = 10^{-38} Z^2 N_e N_i T^{-1/2} e^{\frac{-h\nu}{k_B T}} g_{ff}(c.g.s) \qquad (2.8)$$

The thermal bremsstrahlung is represented with a flat spectrum in a log-log plot and will have a cut-off when $\hbar\omega \approx k_B T$. This is true for an optically thin plasma. Absorption of photons can be estimated from Kirchoff's law, then

$$\alpha_\nu = 3.8 \cdot 10^8 T^{-1/2} Z^2 N_e N_i \nu^{-3}(1 - e^{\frac{-h\nu}{k_B T}})g_{ff}. \qquad (2.8)$$

For $h\nu << kT$, we are in the Rayleigh-Jeans regime and eq. (2.8) becomes

$$\alpha_\nu = 0.018 T^{-3/2} Z^2 N_i N_e \nu^{-2} g_{ff}. \qquad (2.9)$$

The free-free emission spectrum from a plasma with temperature T is given in Fig. 3.

The bremsstrahlung emission from a relativistic electron colliding with an ion is (Rybicki and Lightman, 1979)

$$\frac{dW}{d\omega} = \frac{8Z^2 e^6}{3\pi \beta^2 c^5 m_e^2}(\frac{b\omega}{\gamma^2 c})^2 K_1(\frac{b\omega}{\gamma^2 c}) \qquad (2.10)$$

where $K_1(x)$ is the modified Bessel function of order one, $\gamma = (1 - (v/c)^2)^{-1/2}$ is the relativistic factor. Several approximations of this formula are discussed in the literature; the most useful expression for the frequency integrated power $(erg cm^{-3} sec^{-1})$ of a thermal relativistic plasma is (Novikov and Thorne, 1973)

$$\frac{dW}{dV dt} = 1.4 \cdot 10^{-27} T^{1/2} Z^2 N_e N_i g(1 + 4.4 \cdot 10^{-10} T). \qquad (2.11)$$

For other types of non-thermal distributions we can use eq. (2.6) to estimate the emission coefficient. It is important to notice that if a small number of non-thermal electrons radiate, the emission coefficient will probably be controlled by them. The absorption coefficient on the other hand should be estimated from the ambient Maxwellian plasma. In special cases, when the non-thermal particles reach relatively large numbers ($n_b/n_0 \approx 1$, where n_b is the density of the energetic particles) the **self absorption** from the energetic particles should be considered.

2.2 Cyclotron and Synchrotron emission

The presence of magnetic field inside the plasma will force the ions and electrons to gyrate around the magnetic field lines. The emission is primarily due to electrons gyrating around the magnetic field. The equation of motion is

$$\frac{d\mathbf{p}}{dt} = \left(\frac{e}{c}\right)(\mathbf{v} \mathbf{x} \mathbf{B}) \tag{2.12}$$

where $\mathbf{p} = \gamma m_e \mathbf{v}$, m_e is the electron rest mass. Depending on the energy of the electron, the spectrum changes dramatically. The emitted radiation is called cyclotron emission if the velocity of the particle is non-relativistic or synchrotron emission if the velocity is close to the speed of light.

We assume that the energy lost per cycle is so small that the energy of the particle remains almost constant (note that the velocity is perpendicular to the force $(e/c)(\mathbf{v} \mathbf{x} \mathbf{B})$ acting on the electron). Solving eq. (2.12) we obtain the velocity of the particle and its position

$$v = e_x v_\perp \cos(\omega_0 t) + e_y v_\perp \sin(\omega_0 t) + e_z v_\parallel \tag{2.13}$$

$$r = e_x(\frac{v_\perp}{\omega_0})\sin(\omega_0 t) - e_y(\frac{v_\perp}{\omega_0})\cos(\omega_0 t) + e_z v_\parallel t \tag{2.14}$$

where e_x, e_y, e_z are the unit vectors along the axis x,y,z, $\omega_0 = -\frac{eB}{\gamma m_e c}$. Without loss of generality we assume that the observer is located in the x-z plane. Using eq. (1.9) we obtain the emissivity (Bekefi, 1965)

$$\eta_\omega(\omega, v, \theta) = \frac{e^2 \omega^2}{2\pi c} \left[\sum_1^\infty \left(\frac{\cos\theta - \beta_\parallel}{\sin\theta} \right)^2 J_m^2(x) + \beta_\perp^2 J_m'^2(x) \right] \delta(y) \tag{2.15}$$

where $x = \frac{\omega}{\omega_0}\beta_\perp \sin\theta$, $y = m\omega_0 - \omega(1 - \beta_\parallel \cos\theta)$, $\beta_\perp = v_\perp/c$, $\beta_\parallel = v_\parallel/c$, J_m is the Bessel function and $J_m' = \frac{dJ_m}{dx}$. Harmonics with m=0,-1,-2,-3 ... were eliminated by the delta function $\delta(y)$. The total emission in harmonic m is obtained by integrating a single term over frequency ω and solid angle $d\Omega = 2\pi \sin\theta d\theta$,

$$\eta_m^T = \frac{2e^2\omega_b^2}{c}\frac{1 - \beta_0^2}{\beta_0}[m\beta_0^2 J_{2m}'(2m\beta_0) - m^2(1 - \beta_0^2)\int_0^{\beta_0} J_{2m}(2mt)dt] \tag{2.16}$$

where $\beta_0 = \beta_\perp/\sqrt{1 - \beta_\parallel^2}$ and $\omega_b = -\frac{eB}{m_e c}$. Finally the total power emitted is obtained by summing over all harmonics

$$\eta^T = \frac{2e^2\omega_b^2}{3c}\left[(1 - \beta_\parallel^2)(\frac{\epsilon}{m_0 c^2})^2 - 1 \right]. \tag{2.17}$$

where ϵ is the total energy of the electron. Several important conclusions can be drawn from eq. (2.17): (1) We can estimate the total emission lost per cycle $\Delta W/\Delta t$, when $\Delta t = 2\pi/\omega_0$ (assuming $\beta_\parallel = 0$)

$$\frac{\Delta W}{\epsilon} = \frac{4\pi e^2 \omega_b}{3c}\frac{1}{m_0 c^2}\left[\left(\frac{\epsilon}{m_0 c^2}\right)^2 - 1 \right], \tag{2.18}$$

even for extreme values of ϵ and B the loss is totally insignificant; (2) We can also obtain the time needed by a non-relativistic electron or mildly relativistic electron ($\epsilon << m_0 c^2$) to radiate away its energy

$$\tau = \frac{2.58 \cdot 10^8}{B^2} \quad seconds \qquad (2.19)$$

which is several hours in normal stars ($B \approx 100 - 1000G$). For highly relativistic electrons the time constant for energy loss depends on the initial energy

$$\tau = \frac{8.8 \cdot 10^8}{B^2} \frac{m_0 c^2}{\epsilon(t=0)} \quad seconds$$

that is, synchrotron losses can be important for the duration of transient events and continuous acceleration may be needed when long lasting events are observed. When the energy of the electron is such that $m\beta << 1$ (for m=1,2,...), the Bessel function can be replaced by the asymptotic series expansion for small arguments. Thus

$$\eta_m^T = \frac{2e^2 \omega_b^2}{c} \frac{(m+1)(m^{2m+1})}{(2m+1)!} \beta^{2m}. \qquad (2.20)$$

The spectrum consist of a series of discrete lines of rapidly decreasing intensity. The lines are separated by ω_b (see Fig. (6.2) in Bekefi, 1965).

A number of effects will influence the width of the cyclotron lines. The most important factors are: (1) the energy loss by radiation will impose a frequency width $1/\tau$, where τ is given in eq. (2.19); (2) collisions will force the line width to be of the order of the collision frequency; (3) the random motion of particles along the line of force; (4) the relativistic change of the mass will influence the cyclotron frequency, and (5) non-uniformity of the ambient magnetic field inside the radiation source.

The radiation emitted from highly relativistic electrons is beamed along the forward direction into a narrow cone of angle $2\cos^{-1}(\beta)$, as seen by the observer in the distant field of the electron.

The power emitted in a given harmonic is found from eq. (2.16) when $\beta \to 1$, $\gamma >> 1$ and $\beta_{\parallel} = 0$

$$\eta_m^T = \frac{2e^2 \omega_b^2}{c} \frac{m}{\gamma^2} \left[J_{2m}'(2m\beta) - \frac{m}{\gamma^2} \int_0^\beta J_{2m}(2mt)dt \right] \qquad (2.21)$$

The emission is predominant at high harmonics, so if in addition to $\gamma >> 1$ we impose the conditions $m >> 1$ then (Bekefi,1965):

$$\eta_m^T = \frac{e^2 \omega_b^2}{\sqrt{3}\pi c} \frac{m}{\gamma^4} \int_{\frac{2m}{3\gamma^3}}^\infty K_{5/3}(t)dt \qquad (2.22)$$

where K is the Hankel function. The successive harmonics are now closely spaced ($\Delta\omega \approx \omega_b/\gamma$). Now if we replace m by ω/ω_0

$$\eta_\omega^T = \frac{\sqrt{3}e^2\omega_b}{2\pi c}\left[\frac{\omega}{\omega_c}\int_{\frac{\omega}{\omega_c}}^{\infty}K_{5/3}(t)dt\right] \tag{2.23}$$

Here, the parameter $\omega_c = (3/2)\omega_b\gamma^2 = 2.10^7 B(Gauss)[\frac{E}{m_0c^2}]rad/sec$. The spectrum is plotted in Fig. 4; it rises gently at low frequencies up to $\omega \approx 0.3\omega_c$ and the drops sharply at $\omega \gg \omega_c$. The spectrum can be approximated by the formulae

$$\eta_\omega^T = \frac{\sqrt{3}\Gamma(2/3)e^2\omega_b}{2^{1/3}\pi c}(\frac{\omega}{\omega_c})^{1/3} \tag{2.24}$$

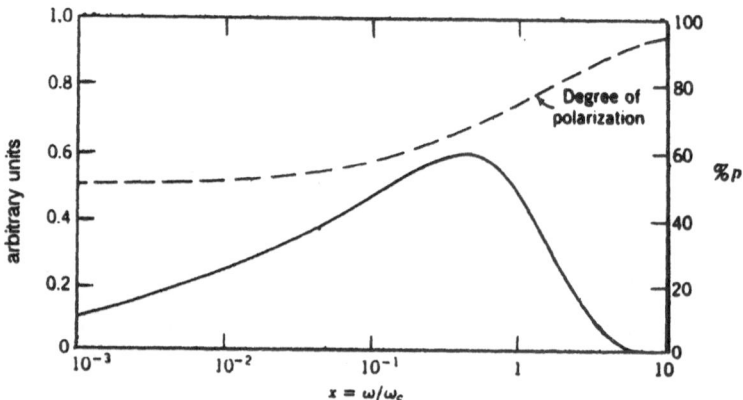

Fig. 4. Synchrotron spectrum from a relativistic electron

when $\omega \ll \omega_c$ and

$$\eta_\omega^T = \frac{\sqrt{3}e^2\omega_b}{2^{3/2}\pi^{1/2}c}(\frac{\omega}{\omega_c})^{1/2}e^{-\frac{\omega}{\omega_c}} \tag{2.25}$$

for $\omega \gg \omega_c$, where $\Gamma(x)$ is the Gamma function.

In all calculations presented above we neglected the motion of the electron along the magnetic field ($\beta_\parallel = 0$). In other words the results obtained are correct for an observer moving with $v_\parallel$, for a stationary observer we should replace $\omega_b \to \omega_b\sqrt{1-\beta_\parallel}$ in eqs (2.24) and (2.25).

The polarisation of the emitted radiation can be estimated from the radiation propagated at right angles with respect to the magnetic field ($\theta = \pi/2$). The emissivity given in eq. (2.15) is made up of two factors: one with $E \parallel B$ which is called ordinary wave (O-mode)

$$\eta_\omega^{(o)}(\pi/2) = \frac{e^2\omega^2}{2c} \sum_1^\infty \beta_\parallel^2 J_m^2(m\beta_\perp)\delta(m\omega_0 - \omega) \qquad (2.26)$$

and the other with $E \perp B$ which is called extraordinary wave (X-mode)

$$\eta_\omega^{(x)}(\pi/2) = \frac{e^2\omega^2}{2c} \sum_1^\infty \beta_\perp^2 J_m'^2(m\beta_\perp)\delta(m\omega_0 - \omega) \qquad (2.27)$$

Since $p_\parallel = p\cos\Theta, p_\perp = p\sin\Theta$ where Θ is the pitch angle, we can estimate the average emissivity

$$< \eta_\omega^{(o,x)} > = \frac{1}{4\pi} \int_0^\pi \eta_\omega^{(o,x)} 2\pi \sin\Theta d\Theta \qquad (2.28)$$

and the degree of polarisation

$$\wp = \left| \frac{\eta_\omega^x - \eta_\omega^o}{\eta_\omega^x - \eta_\omega^o} \right| = \left| \frac{K_{2/3}(\frac{\omega}{\omega_c})}{\int_{\frac{\omega}{\omega_c}}^\infty K_{5/3}(t)dt} \right| \qquad (2.29)$$

for $\omega << \omega_c$, $\wp \to 0.5$ and for $\omega >> \omega_c$, $\wp \to 1$. When we can assume that the ensemble of electrons is present with isotropic distribution $f(p)$, then we can estimate the emission coefficient

$$j_\omega^{(o,x)} = \int_0^\infty < \eta_\omega^{(o,x)} > f(p)4\pi p^2 dp. \qquad (2.30)$$

Since power law distributions for electrons ($f(p) \approx p^{-2r}$) are quite common in normal stars, we can estimate

$$j_\omega = j_\omega^x + j_\omega^o \approx \omega^{-\frac{(r-1)}{2}}. \qquad (2.31)$$

and the polarisation

$$\wp = \frac{r+1}{r+7/3} \qquad (2.32)$$

The value of r=1.6-5.0 and $\wp$ lies between $66 - 82\%$.

Energetic particles in normal stars have energies between $20keV - 500keV$. This energy range is usually called mildly relativistic and the approximations made above are not valid. When the energy is not too high, the total power emitted can be obtained from the first terms of a series expansion of eq. (2.17)

$$\eta^T \approx \frac{2e^2\omega_b^2}{3c}\beta_\perp^2(1 + \beta^2 + ...) \qquad (2.33)$$

Comparing η^T with the emission intensity in the fundamental frequency we conclude that almost 95% of the radiation is in higher harmonics.

Assuming that the plasma is in thermal equilibrium we can estimate the emission and absorption coefficient easily (Bekefi, 1965)

$$j_\omega^{(o,x)}(\pi/2) = \left(\frac{\omega_e^2}{\omega_b c}\right)\left(\frac{\omega^2 k_B T}{8\pi^2 c^3}\right) \sum_{m=1}^{\infty} \Phi_m \tag{2.34}$$

$$\alpha_\omega^{(o,x)}(\pi/2) = \left(\frac{\omega_e^2}{\omega_b c}\right) \sum_m \Phi_m \tag{2.35}$$

Here $\omega_e = (4\pi n_0 e^2/m_e)^{1/2}$, the absorption coefficient is estimated from the emission through Kirchoff's law, the function

$$\Phi_m = \sqrt{2\pi}\frac{\mu^{5/2}}{x^4}m^2\sqrt{m^2 - x^2}e^{-\mu(\frac{m}{x}-1)}A_m^{(o,x)}(m/x)$$

with $x = \frac{\omega}{\omega_b}, \mu = \frac{m_e c^2}{k_B T}$ and furthermore

$$A_m^{(o,x)} = \frac{(m\beta)^{2m}}{(2m+1)!}\left(\frac{\beta^2}{(2m+3)};1\right) \tag{2.36}$$

for non-relativistic particles ($m\beta << 1$) and

$$A_m^{(o,x)} = \frac{e^{2m/\gamma}}{\sqrt{(16\pi m^3\gamma)}}\left(\frac{\gamma-1}{\gamma+1}\right)^m\left[\frac{\gamma(\gamma^2-1)}{2m};1\right] \tag{2.37}$$

for relativistic electrons ($\gamma^3 << m$). In Fig. 5 we show the plot of Φ for the first twenty harmonics for a plasma with $\mu = 10(50 keV)$.

The absorption coefficient (which is called self-absorption here) is derived from eq. (2.36) and the emission from a uniform slab of plasma with thickness L is

$$I_\omega^{(o,x)} = B_o(\omega, T)\left[1 - e^{-\alpha_\omega L}\right]. \tag{2.38}$$

It is obvious that when $\alpha_\omega L >> 1$ the synchrotron emission will be self-absorbed and the observer will see a black body spectrum up to a characteristic harmonic m^*. The plasma will be optically thin only if $\alpha_\omega^{(o,x)}L = \Lambda\sum\Phi < 1$, where $\Lambda = \omega_e^2 L/(\omega_b c)$. For $L = 10^8 cm$, $n = 10^9 cm^{-3}$, $T = 50 keV$ and $B = 1000G$ we find $\Lambda \approx 10^5$, which means that when $\sum\Phi < 10^{-5}$ the $\alpha_\omega^{(o,x)}L$ will be smaller than unity. Using Fig. 5 we conclude that $m > m^* = 14$ for optically thin emission. In summary, only high harmonics will escape from a normal star when the source is heated to very high temperatures (e.g., during flares).

A detail study of the emission spectrum for other types of distributions (e.g. loss cone and runaway distributions) have been presented in the literature (see Winskie and Boyd, 1983; Winskie et al., 1983). The spectra are very sensitive to the angular shape of the distribution and these results are useful only when detailed modelling of the velocity distribution is available. In normal stars only two types of distributions are used: a Maxwellian when we model a hot plasma or a power law when we model a population of high energy particles. The number of free-parameters is high in most stellar sources so it is difficult to distinguish the shape of the distribution using only radio observations (radio emission in normal stars is partially due to synchrotron emission). A combination of X-ray

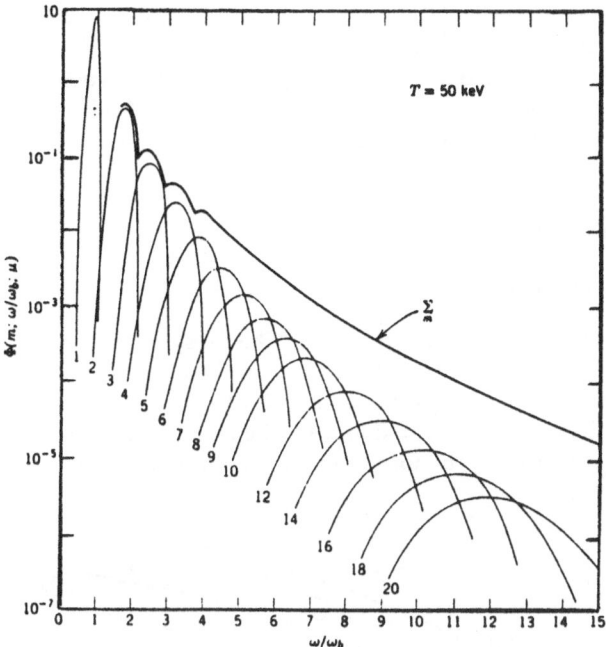

Fig. 5. Spectrum of Synchrotron radiation from a thermal plasma (kT=50 keV)

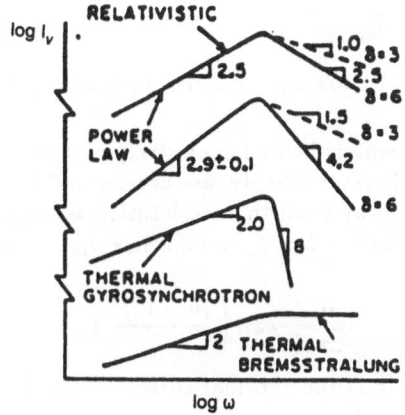

Fig. 6. Schematic spectra for thermal and power law distributions

144

and radio observations from the same event will be the best way to narrow down the choice of the shape of the distribution function.

Dulk (1985) presented a detailed comparison of the bremsstrahlung and synchrotron spectra for thermal and power law ($f \propto E^{-\delta}$) distributions (see Fig. 6). The slope of the spectrum is indicated for $\delta = 3$ and $\delta = 6$.

3 Collective plasma emission

3.1 Plasma radiation

Plasmas in nature are not always stable. The presence of energetic electrons or ions is a source of free energy which can drive several plasma instabilities. We discuss here the simplest unstable system that can occur in normal stars when energetic particles are accelerated locally: the beam-plasma interaction. In Fig.7 we present a scenario for the formation of a beam-plasma distribution function.

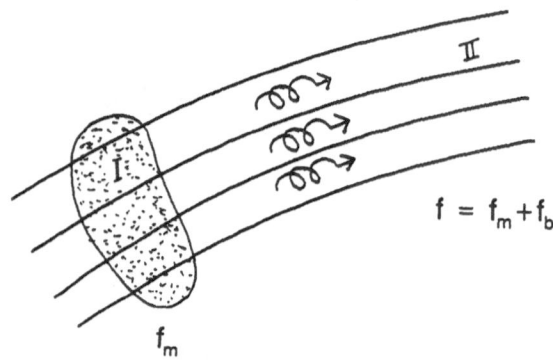

Fig. 7. Formation of a beam-plasma distribution in space plasmas

In the volume I the plasma is heated explosively and electrons with velocity $v \gg V_j$, where V_j is the thermal velocity, are accelerated to high energies. The energetic particles stream away from the acceleration volume and in volume II form a beam-plasma distribution $f = f_m + f_b$, where f_m is the local Maxwellian distribution,

$$f_b = \frac{n_b}{\sqrt{2\pi}V_{eb}} exp\left(\frac{(v-V_b)^2}{2V_{eb}^2}\right),\qquad(3.1)$$

V_b is the beam velocity, V_{eb} is the thermal spread of the beam and n_b is the density of the beam particles. This type of systems are rather common in stellar atmospheres above evolving active regions. In the Sun, the beam-plasma interaction is thought to be responsible for a very common radio burst, called type III burst.

We present a qualitative description of the evolution of such a system in a non- magnetised plasma (see Nicholson, 1983). Three types of waves can be

145

excited in a non-magnetised plasma. (a) The high-frequency electrostatic waves, called plasma or Langmuir waves. Their dispersion relation is

$$\omega_L^2 = \omega_e^2 + 3k^2 V_e^2 \tag{3.2}$$

(b) The low-frequency electrostatic waves, called ion acoustic waves for which

$$\omega_s^2 = k^2 c_s^2 + \frac{k^2 \gamma_e (k_B T_e/m_i)}{1 + \gamma_e k^2 \lambda_d^2} \tag{3.3}$$

Here $c_s = ((\gamma_e k_B T_e + \gamma_i k_B T_i)/m_i)^{1/2}$ is the ion sound speed, γ_e, γ_i, are the ratios of specific heats and λ_d is the Debye length. (c) The high-frequency electromagnetic waves, for which

$$\omega_{em}^2 = \omega_e^2 + k^2 c^2. \tag{3.4}$$

In Fig. 8 we present the dispersion relations for the normal modes of the non-magnetised plasma.

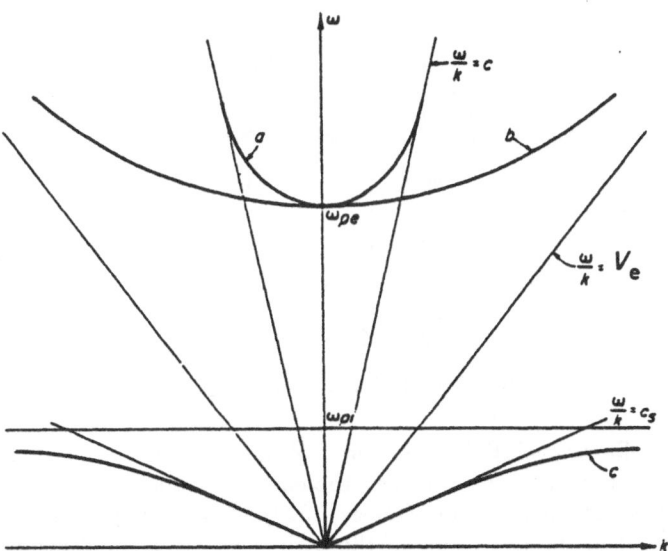

Fig. 8. The normal modes of a non-magnetised plasma, (a) electromagnetic wave, (b) plasma wave, (c) ion acoustic wave

It is important to mention that the electrostatic waves cannot propagate outside the plasma (e.g., when $\omega_e \rightarrow 0, T_e \rightarrow 0$). Only for the high-frequency electromagnetic wave the dispersion relation inside the plasma joins smoothly with the vacuum dispersion relation ($\omega_{em} = kc$). The plasma waves can escape from the plasma only when they are coupled non-linearly with the electromagnetic wave through a wave-wave interaction. The plasma emission is an *indirect* process. When the electromagnetic wave is excited from an unstable distribution, then the radiation will escape immediately and it is called *direct emission*. In the next section we will briefly discuss a direct process: the electron cyclotron maser instability.

The presence of a beam of mildly relativistic electrons will excite plasma waves with a linear growth rate (Nicholson, 1983)

$$\gamma_b = \frac{\pi}{2}\frac{n_b}{n_0}\left(\frac{\omega_e}{k}\right)^2\left(\frac{\partial f_b}{\partial v}\right)\omega_e \tag{3.5}$$

The electric field associated with this instability is

$$E = \left[E_0 e^{\gamma_b t}\right]e^{i(\boldsymbol{k}\cdot\boldsymbol{r}-\omega_L t)} \tag{3.6}$$

where $E_0 e^{\gamma_b t}$ is the amplitude of the wave and ω_L is given by eq. (3.2).

In Fig. 9 we present the evolution of a beam plasma system.

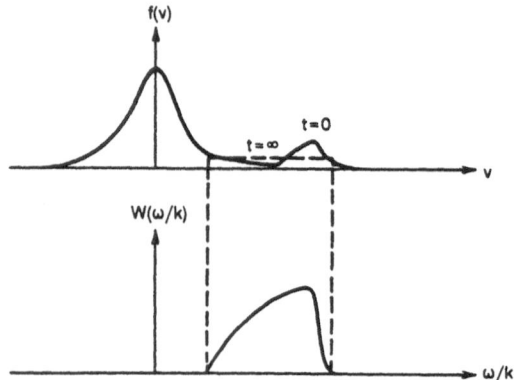

Fig. 9. The evolution of a beam plasma system

The non-linear evolution of the beam plasma system is not trivial. There are two separate ways to approach the problem. (1) In the framework of the *weak turbulence theory* (Sagdeev and Galeev, 1969) we can use standard perturbation techniques when the amplitude of the excited waves is small compared to the thermal energy of the ambient plasma. (2) The amplitude of the wave is so high that they modify of the linear dispersion relation and the perturbation analysis is not valid. The analysis followed in this case is not standard and is called *strong turbulence theory*. A full description of the weak turbulence theory or the strong turbulence theory is outside the scope of these lectures. We will outline briefly the main conclusions from the application of each theory to the beam plasma system.

The wave particle interaction in the framework of the weak turbulence theory is handle by a system of equations known as Quasilinear equations (Nicholson, 1983). The evolution of the velocity distribution is shown in Fig. 9. The instability saturates due to the change of the velocity distribution of the beam particles

(f_b). The beam velocity distribution will become flat when $t \to \infty$ and the linear growth rate (eq (3.5)) approaches zero.

The beam plasma system follows a different evolution when the wave energy exceed a threshold (see Goldman, 1984). The growing electrostatic waves change the linear dispersion relation of the plasma waves (forming solitons in the electric field and cavitons in the density of the ambient plasma). The wave number of the beam excited wave will be forced to become larger (in order to be a mode on the new dispersion relation) and the phase velocity of the beam driven waves will approach the thermal speed. The system is now non-linearly stable since the interaction of waves with the beam particles is not possible.

The results from the two approaches are remarkably different. The weak turbulence theory (valid mostly for very weak and not so energetic beams) suggests that the beam quickly change its shape (becoming flat in the velocity space) and the growth of the waves vanish. The beam, according to this scenario, will not propagate very far. The strong turbulence theory, on the other hand, suggests that the beam quickly enters a non-linearly stable state which allow the beam to propagate long distances (Papadopoulos et al., 1974). A useful conclusion from this analysis is that strong ($n_b/n_0 > 10^{-4}$) and energetic beams (several keV) will easily cross the corona and propagate in the interplanetary space since they are non-linearly stable.

Several scenarios for the evolution of beam plasma systems that are not so strict show that the picture is never so clear. In most stellar applications the beam energetics are such that a mixed evolution is most probable (see Hillaris, 1988).

The next question that is relevant to our lectures is the radiation emitted from a beam of particles. The weak turbulence theory suggest that the excited waves will resonate with the other two normal modes of the non-magnetised ambient plasma (ion acoustic and electromagnetic waves) and transfer energy to the electromagnetic wave. The three wave interaction

$$\omega_L + \omega_s \to \omega_{em} \qquad (3.7)$$

and

$$k_L + k_s \to k_{em} \qquad (3.8)$$

(multiplying Eqs (3.7) and (3.8) with $\hbar$ we recover the well known relation for the conservation of energy and momentum in Quantum mechanics) is a well known problem that is analysed in many plasma physics textbooks (e.g., Sagdeev and Galeev, 1969). Other interactions are also possible, e.g., the interaction of two plasma waves with an electromagnetic wave with frequency $\omega_{em} = 2\omega_e$, the decay of a plasma wave to another plasma wave and an ion acoustic wave or to an electromagnetic wave and an ion acoustic wave (Melrose, 1986). We then conclude that the presence of a beam excited spectrum of plasma waves can drive non-linearly several normal modes of the ambient plasma including electromagnetic waves which can escape from the plasma and propagate inside the free space (vacuum) with frequencies $\omega_{em} = \omega_e$ or $2\omega_e$.

We can reach similar results using a fully non-linear $2\frac{1}{2}$ numerical code when the system is driven in the strong turbulence regime. Akimoto et al. (1988) found that solitons emit electromagnetic radiation with frequencies $\omega_e, 2\omega_e$ and even $3\omega_e$ (see Fig.10)

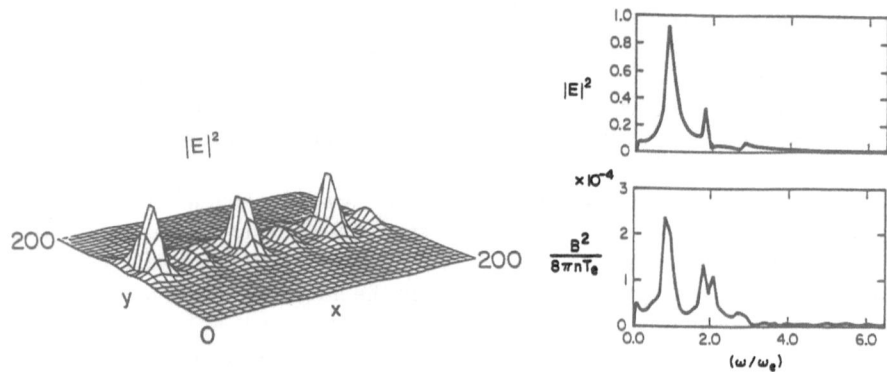

Fig. 10a. The electrostatic field form solitons

Fig. 10b. Electromagnetic radiation from solitons at $\omega_e, 2\omega_e$ and $3\omega_e$

3.2 Electron Cyclotron Maser Instability

The presence of magnetic field in the stellar atmosphere is responsible for many radical changes on the flow of plasmas, e.g., closed field lines are associated with active regions. A collection of loops with low magnetic field in the coronal part and strong field in the foot will trap energetic electrons and form a special type of velocity distribution known as *loss-cone* distribution.

Charged particle gyrating inside the magnetic field conserve their magnetic moment

$$\mu = \frac{1/2mv_\perp^2}{B} \qquad (3.9)$$

where $v_\perp$ is the velocity component vertical to the magnetic field. If the particle starts at the apex of the loop with $v_\perp \gg v_\parallel$ and propagates towards the strong magnetic field, it will reach a point where $v_\perp = v$ and $v_\parallel = 0$. The particle will be reflected at this point and return back to the coronal part of the loop. Assuming now that an isotropic distribution is injected towards the foot of the arcade and that the particles reflected inside the chromosheric part of the loop will lose all their energy to collisions, trapped particles will form a distribution where a cone is missing (see Fig. 11)

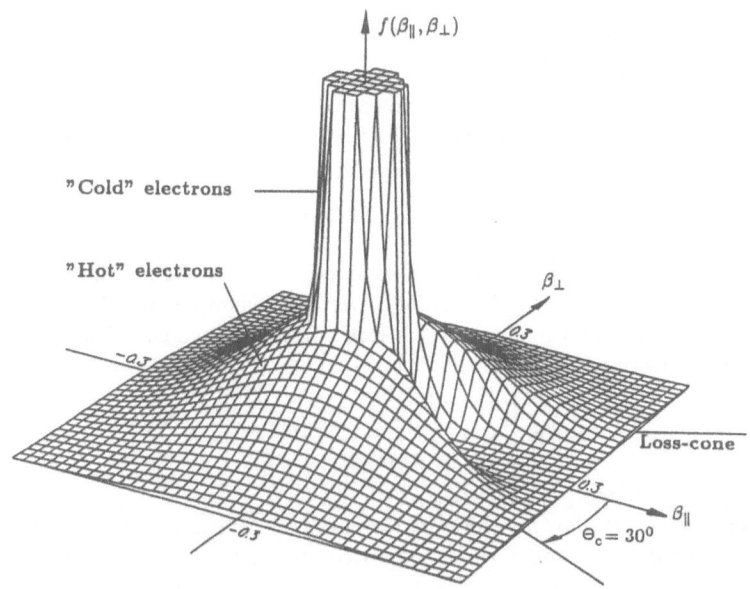

Fig. 11. A loss cone distribution

We can estimate the angle of the cone from the relation

$$\frac{1/2mv_{\perp 1}^2}{B_{top}} = \frac{1/2mv_{\perp 2}^2}{B_{foot}} \tag{3.10}$$

with $B_{top} \ll B_{foot}$. Since $v_\perp = v\sin\Theta$, where Θ is the pitch angle of the charged particle, and v is the total velocity of the particle which is constant when a particle is moving inside a magnetic field, we can replace eq. (3.10) with

$$\frac{\sin^2\Theta_1}{B_1} = \frac{\sin^2\Theta_2}{B_2} \tag{3.11}$$

or

$$\sin\Theta_1 = (B_1/B_2)^{1/2}\sin\Theta_2. \tag{3.12}$$

For $\Theta_2 = 0$ $\sin(\Theta_{cr}) = (B_1/B_2)^{1/2}$. Electrons with pitch angles smaller than Θ_{cr} at the apex of the loop will precipitate at the upper chromosphere and collisionally lose all their energy. Electrons with pitch angles larger than Θ_{cr} will be reflected back in the coronal part of the loop. Θ_{cr} is the loss cone angle if $B_2 = B_{foot}$, where B_{foot} is the value of the magnetic field at the transition region.

We can model a loss cone velocity distribution with a distribution function

$$f_b(v_\perp, v_\parallel) = \frac{n_b}{(2\pi)^{2/3}V_{eb}^3}exp\left[-\frac{(v_\parallel - V_b)^2}{2V_{eb}^2} - \frac{v_\perp^2}{2V_{eb}^2}\right]H(v_\perp - \frac{v_\parallel}{\sigma}) \tag{3.13}$$

where H is a step function which takes into account the loss cone anisotropy, σ is the mirror ratio. Mildly relativistic electrons forming loss-cone type distributions are unstable to electromagnetic cyclotron waves with frequency $\omega = n\omega_0$ (Wu, 1985). Electrons resonate with the cyclotron wave $\omega_0 = \frac{eB}{m_e c}(1 - v/c)^{-1/2} = \omega_b(1 - v^2/(2c^2))$ when they satisfy the condition

$$\omega_r - \omega_0 - k_\parallel v_\parallel = 0. \tag{3.14}$$

If the velocity distribution has a positive slope in this velocity regime electromagnetic cyclotron waves will be excited. The most favourable mode is the first harmonic extraordinary wave, when the frequency is above the X-mode cut-off frequency (Nicholson, 1983)

$$\omega_x = (1/2)\omega_0 + (1/2)(\omega_0^2 + 4\omega_e^2)^{1/2} \approx \omega_0 + \frac{\omega_e^2}{\omega_0} \tag{3.15}$$

for strongly magnetised plasma with $\omega_e << \omega_0$.

The dispersion relation for the first harmonic extraordinary wave $\omega \approx \omega_0 \approx kc >> \omega_e$ propagating almost vertically to the ambient magnetic field is (Wu, 1985)

$$1 - \frac{c^2 k^2}{\omega^2} + \frac{\omega_e^2}{\omega^2}\int d^3v \left(\omega_b \frac{\partial f}{\partial v_\perp} + k_\parallel v_\perp \frac{\partial f}{\partial v_\parallel}\right) \frac{v_\perp (J_1'(b))^2}{(\omega - \omega_b/\gamma - k_\parallel v_\parallel)} = 0 \tag{3.16}$$

and for the ordinary mode

$$1 - \frac{c^2 k_\perp^2}{\omega^2} + \frac{\omega_e^2}{\omega^2}\int d^3v \left(\omega_b \frac{\partial f}{\partial v_\perp} + k_\parallel v_\perp \frac{\partial f}{\partial v_\parallel}\right) \frac{v_\parallel^2 J_1^2(b)}{v_\perp(\omega - \omega_b/\gamma - k_\parallel v_\parallel)} = 0, \tag{3.17}$$

where $J' = dJ/db, b = k_\perp v_\perp/\omega_b, k_\perp/k_\parallel >> 1$. Assuming that the particles are mildly relativistic ($v/c << 1$) we can make several assumptions. (1) We can expand the relativistic correction factor $\gamma = (1 - v^2/c^2)^{-1/2} \approx (1 - v^2/2c^2)$, (2) b is small and we can replace the Bessel functions $J_1^2(b)$ with $v_\perp^2/4c^2$ and $(J_1'(b))^2$ with $1/4$, (3) the term $\partial f/\partial v_\parallel$ is neglected when $\partial f/\partial v_\perp \approx \partial f/\partial v_\parallel$ since $k_\parallel \approx 0$, and finally (4) we ignore the presence of the ambient plasma. Then the growth rate for the first harmonic X-mode is

$$\gamma_x = \frac{\pi^2 \omega_e^2}{4\omega_L}\int_{-\infty}^{\infty} dv_\parallel \int_0^{\infty} dv_\perp v_\perp^2 \delta\left[\omega_L - \omega_b\left(1 - \frac{v^2}{2c^2}\right) - k_\parallel v_\parallel\right]\omega_b \frac{\partial f}{\partial v_\perp}, \tag{3.18}$$

and for the O-mode,

$$\gamma_o = \frac{\pi^2 \omega_e^2}{4\omega_L}\int_{-\infty}^{\infty} dv_\parallel \int_0^{\infty} dv_\perp (\frac{v_\perp^2}{c^2})v_\parallel^2 \delta\left[\omega_L - \omega_b\left(1 - \frac{v^2}{2c^2}\right) - k_\parallel v_\parallel\right]\omega_b \frac{\partial f}{\partial v_\perp}, \tag{3.19}$$

It is obvious that the growth rate for the X-mode is higher and the radiation emitted in a narrow cone around 90^0.

Including the ambient plasma in the velocity distribution and using the complete dispersion we can estimate the growth rate of the first harmonics (see Fig.

151

12). It is obvious that as the plasma density increases the linear growth rate
of the first and higher harmonics of the extraordinary and the ordinary mode
changes dramatically.

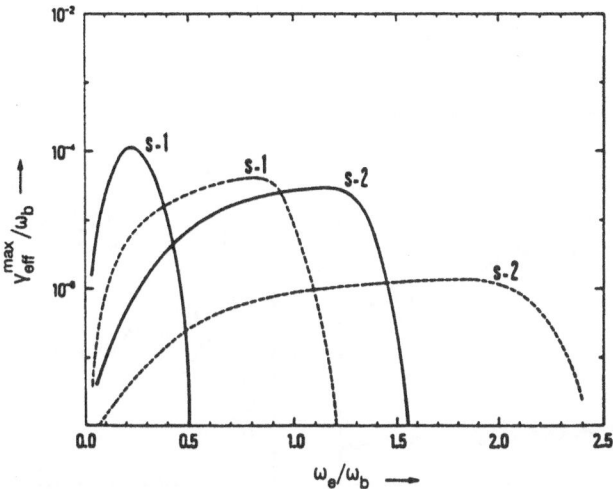

Fig. 12. The maximum growth rate as a function of ω_e/ω_0. The first and second
harmonic for the X-mode (solid line) and O-mode (dashed line) are shown

In fact, for $\omega_e/\omega_b < 0.3$, the fundamental X-mode has the highest growth
rate, for $0.3 < \omega_e/\omega_0 < 1$ the fundamental O-mode and the second harmonic X-
mode dominate (Sharma and Vlahos, 1984). As the ratio ω_e/ω_b increases above
unity, no direct emission is expected (the electrostatic modes have the largest
growth rate). It is easy to show that the mode with the largest growth rate
will be the dominant mode and will saturate the instability before the other
modes grow to any appreciable level (Sharma et al., 1982). The effects of finite
plasma temperature and superthermal tails on the ambient distribution on the
evolution of the Electron Cyclotron Maser (ECM) have been studied by many
authors (Sharma and Vlahos, 1984; Winglee, 1985; Vlahos and Sharma, 1985).
These results are important for the application of the ECM in solar or stellar
flares since the loss-cone particles and the super hot and dense plasma almost
co-exist.

The linear theory discussed so far suggest only the frequency range and the
wave numbers that will be excited by the loss cone distribution. The exponential
growth lasts only 50-100 growth times, $t_g \approx \gamma_j^{-1}$ (where γ_j is the growth rate
of the mode j). The saturation time for typical solar or stellar parameters is
$\approx 10^{-6} sec$. The evolution of the burst follows very closely the characteristics of
the injection function of the energetic particles.

The amplification of the Electron Cyclotron Maser is due to the positive
slope in the direction perpendicular to the magnetic field. One possible way
to saturate the instability is for the growing waves to flatten the distribution

in the direction perpendicular to the magnetic field (using the standard quasi-linear theory mentioned above). Numerical simulations by Wagner et al. (1984), Pritchet (1984), Lin et al. (1984) demonstrated that only $1-5\%$ (Fig.13) of the energy carried by the energetic particles will be radiated away. This efficiency is enough for producing extremely high brightness temperature ($T_b \approx 10^{16} - 10^{17} K$).

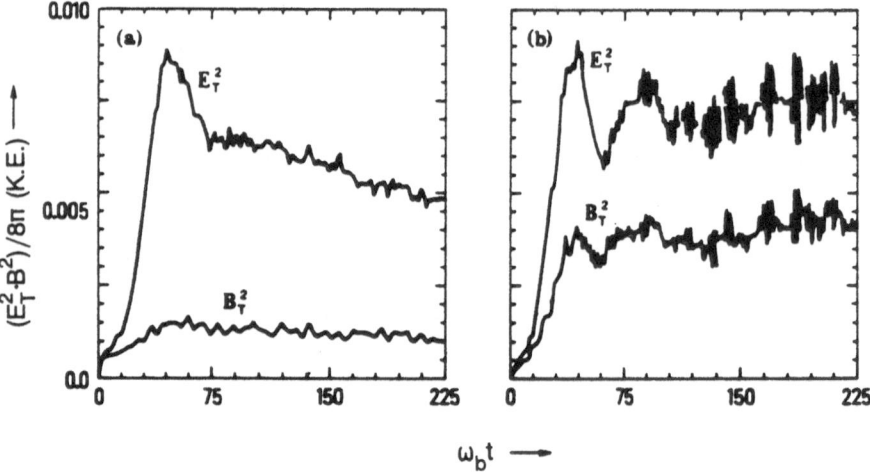

Fig. 13. The time histories of the transverse electric (E_T^2) and magnetic (B_T^2) energy normalised to the initial electron kinetic energy

4 Observations

Magnetic-field dynamics in the convection zone and solar atmosphere is crucial for the high-energy emission since: (1) the energy release processes and particle energisation will form and sustain the injection distribution, (2) the magnetic-field topology will guide the energetic particles away from the energy release (energetic particle transport) and re-shape the initial distribution. Injected distributions evolve differently in open or closed magnetic field lines. In the stellar atmospheres two types of magnetic topologies exist: (1) closed field lines associated mainly with the active regions, (2) open field lines associated mainly with coronal holes. In Fig. 14 we present a cascade of scales that naturally appear during the evolution of magnetic fields.

We will focus our attention on active regions where most of the high-energy electrons are accelerated. A variety of observations have been recorded and it is impossible to review all of them here; instead, we plan to present only a small subset of the available data that are related to the processes presented above.

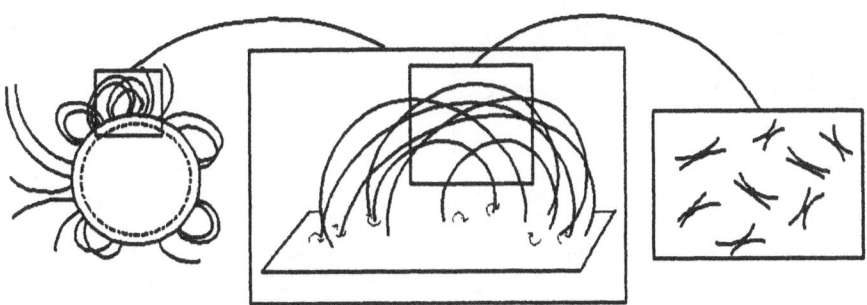

Fig. 14. Multi-scale evolution of the active region magnetic field

4.1 Hard X-ray emission

In Fig. 15a we present a series of solar bursts in the energy range $40keV < E < 25MeV$. The spectrum at the peak of the intensity is given in Fig. 15b. The existing data can be interpreted as follows:

A power law velocity distribution is formed impulsively inside a loop or in the boundaries of a collection of loops. This distribution is called the injection distribution. Energetic electrons trapped and precipitated in the low atmosphere will collide with the ambient plasma and emit X-rays. This model can explain the spectra and the time evolution of the emitted X-ray radiation (Vilmer, 1987). Bremsstrahlung emission is the main radiation process that can interpret successfully the X-ray data. Many questions remain open on the critical energy where the distribution changes from thermal to power law, and the energy where the slope of the distribution changes (usually at very high energies). We can now ask if this type of injection distribution can explain the radio data. We will discuss this topic in the next section.

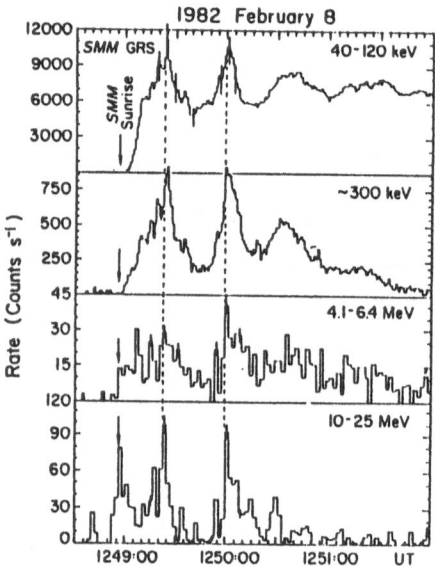

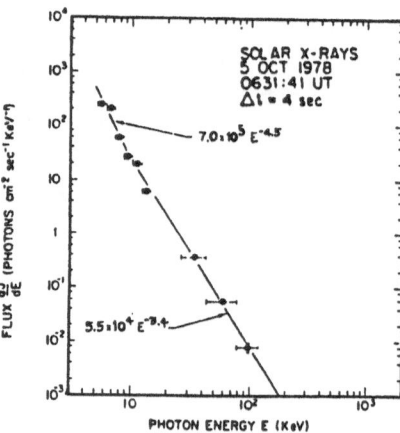

Fig. 15a. The observed time histories in 4 energy bands

Fig. 15b. Typical X-ray spectra

4.2 Radio Observations

Trapped energetic electrons gyrate in the strong magnetic field of the low lying arcades and emit synchrotron radiation that is responsible for the microwave bursts. Microwave and X-ray emission are almost co-spatial and have the same time evolution (Kundu and Vlahos, 1982). Unfortunately only a few attempts have been made for simultaneous modelling of both signatures.

In section 3.2 we proved that trapped energetic electrons form loss cone velocity distributions that are unstable to the ECM instability when $\omega_0 >> \omega_e$. A very intense (brightness temperature $\approx 10^{15} K$), highly polarised and spiky radio burst was observed (Slottje, 1978) which is believed to be the signature of this instability. In Fig. 16 we show an example of a microwave burst recorded with a high time resolution, and we can make three important points: (1) The spiky component is added on top of the normal microwave burst, (2) the duration of each spike is a few milliseconds and its polarisation is very high (almost 100%) of the spike, (3) the spiky emission stops at the peak of the normal burst, probably due to the changes of the ambient plasma. All these characteristics can be explained by the ECM theory (Vlahos, 1987; Melrose, 1991).

Electrons accelerated in the boundary of arcades or drifted outside the closed field lines will stream along the open field lines forming field aligned beams of energetic electrons. These beams excite a number of bursts that have been observed, e.g., type III bursts. In section 3.1 we analysed the emission from

155

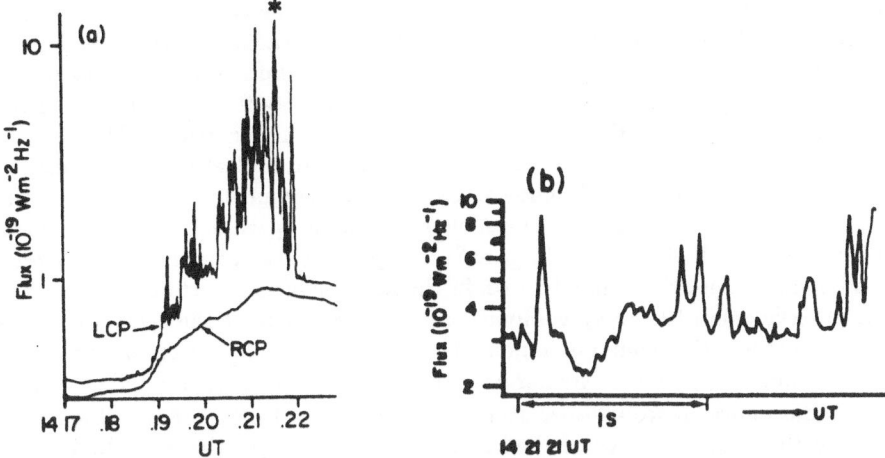

Fig. 16. (a) Flux density of left (LCP) and right (RCP) circularly polarised components of the solar burst of 1978 April 11 observed by Slottje (1978). (b) An expansion of the portion shown by an asterisk in (a) show clearly that the spikes are irregular in time with no obvious pulsation frequency.

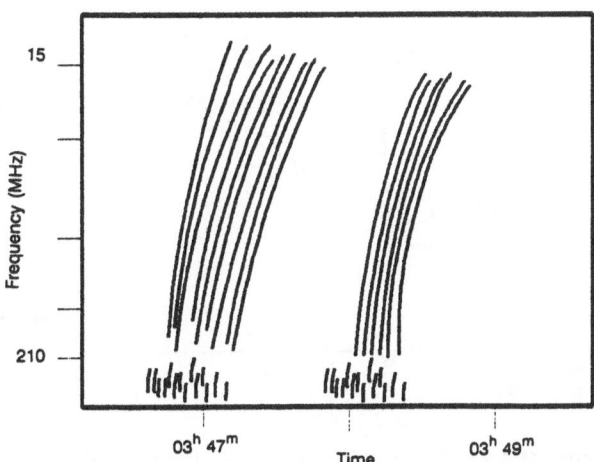

Fig. 17. A schematic presentation of a group of type III bursts and the corresponding dcm spikes. The vertical scale is logarithmic.

a beam-plasma system. We proved that the beam driven plasma waves will resonate with the electromagnetic waves and emit radiation with frequencies ω_e and $2\omega_e$. The plasma frequency is a function of the ambient density ($\omega_e \propto \sqrt{n_0}$) and since the density decreases as we move away from the surface of the Sun, we should expect the emission to drift to smaller frequencies. The slope of the emission in the frequency vs time diagram is an indication of the velocity of the beam if the $n_0(z)$ is known. Using such arguments, we estimated the $V_b = 10^{10} cm/sec$. In Fig. 17 we present the dynamic spectrum of type III burst. Energetic particles trapped in large arcades will form loss cone (in weak magnetic fields) and beam driven instabilities. In the high frequency part of the spectrum we observe again the spiky component in the dcm part of the spectrum. This component is rather common in most flares and a close association of dcm spikes and a group of type III bursts has been reported (Aschwanden and Gudel, 1992).

Radio emission from other stars (see review by Dulk, 1985) follows very closely the characteristics of the emission observed from the Sun. In Fig. 18 we present an example of highly polarised very intense radio burst recorded at Arecibo from the dMe star AD Leo, which is interpreted as the signature of the ECM instability.

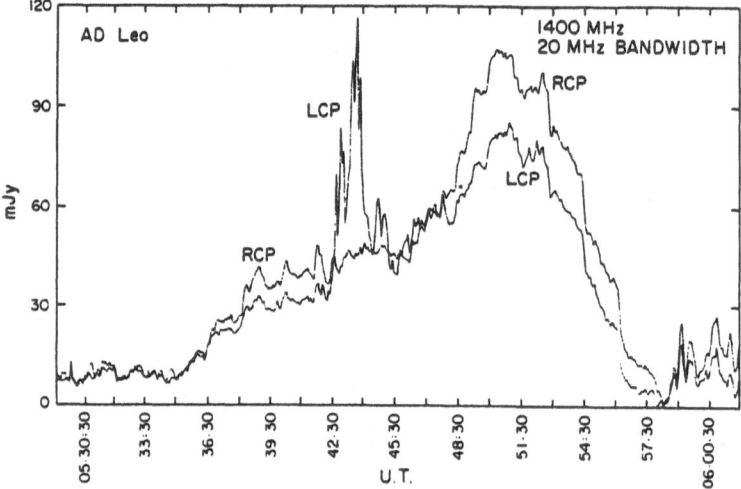

Fig. 18. A radio burst at 1.4 GHz, recorded at Arecibo from the dMe star AD Leo. A purely LH-polarised burst. Even with a time resolution of 200 ms some spikes are not resolved (from Lang et.al., 1983).

In summary, using simple models for the magnetic-field structure and the injection distribution we have succeed to interpret successfully the observed radio and X-ray data. A major setback to our analysis is the lack of simultaneous multi-wavelength data. It is true that many open questions still exist but they are mostly related to the dynamic evolution of the burst, the polarisation etc., that require more careful modelling. Models for the evolution and propagation of radiation are difficult to construct since they require much more detailed knowledge of the environment.

5 Summary and Conclusions

We have made an attempt in these lectures to present a global view of the high-energy emission from normal stars. We focussed only on the emission associated with high-energy electrons and avoided completely the emission of the high-energy ions, e.g., nuclear lines etc. We showed that the energy flow inside a normal star follow several stages.

Energy released inside the stellar core $\longrightarrow$ propagation inside the radiation zone $\longrightarrow$ drive turbulence in the convection zone and generate magnetic fields (dynamo) $\longrightarrow$ escape of the magnetic field from the active region (buoyancy) and formation of active regions $\longrightarrow$ random photospheric motions and/or emergence of new flux generate neutral sheets and local explosions $\longrightarrow$ acceleration of particles, heating and local flows $\longrightarrow$ interaction of energetic particles with the ambient plasma and the magnetic field generate high energy emission through processes such as (1) bremsstrahlung, (2) synchrotron radiation, (3) collective plasma emission, and (4) nuclear interactions (not discussed here). Analysing the observations (ground based or from satellites) we can infer the shape of the injection distribution and the topology of the magnetic field.

The picture that has emerged over the years is as follows:

1. Energetic electrons (Energy $> 100 keV$) are either trapped and precipitate in the low atmosphere, or stream in open field lines to generate:

Hard X-ray emission	trapped or precipitate	bremsstrahlung
microwave bursts	trapped electrons	Synchrotron
Microwave spikes	trapped (loss cone)	ECM
type III bursts	streaming	beam instabilities

2. The injected distribution of energetic electrons above $100keV$ has the following characteristics:

(a) power law in energy $f \propto E^{-\delta}$, where $\delta = 3 - 10$; (b) acceleration time 10 milliseconds to minutes, (c) total number of accelerated electrons is $10^{37} - 10^{38}$

These characteristics are mostly derived from Solar data that are more detailed, but observations from other stars follow similar patterns: the scenario presented in these lectures is rather universal and covers the evolution of most normal stars.

Acknowledgements

I would like to thank Prof. J. van Paradijs, the director of the school, and Prof. M.H. Maitzen, the local organizer, for providing an excellent atmosphere for scientific discussions during the school. Several questions and comments made by the students attending the school have helped me to improve these lecture notes.

References

Akimoto, K., Rowland, H.L. and Papadopoulos, K. (1988): Phys. Fluids **31**, 2185.
Anastasiadis, A. and Vlahos, L. (1991): Astr. Astroph. **245**, 271.
Aschwanden, M.J. and Guedel, M. (1992): Astroph. J. **401**, 736.
Athay, G. (1972):"Radiation Transport in Spectral Lines" (Reidel, Dordrecht).
Bekefi, G. (1966) "Radiation Processes in Plasmas" (Wiley, New York).
Bowers, R. and Deeming, T. (1984): "Astrophysics, Vol. I" (Jones and Bartlett Publishers, Inc, Boston)
Dulk, G. (1985): Ann. Rev. Astr. Astroph. **23**, 169.
Goldman, M.N. (1984): Rev. Mod. Phys. **56**, 709.
Griem, H.R. (1974): "Spectral Line Broadening by Plasmas" (Academic Press, New York)
Heyvaerts, J. (1989): 'Coronal Heating', in IAU Symp. 142, ed. E.R. Priest and V. Krishan, (Kluwer, Dordrecht), p. 207.
Hillaris, A., Alissandrakis, C. and Vlahos, L. (1988): Astr. Astroph. **195**, 301.
Karzas, W. and Latter, R. (1961): Astroph. J., Suppl. **6**, 167.
Kundu, M. R. and Vlahos, L. (1982): Space Sci. Rev. **32**, 405.
Lang, K.R., Bookbinder, J., Golub, L. and Davis, M. (1983): Astrophys. J., (Lett) **272**, L15.
Lin, A.T., Chang, W.W. and Lin, C.C. (1984): Phys. Fluids **27**, 1054.
Martin, S. (1990): "Solar Magnetic Fields" in J. Italian Astr. Soc. **61(2)**, 293.
Melrose, D.B. (1986): "Instabilities in Space and Laboratory Plasmas" (Cambridge University Press, Cambridge).
Melrose, D. (1991): Ann. Rev. Astr. Astroph., **29**, 31.
Nicholson, R. (1983): "Introduction to Plasma Physics Theory" (Wiley, New York).
Novikov, I.D. and Thorne, K.S. (1973): in "Black Holes", Eds. C. De Witt and B. De Witt (Gordon and Breach, New York).
Papadopoulos, K., Goldstein, M.L. and Smith, R.A. (1974): Astroph. J., **190**, 175.
Parker, E.N. (1972): "Cosmic Magnetic Fields" (Clarendon, Oxford).
Priest, E.R. (1990): "Reconnections on the Sun" in IAU Symp. 142, eds E.R. Priest and V. Krishan, (Kluwer, Dordrecht), p. 271.
Pritchet, P.L. (1984): Phys. Fluids **27**, 2393.
Raoult, A., Vlahos, L. and Mangeney, A. (1990): Astr. Astroph. **233**, 229.
Ramaty, R. and Murphy, R.J. (1987): Space Sci. Rev. **45**, 213.
Rybicki, G. B. and Lightman, A.P. (1979): "Radiative Processes in Astrophysics" (Wiley, New York).
Sagdeev, R.Z. and Galeev, A.A. (1969): "Nonlinear Plasma Theory" (W.A. Benjamin, New York)

Sharma, R.R. and Vlahos, L. (1984): Astroph. J. **280**, 405.

Sharma, R.R., Vlahos, L. and Papadopoulos, K. (1982): Astr. Astroph. **112**, 377.

Shu, F. (1982):" The Physical Universe" (University Science Books, California).

Slottje, C. (1978): Nature **275**, 520.

Sobelman, I.I. (1979): "Atomic Spectra and Radiative Transitions" (Springer-Verlag, Heidelberg).

Sobolev, V.V. (1963): "A Treatise in Radiative Transfer" (van Nostrand, Princeton).

Stix, M. (1989): "The Sun" (Springer-Verlag, Berlin).

Vas'kov, V.V., Gurevich, A.V. and Dimant, Ya.S. (1983): Sov. Phys. JEPT **52**, 310.

Vilmer, N. (1987): Solar Phys. **111**, 207.

Vlahos, L. (1987): Solar Phys. **111**, 155.

Vlahos, L. (1989): Solar Phys. **121**, 431.

Vlahos, L. and Sharma, R.R. (1985): Astroph. J. **290**, 347.

Wagner, J.S., Lee,L.C., Wu, C.S. and Tajima, T. (1984): Radio Sci. **19**, 509.

Winglee, R.M. (1985): Astroph. J. **291**, 160.

Winskie, D. and Boyd, D.A. (1983): Phys. Fluids **26**, 755.

Winskie, D., Peter, Th. and Boyd, D. (1983): Phys. Fluids **26**, 3497.

Wu, C.S. (1985): Space Sci. Rev. **41**, 215.

Accretion in Close Binaries

A.R. King

Astronomy Group, The University, Leicester LE1 7RH, U.K.

1. Introduction

Virtually all binaries with periods of less then a few years will interact at some stage of their evolution, because the radius of one of the stars will become comparable with the binary separation. The interaction usually involves the transfer of mass from one star to another. As we shall shortly see, accretion of this matter by the other star can be an extremely energetic process if the latter star is compact, i.e. a white dwarf, neutron star or black hole. Accordingly, the investigation of close binaries is a particularly fascinating field, characterized by the interplay of stellar evolution theory and the study of accretion processes. A general reference for much of the material presented here is the book by Frank *et al.* (1992).

1.1. Energy Yields

Nuclear Burning

The most efficient way of extracting energy from matter in thermonuclear processes is via the conversion of hydrogen to helium: other processes yield considerably less. The maximum energy which can be liberated from a mass ΔM by thermonuclear processes is

$$\Delta E_{\text{nuc}} = 0.007 \Delta M c^2$$

$$= 6 \times 10^{18} \quad \text{erg g}^{-1} \tag{1}$$

Accretion on to a Compact Object

For an object of mass M with a hard surface of radius R_* the energy yield is

$$\Delta E_{\text{acc}} = \frac{GM}{R_*} \Delta M \tag{2}$$

so that the efficiency evidently rises the more *compact* the accreting star is, i.e. the larger the ratio M/R_*. For a white dwarf ($M \sim M_\odot$, $R_* \sim 10^9$ cm) we find

$$\Delta E_{\text{acc}} \sim 10^{17} \ \text{erg g}^{-1} \tag{3},$$

while for a neutron star ($M \sim M_\odot$, $R_* \sim 10^6$ cm) we have

$$\Delta E_{\text{acc}} \sim 10^{20} \ \text{erg g}^{-1} \tag{4}.$$

A black hole might have a somewhat higher mass (few $M_\odot$), but does not have a hard surface, so that we expect an estimate like (4) again.

We conclude that accretion on to neutron stars and black holes is the most energetically efficient process, being some $20\times$ the maximum nuclear burning efficiency. In principle nuclear burning of matter on the surface of a white dwarf should yield more energy than released as it accretes. However the nuclear processes are frequently explosive (*novae*), so that here too accretion is the dominant energy source most of the time. The instantaneous luminosity from accretion on to a star with a hard surface is

$$L_{\text{acc}} = \frac{GM\dot{M}}{R_*} \tag{5}.$$

1.2. Radiation Spectrum

We expect the radiation emerging from accretion processes to be characterized by two temperatures:

$$T_{\text{eff}} = \left(\frac{L_{\text{acc}}}{4\pi R_*^2 \sigma} \right)^{1/4} \tag{6}$$

and

$$T_V = \frac{GM m_{\text{p}}}{k R_*} \tag{7},$$

rather like the effective and central temperatures of a star. T_{eff} corresponds to cases where the accretion energy is thermalized before being radiated, and T_V to cases where the infall energy (mainly in protons of mass m_{p}) is given up directly, e.g. in shocks. We thus get an idea of the likely emission from accretion processes by computing kT_{eff} and kT_V.

For neutron stars and black holes, with 10^{36} erg s$^{-1} \lesssim L_{\text{acc}} \lesssim 10^{38}$ erg s^{-1} we find $T_{\text{eff}} \sim 10^7$ K, $kT_{\text{eff}} \sim 1$ keV, $T_V \sim 10^{12}$ K, $kT_V \sim 100$ MeV, so that we can expect X-rays and possibly gamma rays.

For accretion on to white dwarfs, $L_{acc} \sim 10^5$ erg s^{-1} gives $T_{eff} \sim 10^5$ K, $kT_{eff} \sim 10$ eV, $T_V \sim 10^9$ K, $kT_V \sim 100$ keV, implying UV, soft X-ray and possibly hard X-ray emission.

We can thus make the identifications

$$accreting \quad neutron \quad stars, \quad black \quad holes \equiv X-ray \quad binaries \quad (XRBs)$$

$$accreting \quad white \quad dwarfs \equiv cataclysmic \quad variables \quad (CVs)$$

Thermonuclear Events

For accreting compact stars we can easily identify the thermonuclear events as
  neutron stars, black holes: (most) X-ray bursts
  white dwarfs: novae.
Since $\Delta E_{acc} < \Delta E_{nuc}$ for white dwarfs but not neutron stars, we expect mass ejection to occur in novae, but not in X-ray bursts.

1.3. Rotational Energy

A star can store energy in the form of rotation, particularly if it has been spun up by accreting angular momentum (and mass) from a binary companion. Here again the important objects are compact. For the spin angular velocity Ω cannot be so large that centrifugal force exceeds gravity at the equator of the star:

$$R_* \Omega^2 \lesssim \frac{GM}{R_*^2} \tag{8}$$

i.e.

$$\Omega^2 \lesssim \frac{GM}{R_*^3} \tag{9}.$$

Thus the maximum rotational energy which can be stored is

$$E_{rot}(\max) = \frac{I\Omega^2}{2} \tag{10}$$

where the moment of inertia I is given by $k^2 M R_*^2$, with $k \lesssim 1$ the radius of gyration. Hence

$$E_{rot}(\max) \simeq \frac{k^2}{2} \frac{GM^2}{R_*}, \tag{11}$$

which is of the order of the binding energy.

The most spectacular example of this is provided by the radio pulsars, which are rapidly rotating neutron stars. Some of them have periods of milliseconds, implying that they are near the limit (11). Some of the millisecond pulsars are observed to have close companions, while even the presently isolated ones may have been spun up by a companion star which has now disappeared.

2. Accreting Binaries

2.1. The Sizes of Binaries

We shall consider close binaries in which a compact primary star of mass M_1 accretes from a secondary or donor star of mass M_2; we write the total binary mass as $M = M_1 + M_2$. In almost all cases one can measure the binary period P. Binaries are relatively easy to study because this directly gives an idea of the size of the binary: the orbital semi-major axis (stellar separation for a circular orbit) a is given by Kepler's third law as

$$a^3 = \frac{GMP^2}{4\pi^2} \qquad (12).$$

It is convenient to write $m_1 = M_1/M_\odot$, $m_2 = M_2/M_\odot$, $m = M/M_\odot$, and $P_{\rm hr}, P_{\rm day}$ for the binary period measured in hours or days respectively. Then

$$a \simeq 0.5 R_\odot m^{1/3} P_{\rm hr}^{2/3} \simeq 3.5 \times 10^{10} m^{1/3} P_{\rm hr}^{2/3} \quad {\rm cm}, \qquad (13)$$

giving the typical size of CVs and some low–mass X–ray binaries (LMXBs), which have periods of a few hours. For massive X–ray binaries

$$a \simeq 40 R_\odot (m/10)^{1/3} (P_{\rm day}/10)^{2/3} \simeq 2.9 \times 10^{12} (m/10)^{1/3} (P_{\rm day}/10)^{2/3} \quad {\rm cm} \ (14).$$

These estimates show that in CVs and LMXBs with $P = $ a few hours, a lower main–sequence secondary star ($R_2 \lesssim R_\odot$) will occupy a large fraction of the binary orbit, while in massive XRBs with $P = $ days a giant or supergiant donor ($R_2 \sim$ several $R_\odot$) will do this. From (5) we know that the typical luminosities $L_{\rm acc} \sim 10^{33} - 10^{34}$ erg s^{-1} and $10^{36} - 10^{38}$ erg s^{-1} in CVs and X–ray binaries, respectively, require accretion rates $\dot{M} = -\dot{M}_2$ of order $10^{16} - 10^{17}$ g s^{-1} ($10^{-10} - 10^{-9}$ M$_\odot$ y^{-1}) for CVs and $10^{16} - 10^{18}$ g s^{-1} ($10^{-10} - 10^{-8}$ M$_\odot$ y^{-1}) for X–ray binaries. In Sections 5 and 6 we will discuss how such rates of mass transfer can arise.

2.2. The Eddington Limit

In all cases we know that there is a limit to the steady accretion luminosity; if this becomes too large its radiation pressure will inhibit further accretion. Consider matter accreting on to a star of mass M_1, at a distance r from its centre. If the luminosity is L the electrons in this matter will each feel a radiation pressure force

$$F_{\rm rad} = \frac{L\sigma_{\rm T}}{4\pi r^2 c}$$

outwards, where $\sigma_{\rm T}$ is the Thomson cross–section. However the electrons are unable to separate entirely from the positive ions in the flow because overall electrical neutrality will be maintained by Coulomb forces. For normal cosmic matter (mostly hydrogen) there will be about one proton per electron; hence the gravitational force opposing F_{rad} will be

$$F_{grav} = \frac{GM_1}{r^2}(m_p + m_e) \simeq \frac{GM_1 m_p}{r^2}. \tag{15}$$

Since both F_{rad} and F_{grav} scale as r^{-2}, steady accretion is evidently only possible if

$$L < L_{Edd} \simeq \frac{4\pi GM_1 m_p c}{\sigma_T} \simeq 10^{38} m_1 \text{ erg s}^{-1}. \tag{16}$$

It is worth making three points about L_{Edd}.

(a) The restriction $L < L_{Edd}$ only holds in a steady state (it is violated by many orders of magnitude in a supernova explosion for example).

(b) The precise value of L_{Edd} depends on the chemical composition, i.e. the mean mass per electron.

(c) The limit clearly applies also to the intrinsic stellar luminosity also: combined with the upper main–sequence mass–luminosity relation $L \propto M^3$ it leads to an upper mass limit $M \lesssim 60 M_\odot$ for main–sequence stars.

3. Tidal Effects

In most close binaries the orbit is circular and any extended stellar component (typically the secondary) rotates synchronously. The first of these results follows from a basic principle of Kepler orbits which we shall invoke several times in these notes:

the orbit of lowest energy for a given angular momentum is a circle.

Thus if the binary orbit is eccentric (and the rotation of the secondary necessarily asynchronous), the varying stellar separation will cause tides to be raised in this star. This flexing must lead to dissipation (via e.g. viscosity) and consequent energy loss, usually on a timescale much shorter than angular momentum can be lost from the binary orbit. Thus the binary will reduce its energy to the minimum allowed by its angular momentum, thus circularizing the orbit. If the secondary is not synchronous, tides will cause further dissipation until it exactly corotates with the binary. In this state everything rotates rigidly in space and there is no further dissipation. Even if angular momentum is gradually lost from the binary orbit, the timescales for circularization and synchronization are sufficiently short that it is generally always a good approximation to regard the binary orbit as circular and the secondary as synchronously rotating. Note that this is not true of the compact primary star, where tides are ineffective.

A further simplification is that we can usually regard both stars as mass points for dynamical purposes. Any gas flow in the binary can then be treated as test particle motion under the combined gravity of the two rotating mass points: this is known as the restricted 3–body problem.

4. The Roche Potential

The restricted 3–body problem is usually treated by adopting a reference frame rotating with the binary; neglecting Coriolis forces we can combine the gravitational and centrifugal forces per unit mass as $-\nabla\Phi_R$, where Φ_R, the Roche potential, is given by

$$\Phi_R = -\frac{GM_1}{|r - r_1|} - \frac{GM_2}{|r - r_2|} - \frac{1}{2}(\omega \wedge r)^2 \qquad (17)$$

where ω is the binary rotation vector, with $\omega = 2\pi/P$. The great advantage of Φ_R is that, as usual, we can get considerable insight into gas motions simply by studying the equipotential surfaces rather than having to solve the equations of motion. In particular, the sections of these equipotentials in the orbital plane, shown in Fig.1, give direct insight into when mass transfer occurs.

The scale of the equipotential pattern is set by the binary separation a, and the precise shapes by the mass ratio $q = M_2/M_1$. The stationary points of Φ_R satisfy $\nabla\Phi_R = 0$ and are called the Lagrange points. The most important of these is the inner Lagrange point L_1, which is a saddle point. It plays the role of a high mountain pass between the two deep valleys around each of the mass points M_1, M_2, which are called the Roche lobes (the two parts of the figure–of–eight equipotential 2 of Fig.1).

5. Roche Lobe Overflow

5.1. Occurrence

The mass transfer process in close binaries is determined by the relation of the sizes of the stars to their Roche lobes. Three generic situations are possible.

(a) Detached. Both stars are smaller than their Roche lobes: mass has to climb a considerable potential barrier to leave either of the stars. Mass transfer can only occur if one of the stars has a strong wind, and is likely to be inefficient (see Section 6).

(b) Semi–detached. One of the stars fills its Roche lobe: because L_1 is a saddle point the gas here pours through into the other star's Roche lobe and is all captured by this star. Because the mass is removed on a dynamical timescale a star can never get much bigger than its Roche lobe (see Sect 5.2).

(c) Contact. Both stars fill their Roche lobes. What happens in this case is rather unclear, but in any case this case does not occur for compact stars.

Case (a) includes all wide binaries, as well as some high–mass X–ray binaries. Case (b) covers contact binaries such as the W UMa systems. The most interesting case applying to CVs, LMXBs, and some high–mass X–ray binaries is the semi–detached case (b). Long–lived efficient mass transfer will ensue if the extended star in a compact binary fills its Roche lobe. This can occur in one of two ways: the star may expand for reasons connected with its internal structure, such as the expansion towards the giant branch at the end of main–sequence

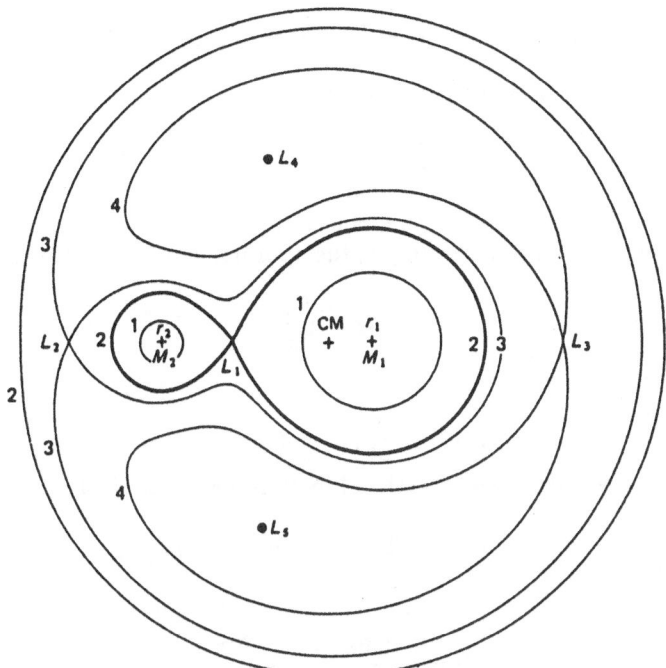

Fig. 1. Sections in the orbital plane of the Roche equipotentials Φ_R = constant, for a binary with mass ratio $q = M_2/M_1 = 0.25$. Stationary points of the potential are marked $L_1 - L_5$ and the centre of mass as CM. The equipotentials are labelled $1 - 4$ in order of increasing Φ_R. The inner Lagrange point L_1 is a saddle point, acting as a 'pass' between the two Roche lobes, the two parts of the figure–of–eight equipotential 2. The full equipotential surfaces are approximately given by rotating the figure around the line of centres $M_1 - M_2$

evolution; or the binary may shrink under angular momentum loss, reducing a and hence the size of the Roche equipotentials. We shall examine these cases in detail later

To make our considerations precise we need an expression for the secondary's Roche lobe; as a suitable measure we take the radius R_2 of a sphere occupying the same volume. By the remarks in Sect. 4 above we expect R_2/a to be purely a function of the mass ratio q. For $q \lesssim 1$, the important case for CVs and LMXBs, a tolerable approximation is that of Paczyński:

$$\frac{R_2}{a} = 0.462\left(\frac{q}{1+q}\right)^{1/3} = 0.462\left(\frac{M_2}{M}\right)^{1/3} \tag{18}$$

This form gives a simple relation between the period P and the mean density $\bar{\rho}$ of a secondary which fills the Roche lobe:

$$\bar{\rho} = \frac{3M_2}{4\pi R_2^3} = \frac{3M_2}{4\pi a^3}\left(\frac{a}{R_2}\right)^3 \propto \frac{M_2}{a^3}\frac{M}{M_2} \propto P^{-2} \tag{19}$$

where we have used Kepler's 3rd law (12). Numerically one finds

$$\bar{\rho} = \frac{115}{P_{\text{hr}}^2}\ \text{g cm}^{-3} \tag{20}$$

Thus assuming a structure for the lobe–filling star fixes it uniquely: if it is on the lower main sequence for example, the approximate mass–radius relation

$$m_2 = \frac{R_2}{R_\odot} \tag{21}$$

combines with (20) to give

$$m_2 = 0.11 P_{\text{hr}},\ R_2 = 0.11 P_{\text{hr}} R_\odot. \tag{22}$$

A degenerate secondary with fractional hydrogen content X obeys

$$R_2 \simeq 10^9 (1+X)^{5/3} m_2^{-1/3}\ \text{cm}, \tag{23}$$

giving

$$m_2 = \frac{0.015(1+X)^{5/2}}{P_{\text{hr}}} \tag{24}$$

5.2. The Degree of Roche Lobe Filling

Clearly a star does not have a sharp edge, and the phrase "filling the Roche lobe" is necessarily imprecise. If ΔR is the difference $R_L - R_*$, where R_L is the position of the Roche lobe (say near L_1) and R_* is some reference level (e.g. the photosphere) in the star, a study of the gas flow near L_1 shows (e.g. Ritter, 1988) that the *instantaneous* mass overflow rate through L_1 is

$$-\dot{M}_2 \simeq \frac{CM_2}{t_{\text{dyn}}} e^{-\Delta R/H} \tag{25}$$

where t_{dyn} is the star's dynamical time, H its scaleheight near L_1, and C a constant depending on the choice of reference level. We see that unless the argument of the exponential is very large and negative the star would lose most of its mass on a very short timescale (of order the binary period). This is not surprising: if ΔR were small, the star's atmosphere would significantly exceed its Roche lobe, and its mass would simply "fall off" towards its companion. Thus ΔR is always small compared with R_2 and a star can never be much bigger than its Roche lobe.

Since the quantity $\Delta R/H$ is large, $-\dot{M}_2$ is extremely sensitive to ΔR: changing ΔR by of order a scaleheight changes $-\dot{M}_2$ by large factors. Reversing the argument, we see that ΔR is very *insensitive* to the value of the mass transfer rate $-\dot{M}_2$, determined in the long term by evolutionary expansion of the star or angular momentum losses. Hence the star can accomodate itself to any imposed mass transfer rate by small adjustments of ΔR; for most purposes we can regard its radius as exactly equal to that of its Roche lobe. Thus if mass transfer is driven by angular momentum loss ($\dot{R}_L < 0$), R_* will be forced to shrink at least as quickly to prevent ΔR decreasing too much, i.e. to keep the mass transfer stable. Conversely, if mass transfer is driven by evolutionary expansion of the star ($\dot{R}_* > 0$), stable mass transfer will occur if the changes of a and q resulting from the overflow are sufficient to move the Roche lobe out at least as quickly.

5.3. Driving Mass Transfer

Mass transfer in a binary changes its orbital elements and hence the Roche lobe radius: for long–lived mass transfer the driving mechanism must be able to keep the star in contact with the lobe. Generally tides are able to keep the binary orbit circular, so the behaviour of M_2, R_2 etc is determined by that of the orbital angular momentum

$$J = M_1 M_2 \left(\frac{Ga}{M} \right)^{1/2}. \tag{26}$$

We assume that all the mass lost from star 2 is accreted by star 1 (i.e. $\dot{M} = \dot{M}_1 + \dot{M}_2 = 0$). Then logarithmic differentiation of eq. (26) gives

$$\frac{\dot{a}}{a} = \frac{2}{3} \frac{\dot{P}}{P} = \frac{2\dot{J}}{J} + \frac{2(-\dot{M}_2)}{M_2} \left(1 - \frac{M_2}{M_1} \right) \tag{27}$$

showing how the separation and binary period change on mass transfer. Now logarithmically differentiating (18) in the form $R_2 \propto M_2^{1/3} a$ and combining with (27) gives

$$\frac{\dot{R}_2}{R_2} = \frac{2\dot{J}}{J} + \frac{2(-\dot{M}_2)}{M_2} \left(\frac{5}{6} - \frac{M_2}{M_1} \right). \tag{28}$$

In *conservative* mass transfer, i.e. $\dot{M} = \dot{J} = 0$, we see that the sign of $\dot{a}$ (and $\dot{P}$) is the same as that of the quantity $1 - q$ ($q = M_2/M_1$), while the sign of $\dot{R}_2$ is the same as that of $5/6 - q$. If $q > 5/6$ mass transfer causes the Roche lobe to shrink; this will cause further mass transfer unless the star shrinks still more rapidly. Frequently this cannot happen, for example if the star has a deep convective envelope, and the case $q > 5/6$ is unstable in the sense described in Sect. 5.2. Conversely, the case $q < 5/6$ is usually stable; however in this case $\dot{R}_2 > 0$, so the lobe expands: mass transfer can only be sustained if the star expands, usually as a result of nuclear evolution, say from the main sequence to the giant branch.

For periods $P \lesssim$ day the binary is not wide enough to contain an evolved star, so mass transfer must be non–conservative ($\dot{J} < 0$). If we know a mass-radius relation for the lobe–filling star, (28) immediately tells the average mass transfer rate driven by whatever angular momentum loss process operates. For example if the mass–losing star is close to the main sequence, so that $R_2 \propto M_2$, (28) becomes

$$-\frac{\dot{M}_2}{M_2} = \frac{-\dot{J}/J}{4/3 - q},$$

(29)

while for a degenerate star (23) gives $\dot{R}_2/R_2 = -\dot{M}_2/3M_2$, so that (28) becomes

$$-\frac{\dot{M}_2}{M_2} = \frac{-\dot{J}/J}{2/3 - q}.$$

(30)

we see that in both cases mass is transferred on the angular–momentum–loss timescale; the denominators $4/3 - q, 2/3 - q$ are fairly close to 1 except near the mass transfer stability limits $q = 4/3, 2/3$, so that $-\dot{M}_2/M_2 \sim -\dot{J}/J$ on both cases.

We shall consider various forms of angular momentum loss processes in these notes, but one which is present in all binaries (though only effective at short periods) is *gravitational quadrupole radiation*. This general relativistic effect extracts angular momentum at the logarithmic rate

$$\frac{\dot{J}}{J}\,|_{\mathrm{GR}} = -\frac{32}{5}\frac{G^3}{c^5}\frac{M_1 M_2 M}{a^4}$$

(31)

(e.g. Landau and Lifschitz, 1958). For the main–sequence or degenerate secondaries considered above, we can use the Roche lobe relation $a \propto R_2 M_2^{-1/3}$ together with the mass–radius relations $R_2 \propto M_2, R_2 \propto M_2^{-1/3}$ and the resulting mass–period relations $M_2 \propto P, M_2 \propto P^{-1}$ (22, 24) to get (e.g. King, 1988)

$$-\dot{M}_2(\mathrm{MS}, \mathrm{GR}) \simeq 10^{-10}\left(\frac{P_{\mathrm{hr}}}{2}\right)^{-2/3}\;\; M_\odot \mathrm{y}^{-1}$$

(32)

and

$$-\dot{M}_2(\mathrm{degen}, \mathrm{GR}) \simeq 1.6 \times 10^{-12}\left(\frac{P_{\mathrm{hr}}}{2}\right)^{-14/3}\;\; M_\odot \mathrm{y}^{-1},$$

(33)

with hydrogen content $X = 0.6$ in the latter equation. In principle we could replace P in these equations by M_2 and M_2^{-1} respectively and integrate to find $M_2(t)$. However we must remember that (32) and (33) were derived assuming main–sequence and degenerate–sequence mass–radius relations. The effect of mass loss on these stars may indeed be to cause them to deviate from these ideal relations. For example, if (32) predicts that mass is removed from the star on a timescale shorter than its thermal timescale, the star will not be able to shrink back to the thermal–equilibrium (i.e main–sequence) radius appropriate to its new smaller mass rapidly enough, and will remain progressively more and more oversized for its mass compared with its main–sequence radius. We shall see in Sect 12 that such effects play a major part in the evolution of close binaries.

We note finally that although the *average* mass transfer rate may be determined by evolutionary expansion or angular momentum loss from the binary as in (32, 33), quite large fluctuations around this value will occur if for example the star's surface oscillates; the evolutionary average of $-\dot{M}_2$ will be established only over a timescale such that the Roche lobe moves relative to the star by a distance of order the scaleheight H. As this time can be $\gtrsim 10^5$ y in many binaries it is clear that we should not necessarily expect the observed mass transfer rate in a given system to be close to the evolutionary mean.

6. Stellar Wind Accretion

If the companion star does not fill the Roche lobe, mass transfer can only occur via mass loss with velocities $\gtrsim$ orbital, which are necessarily highly supersonic. There are two main cases involving compact primary stars.

6.1. OB Supergiant X–ray Binaries

These have periods $P \sim 10$ days, with a neutron star or black hole in an orbit which almost skims the surface of a massive OB supergiant. The neutron star systems frequently show pulsed X–ray emission, indicating that the neutron star is strongly magnetized and modulates the X–rays at its rotation period. The OB star loses mass at a rate $\dot{M}_w \sim 10^{-5}$ $M_\odot y^{-1}$, with wind velocities $v_w \sim$ a few 1000 km s^{-1}. This evolutionary stage must be very short–lived ($\sim 10^4$ y) as the OB star is close to filling its Roche lobe: once this happens the binary will be engulfed and turn off as an X–ray source, as the mass–losing star is more massive than the compact primary, and Roche lobe overflow is unstable (see Section 5).

6.2. Be–star X–ray Binaries

Here the neutron star (again frequently magnetic) has a much longer (~ 100 d) orbit; pulse–timing measurements show some of these orbits to have appreciable eccentricities ($e \gtrsim 0.3$). Not surprisingly, the X–ray emission is often episodic, corresponding probably to periastron passage of the neutron star. Mass loss from Be stars is more complicated than from OB supergiants. While they do have high–velocity winds like the latter, the distinctive feature of Be stars seems to be a slow ($v_w \lesssim$ few $\times 10^2$ km s^{-1}) wind in the rotational equatorial plane (e.g. Waters and van Kerkwijk, 1989).

6.3. Mass and Angular Momentum Capture in Wind Accretion

Since wind material by definition is able to escape the mass–losing star quite easily, only a small and rather uncertain fraction is captured by the compact star. The angular momentum capture rate is even harder to estimate as it is sensitive to very minor readjustments of the wind flow near this star.

On simple energetic grounds it is usual to assume that the compact star captures most of the wind matter passing within the cylindrical "accretion radius"

$$r_{\text{acc}} = \frac{2GM_1}{v_{\text{rel}}^2} \tag{34}$$

where

$$v_{\text{rel}}^2 = v_w^2 + v_{\text{orb}}^2 + c_S^2 \tag{35}$$

with v_{orb} and $c_S(<< v_w, v_{\text{orb}})$ the orbital velocity of the compact star and the local sound speed. Since $v_{\text{rel}} >> c_S$ this necessarily requires some kind of bow shock to form in the wind ahead of the compact star. Thus the accretion rate is

$$\dot{M}_1 \simeq \frac{\pi r_{\text{acc}}^2}{\Omega a^2} \dot{M}_w \tag{36}$$

where Ω is the solid angle over which the wind is lost (assumed to contain the orbit of the compact star). For OB supergiant X-ray binaries $\Omega \sim 4\pi$ and $v_{\text{rel}} \simeq v_w$, leading to accretion rates $\dot{M}_1 \lesssim 10^{-8}$ M$_\odot$ y^{-1}. In Be-star X-ray binaries the mass loss in the high-velocity spherical wind is too low to drive significant accretion rates; since $v_w \sim v_{\text{orb}}$ is much lower in the equatorial wind it is probably this which drives the observed accretion rates $\lesssim 10^{-8}$ M$_\odot$ y^{-1}, the lower mass-loss rate $\dot{M}_w$ being compensated by the smaller solid angle Ω.

If the wind material were uniformly distributed before encountering the bow-shock a simple argument (Illarionov and Sunyaev, 1975; Shapiro and Lightman, 1976) shows that the accreted matter would have specific angular momentum

$$j_{\text{capt}} = \eta \frac{r_{\text{acc}}^2}{4} \frac{2\pi}{P_{\text{orb}}} \tag{37}$$

with an efficiency factor $\eta = 1$. However numerical treatments of the flow, while very difficult, generally agree on a rather small value $\eta \simeq 0.05$ (Livio et al. 1986; Blondin et al. 1990 and references therein), because of gravitational focussing. There is some indication that the actual accretion flow may be time-dependent, with the compact star accreting opposite signs of angular momentum alternately.

7. The Fate of the Accreted Matter

The nature of gravitational energy release from the infalling matter depends strongly on its flow geometry. Once the matter is captured, we can to a good approximation neglect the influence of the secondary and regard the gas as flowing in the potential of the primary alone. Clearly the gas will rapidly become highly supersonic as it falls freely under this potential, so at least initially is will follow ballistic trajectories. In the single-star potential these are just Kepler orbits, fixed by the specific energy and angular momentum of the gas.

7.1. Roche Lobe Overflow

Here the near–sonic velocity through L_1 is much smaller than the velocities acquired through free fall, so the initial conditions at L_1 are rapidly forgotten by the gas, which all follows essentially the same near–parabolic (zero–energy) orbit. The distance of closest approach to the primary, $R_{\min}$, is fixed by the angular momentum: it is easy to show that $R_{\min} = \frac{1}{2}R_{\mathrm{circ}}$, where R_{circ} is the radius of a circular Kepler orbit with the same angular momentum: this is called the circularization radius. Evidently if $R_{\min} < R_1$, the radius of the primary, the gas will crash directly into the accreting star; this is thought to occur in some Algol systems. By contrast if $R_{\min} > R_1$ the gas will have to orbit the star, and must lose angular momentum if it is to accrete.

The specific orbital angular momentum fixing $R_{\min}$ and R_{circ} can be estimated quite simply by neglecting the gas velocity through L_1, and regarding the gas as being "dropped" from L_1 with an initial velocity resulting purely from the binary rotation. If b_1 is the distance of L_1 from the centre of the primary this means that the gas specific angular momentum is $b_1^2\omega$; the definition of R_{circ} requires this to be equal to $(GM_1 R_{\mathrm{circ}})^{1/2}$. Using Kepler's law for the binary we have $\omega^2 = GM/a^3$, and hence

$$R_{\mathrm{circ}} = \frac{M}{M_1}\left(\frac{b_1}{a}\right)^3 b_1 \tag{38}$$

Numerical results for Roche geometry can be fitted by the formula

$$\frac{b_1}{a} = (0.500 - 0.227\log q) \tag{39}$$

so that we finally get

$$\frac{R_{\mathrm{circ}}}{a} = (1+q)[0.500 - 0.227\log q]^4. \tag{40}$$

One can evaluate this equation for a range of parameters, and a simple result emerges. For $q = 1.0$ we find $R_{\min} \simeq a/16$, and for $q = 0.1$ we get $R_{\min} \simeq a/8$. Using (13) we see that in all realistic cases $R_{\min}$ is larger than the size of any compact object, which cannot exceed the radius of a low–mass white dwarf ($\sim 10^9$ cm). Hence in general Roche lobe overflow results in the infalling gas missing the accreting compact star. Since the gas is trapped in this star's Roche lobe and must move alomst in the orbital plane it is clear that the gas trajectories will intersect themselves. These collisions of supersonic gas streams must produce significant dissipation, so that the gas will lose energy ultimately to radiation. Generally speaking the timescale for energy loss is much shorter than that for any redistribution of angular momentum within the gas, which therefore loses energy at nearly constant angular momentum. We may thus invoke the principle enunciated in Sect. 3. and assert that the gas will initially try to follow a circular orbit consistent with its original angular momentum, i.e. at R_{circ}. Evidently this ring of matter must try to rid itself of angular momentum in order to accrete. As the accompanying energy loss will in practice usually be much more rapid, the matter will be gently lowered on to the central star through a sequence of circular orbits. This constitutes an *accretion disc*: we see that such discs are a general consequence of Roche lobe overflow on to a compact star. The one exception to this general conclusion is the case where the central star has a strong magnetic moment. In practice this means a white dwarf, which can have moments μ as high as 10^{34} G cm^3. For binaries of sufficiently small separation (short periods, $\lesssim 4$ hr) the field can prevent disc formation entirely. Systems of this type are known (the AM Herculis stars), and form a fascinating subgroup of their own.

7.2. Stellar Wind Accretion

The clearcut picture of disc formation in Roche lobe overflow does not extend to disc accretion. We have seen already (Sect. 6.3) that the mass and angular momentum capture in this case are quite uncertain. If we adopt the simple estimate (37) of the captured angular momentum and try to compute a circularization radius we find

$$R_{\text{circ}} = \frac{4\pi^2 G^3 M_1^3}{P_{\text{orb}}^2 v_{\text{rel}}^8} \tag{41}$$

This is to be compared with the obstacle presented by the accreting object. In many cases of wind accretion this is a strongly magnetic neutron star, so if the estimate (41) were not already uncertain enough (a 10% change in v_{rel} alters R_{circ} by a factor 2) we have to add the uncertainties of how the magnetic field interacts with the flow. In any case, the estimate (41) is naive, as (37) is unlikely to be accurate. In view of these uncertainties it is hardly surprising that no clear picture of accretion from a stellar has yet emerged. Numerical computations of such flows have so far been confined to 2D or rather low–resolution 3D modelling. These suggest that the average angular momentum of the flow is insufficient to form a disc; however it is possible that the accreting object receives opposite signs of angular momentum at different epochs, leading to transient disc formation (see e.g. Blondin *et al.* 1990; Livio 1991 and references therein).

8. Accretion Discs

We have seen that the formation of an accretion disc is the normal consequence of Roche lobe overflow in close binaries. There are many other astrophysical situations in which matter has to rid itself of angular momentum in order to sink deeper in a potential well, and these too are likely to give rise to discs. Accordingly there is now a vast literature on discs in all contexts, including notably star formation and active galactic nuclei. However the binary case remains the best understood and in some sense is treated as a paradigm. A full review of the subject would require far more space than is available here, so the present Section attempts only a brief introduction. The interested reader is referred to the book by Frank *et al.* (1992) and references therein for more detailed treatments.

The disc is taken as a flattened configuration lying close to the orbital plane $z = 0$ in cylindrical coordinates (R, ϕ, z). By the arguments of Sect. 7 we assume the matter to accrete through a sequence of circular orbits $R = $ constant; frequently these will be Kepler orbits, with azimuthal velocity $v_\phi = R\Omega = (GM_1/R)^{1/2}$. We assume that the "vertical" velocity component v_z is negligible, and that the radial "drift" velocity v_R is $<< |v_\phi|$. Clearly for accretion to occur we require $v_R < 0$ near the central object. We characterize the matter by its surface density

$$\Sigma(R, t) = \int_{-\infty}^{\infty} \rho \, dz \tag{42}$$

where ρ is the local matter density. Armed with these assumptions we can write down conservation equations for mass and angular momentum describing the radial structure of accretion discs. The mass of the annulus between r and $r + \Delta R$ is $2\pi R \Delta R \Sigma$, and mass flows across the curved boundaries of the annulus at local rates $2\pi R \Sigma v_R$. Hence

$$\frac{\partial}{\partial t}(2\pi R \Delta R \Sigma) = -\left[2\pi R \Sigma v_r\right]_R^{R+\Delta R}$$

or

$$R\frac{\partial \Sigma}{\partial t} + \frac{\partial}{\partial R}(R \Sigma v_R) = 0 \tag{43}$$

The angular momentum equation is very similar, with Σ being replaced by $\Sigma R^2 \Omega$ everywhere, except that there must be transport by viscous torques between annuli. If $G(R, t)$ is the viscous torque of an outer annulus on an inner one, we get

$$R\frac{\partial}{\partial t}(\Sigma R^2 \Omega) + \frac{\partial}{\partial R}(R \Sigma v_R R^2 \Omega) = \frac{1}{2\pi}\frac{\partial G}{\partial R} \tag{44}$$

We can exploit the similarity with the mass conservation to eliminate the quantity $R \Sigma v_R$, arriving at

$$R\frac{\partial \Sigma}{\partial t} = -\frac{\partial}{\partial R}\left[\frac{1}{2\pi(R^2\Omega)'}\frac{\partial G}{\partial R}\right] \tag{45}$$

where the prime denotes differentiation with respect to R.

8.1. Viscosity

Given a prescription for G in terms of Σ etc, eqn.(45) will determine $\Sigma(R, t)$. Specifying the viscosity is the crucial part of the accretion disc problem. Without viscosity there can be no disc: this is physically obvious, since there would be no angular momentum transport, and follows from the equations above if we set $G = 0$, which implies $v_R = 0$, i.e. no mass flow. In fact the viscosity has to perform three related functions: circularizing the gas orbits, transporting angular momentum, and dissipating energy. Yet we have no real understanding of this basic driving process, and disc theory has no predictive power except in certain limits; in particular our picture of time–dependent disc evolution lacks a firm basis. This is analogous to the position of stellar structure theory in the 1920s, before thermonuclear energy generation was understood. While the ingenuity of astronomers nevertheless produced a remarkable amount of progress, a full outline of stellar *evolution* in particular had to await this discovery.

The term "viscosity" usually implies the existence of random motions about the mean streaming velocity, and this at least provides a simple picture to fix ideas. If the random motions have a typical scale λ and velocity v', an elementary argument (e.g. Frank *et al.* 1992) shows that they will transport momentum along the velocity gradient of a shearing flow: if we have a linear flow field $v(R)$ orthogonal to the R direction, the resulting force per unit length between adjacent streamlines is

$$\sim \Sigma \lambda v' \frac{\partial v}{\partial R}$$

where Σ is the surface density of the flow. We can express this as saying that the fluid has *kinematic viscosity* [cm s^{-1}] $\nu \sim \lambda v'$.

For a rotating flow the arguments go through similarly; the rate of angular momentum transport orthogonal to a shearing flow $v = R\Omega(R)$ involves the angular velocity gradient Ω'; thus the torque of an outer disc annulus on an inner one is

$$G(R,t) = 2\pi R \nu \Sigma R^2 \Omega' \tag{46}$$

Obviously the torque of an inner annulus on an outer one is just $-G$, and the *net* torque on a ring between radii $R, R + \Delta R$ is

$$\frac{\partial G}{\partial R} \Delta R \tag{47}$$

8.2. Viscous Dissipation

From (47) the rate of working of the viscosity on a ring is (angular velocity × torque) =

$$\Omega \frac{\partial G}{\partial R} \Delta R = \left[\frac{\partial}{\partial R}(\Omega G) - G\Omega' \right] \Delta R.$$

But here the term $\partial(\Omega G)/\partial R$ is not a local term, as is clear from its form as a divergence: in an integral over the whole disc, this term would produce a contribution $[\Omega G]_{\text{inner edge}}^{\text{outer edge}}$ depending only on boundary conditions. We can identify this term as the rate of advection of rotational energy by the viscous torques. The other term $-G\Omega'$ in the square bracket of this equation is clearly local in origin, and gives the rate at which the viscosity dissipates the kinetic energy of the fluid, i.e. turns it into heat, which is then radiated from the disc.

This is the prime observable quantity; since most of the radiation is from the flat faces of the disc, we express it as the dissipation rate per unit face area, $D(R,t) = G\Omega'\Delta r/4\pi R\Delta R$, since the disc has two faces, so that

$$D(R,t) = \frac{1}{2}\nu\Sigma(R\Omega')^2 \tag{48}.$$

We note that $D(R,t) \geq 0$, vanishing only for rigid rotation $\Omega(R) = \text{constant}$.

8.3. The Magnitude of Viscosity

All of the above discussion has avoided any direct identification of what processes might produce the random motions giving rise to ν. It is clear that ordinary "molecular" (atomic) viscosity, where λ is the collisional mean free path and v' the sound speed, is far too small to cause noticeable angular momentum transport on timescales of interest: the resulting Reynolds number

$$\mathcal{R}e = \frac{Rv_\phi}{\lambda v'} \sim \frac{Rv_\phi}{\nu} \tag{49}$$

is of order 10^{14}. This very large value is sometimes taken as an indication that the disc flow is turbulent, so that ν would then be the eddy viscosity corresponding to the largest eddy scales, although there is no proof of this. There are several other candidate mechanisms, e.g. local magnetic fields, or spiral shocks (see Morfill *et al.* 1991 for a recent review of the latter). None of these has yet attained the state of making firm quantitative predictions that can be compared successfully with observational data. The most fruitful approach has been to bound the likely values of λ and v'. If the viscosity is assumed to result from turbulent eddies, we can assert that these are unlikely to exceed the disc semi-thickness H by very much; further, except in restricted regions the motions cannot be very supersonic, as shocks will quickly reduce them to subsonic. Thus we can set

$$\nu = \alpha\lambda c_S \tag{50}$$

and hope that the dimensionless parameter α is $\lesssim 1$. This approach, pioneered by Shakura and Sunyaev (1973) can be applied to other candidate mechanisms with similar results. It is very important to realise that there is no reason at all to suppose α to be constant in time or space; apart from the reasonable expectation that $\alpha \lesssim 1$, (50) gains us nothing. Nevertheless the α–parametrization is useful in allowing numerical estimates, and it is customary to discuss suggested viscosity mechanisms in terms of their effective alphas.

9. Disc Structure

The form (46) inserted in (45) shows that Σ obeys a nonlinear diffusion equation

$$\frac{\partial \Sigma}{\partial t} = \frac{3}{R} \frac{\partial}{\partial R} \left(R^{1/2} \frac{\partial}{\partial R} [\nu \Sigma R^{1/2}] \right) \qquad (51).$$

Given a form of ν and suitable boundary conditions this equation determines $\Sigma(R, t)$, and v_R can be read off from the mass conservation equation as

$$v_R = -\frac{3}{\Sigma R^{1/2}} \frac{\partial}{\partial R} [\nu \Sigma R^{1/2}] \qquad (52).$$

The diffusive behaviour suggested by (51) is clearly visible in simple analytic solutions, e.g. for the unrealistic case $\nu = $ constant: an initial ring of matter spreads to both larger and small R under the viscous torques, the inward-accreting matter losing angular momentum to the small fraction of the mass which moves outwards. The form of the equation shows that matter diffuses locally on the *viscous timescale*

$$t_{\text{visc}} = \frac{l^2}{\nu} \qquad (53)$$

where $l = \Sigma / |\nabla \Sigma|$ is the local density scalelength. Thus a concentrated matter distribution spreads rapidly, while a more uniform one relaxes slowly. It is easy to see from (52) that t_{visc} can be re-expressed as $\sim l/v_R$.

9.1. Steady Discs

In many cases the disc flow is steady on timescale of interest: setting $\partial/\partial t = 0$ the mass conservation equation gives

$$2\pi R \Sigma(-v_R) = \dot{M} \qquad (54)$$

where $\dot{M}$ is the (constant) accretion rate. The diffusion equation (51) has steady solutions of the form

$$\nu \Sigma = A + \frac{B}{R^{1/2}} \qquad (55)$$

where A, B are constants. We see from (52) and (54) that the A–solutions correspond to steady inflow at a rate $\dot{M} = 3\pi A$ and the B–solutions to no flow at all ($v_R = 0$). It is easy to show (e.g. Frank *et al.* 1992) that the solution corresponding to steady accretion on to a non–rotating star of radius R_* is

$$\nu \Sigma = \frac{\dot{M}}{3\pi} \left[1 - \left(\frac{R_*}{R} \right)^{1/2} \right]. \qquad (56)$$

This solution has the property that the maximum of $\Omega(R)$ is the Kepler value very close to R_*; Ω then decreases rapidly towards zero at the stellar surface in a narrow boundary layer. Solutions with $B > 0$ and $A = 0$ correspond to situations in which the disc does not accrete at all but simply extracts rotational energy from the central star (the propellor effect).

Surface Dissipation of Steady Discs

Equation (48) shows that a Keplerian disc has surface dissipation rate

$$D(R) = \frac{9}{8}\nu\Sigma\frac{GM_1}{R^3}. \tag{57}$$

Because this depends only on the combination $\nu\Sigma$ we see from (55) that *in steady discs the surface dissipation is independent of viscosity*. In particular for the solution (56) corresponding to accretion on to a non–rotating star we have

$$D(R) = \frac{3GM_1\dot{M}}{8\pi R^3}\left[1 - \left(\frac{R_*}{R}\right)^{1/2}\right]. \tag{58}$$

Defining the disc effective temperature $T_{\text{eff}}(R)$ through the relation $D(R) = \sigma T_{\text{eff}}^4$ we find

$$T_{\text{eff}}(R) \propto R^{-3/4} \tag{59}$$

for $R >> R_*$. This relation is effectively independent of the inner boundary condition, and so provides a stringent test of steady disc theory. Eclipse mapping of discs in cataclysmic variables (see Horne 1991 for a review) confirms this dependence in cases where we can be sure that the disc is steady and optically thick (so that T_{eff} is easy to estimate from observation). This is important, as a failure here could not be ascribed to our ignorance of viscosity.

Boundary Layers

We have already alluded to the presence of a boundary layer around the accreting star in some cases. The total disc luminosity

$$L_{\text{disc}} = 2\int_{R_*}^{R_{\text{out}}} D(R)2\pi R dR,$$

where $R_{\text{out}} >> R_*$ is the outer disc radius, is only one–half of the accretion luminosity, i.e.

$$L_{\text{disc}} = \frac{GM_1\dot{M}}{2R_*}. \tag{60}$$

This is easy to understand, in that matter in Keplerian rotation near R_* has one–half of the binding energy of matter which has accreted to the non–rotating star. The result suggests that the boundary layer might be as bright as the disc under certain circumstances. EUV observations of non–magnetic cataclysmic variables do not show much evidence for the widespread presence of such boundary layers.

This may be because the accreting white dwarfs are rapidly spun up to near–Kepler velocities, considerably reducing the boundary layer luminosity (King *et al.* 1991).

9.2 Vertical Disc Structure

We have so far simply assumed that the disc is thin in the vertical direction, i.e. $H << R$ at each radius. We should now check this. There is hydrostatic balance in the z–direction between the pressure gradient and the tidal gravitational field:

$$\frac{\partial P}{\partial z} \simeq -\frac{GM_1 z}{R^3}\rho \qquad (61)$$

To get an estimate of H we set $z \sim H$, $\partial P/\partial z \sim P/H$, and express P as ρc_S^2. Then

$$\frac{H}{R} \sim \frac{c_s}{v_K} \qquad (62)$$

where v_K is the Kepler velocity at R. This shows that the disc is thin if and only if the Kepler velocity is highly supersonic. A simple order–of–magnitude analysis of the radial Euler (momentum) equation readily shows that in this case the gas velocity v_ϕ must be very close to v_K, so that the radial momentum balance is supplied by centrifugal force and not pressure. Thus we see that *the disc is thin if and only if it is Keplerian*, which in turn will hold provided that the disc cooling is efficent enough always to keep $c_S << v_K$.

If the disc is thin at each radius, we can treat the vertical structure at fixed R just like that of a one–dimensional star, with viscous heating playing the role of nuclear energy generation in a star. In particular the vertical energy transport through the disc may be by convection or radiation. In the latter case the internal disc temperature $T(z)$ will vary as

$$T \sim \frac{1}{\tau^{1/4}} \qquad (63)$$

where $\tau \propto z$ is the Rosseland mean optical depth, provided that most of the viscous energy generation take place near the disc mid–plane.

10. Disc Stability

We have implicitly assumed above that a steady mass supply leads to steady disc flow. In fact this may not always be possible, and is a suggested cause of dwarf nova outbursts. There is good observational evidence that discs are present in these cataclysmic systems, and that the observed outbursts (luminosity increases by factors ~ 10 in about a day, declining over several days, with intervals of weeks between outbursts) are caused by increased accretion. The key to this type of behaviour is the possibility of two stable types of vertical disc structure (Fig.

2b). At fixed radius R we plot the quantity $\nu\Sigma$ (or T_{eff}^4 or the local mass–flow rate $\dot{M}(R)$) against surface density Σ.

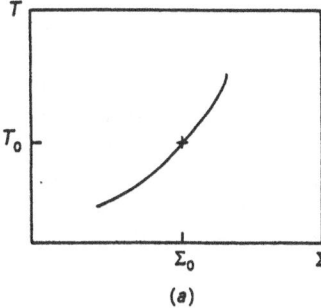
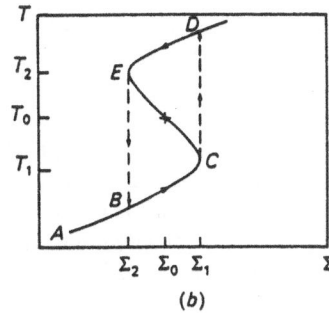

Fig. 2. Effective temperature versus surface density at fixed radius in an accretion disc. In (a) there is a unique stable steady solution (T_0, Σ_0), while in (b) the steady solution lies on an unstable branch of the curve. The systems executes limit cycles $B - C - D - E$

It is easy to show that to the right of the equilibrium curve viscous heating dominates over radiative cooling, with the opposite behaviour to the left of the curve. Hence where the curve has positive slope the structure is stable against thermal perturbations, but no equilibrium is possible on any part of the curve where the slope is negative. If the curve always has a positive slope as in Fig. 2a there is a unique stable solution. However, if the curve has the S shape shown in Fig. 2b there are two stable branches separated by an unstable one. If the external mass transfer rate into the disc corresponds to an equilibrium on this branch, the system will be forced to execute limit cycles around this point: on the lower stable branch mass accumulates in the ring faster than it flows out, as the viscosity is low at these low temperatures. The surface density thus rises until the "knee" at the upper end of the stable branch is reached. Here equilibrium is no longer possible and the annulus jumps to the high–temperature branch at the same density on a local thermal timescale. Here the viscosity is high, and mass flows out of the ring more rapidly than it arrives, reducing the surface density until the ring drops back into the low–temperature equilibrium state. The ring evidently alternates between a long–lived state of low temperature and mass flow, and a short–lived state of high temperature and mass flow, the transitions between the states being fairly rapid. This is clearly qualitatively similar to observed dwarf nova behaviour. The origin of the S–shape can plausibly be ascribed to the rapid increase of hydrogen opacity with temperature near 10^4 K: just below the critical temperature, a local injection of heat raises the opacity, and thus traps more heat leading to a further temperature rise. Conversely, just above the critical temperature slight cooling causes a dramatic opacity decrease, allowing heat to escape more freely.

There are several problems in accepting such disc instability models of dwarf nova outbursts. First, the instability described here is purely local, i.e. at one R.

To achieve a global instability requires that local instability at one radius triggers instability at other radii. By suitable choice of parameters this can be arranged. But in order to obtain wide enough outbursts it is necessary to manipulate the α parameter, raising it on the hot branch above its value on the cool branch. One could argue that there is nothing sacred about holding α fixed. But one might counter that with a free function at one's disposal, the ability to fit dwarf nova outburst light curves is not necessarily very impressive. Of course the use of an α-parametrization is inherently unsatisfactory, and a more physically motivated description of viscosity is urgently needed. Perhaps even more seriously, most of the instability calculations are essentially one-dimensional in space, whereas the physics of the disc is probably fully three-dimensional. In one dimension the instabilities must propagate as fronts through the disc, but the coherence of such fronts is not guaranteed in three dimensions.

11. Discs in Other Binaries

Non-magnetic cataclysmic variables provide the most direct application of accretion disc theory. There is some encouragement for theoreticians in that basic predictions like the $T_{\text{eff}} \propto R^{-3/4}$ do seem to hold up in certain systems. Since comparatively little of disc theory depends crucially on the central boundary conditions one might hope that it would extend straightforwardly to other binaries, in particular the low-mass X-ray binaries (LMXBs), which differ from CVs only in having a neutron star in place of a white dwarf. This is far from being the case; the X-ray light curves of these systems show phenomena which are not easily reconciled with the simple picture described above (see White, 1989; Watson and King, 1991, for reviews). This is a severe challenge, as any model must also explain why these phenomena do not appear to occur in CVs. One can either argue that discs in LMXBs are physically different, or that these same phenomena occur in CVs, but are not observable in the same way. The physical difference posited in the first approach might result from X-ray heating of the disc, although the radiation temperature from this heating is insufficient to support the increased disc scale height sometimes postulated to explain the X-ray light curves (King, 1991). In the second approach one can argue that X-ray observations of LMXBs are observations of the central source and its immediate surroundings, and the CV analogy would be UV light curves, which have not been available hitherto. High time-resolution UV observations with the Hubble Space Telescope should settle this question. The extension of disc theory to deal with magnetized central stars is another difficult area: again, there is little direct evidence supporting a simple extrapolation of the standard picture sketched here.

12. Short–Period Binary Evolution

This Section deal with the evolution of accreting binaries with orbital periods $P \lesssim 12$ hr (see King, 1988 for a more thorough review). There are two main classes: cataclysmic variables (CVs), in which a white dwarf accretes from a main–sequence companion, and low–mass X–ray binaries (LMXBs) in which the white dwarf is replaced by a neutron star or black hole. From (22) we see that the companion star must have mass $M_2 \lesssim 1 M_\odot$, so that evolutionary expansion is unlikely. The mass transfer and evolution of these binaries must be driven by the loss of orbital angular momentum, as discussed in Sect. 5.3. However it is clear that the GR mass transfer rates $\lesssim 10^{-10}, 10^{-11}$ $M_\odot$ yr^{-1} are too low for CVs with periods $\gtrsim 3$ hr, where observation suggests $-\dot{M}_2 \gtrsim 5 \times 10^{-10}$ $M_\odot$ yr^{-1} (although there is a huge dispersion in this figure).

The favoured candidate mechanism for this angular momentum loss at periods $\gtrsim 3$ hr is *magnetic stellar wind braking*, often referred to as magnetic braking (MB). This is the process by which the Sun is thought to have lost most of its original angular momentum. The idea basic is that the magnetic field of the Sun (or the secondary in a close binary) is able to force the stellar wind to corotate out to large radii. The resulting large specific angular momentum means that an insignificant mass loss rate ($\sim 10^{-13}$ $M_\odot$ yr^{-1} for the Sun) can carry off large amounts of angular momentum. Because tides lock the secondary star's rotation to that of the binary (see Sect. 3) this drain of angular momentum from the star ultimately shrinks the binary. Theories of this process (e.g. Mestel and Spruit, 1987) suggest an angular momentum loss rate increasing with period, but contain parameters from dynamo theory for the stellar magnetic fields which cannot be regarded as well established. Empirical estimates from the average spin rates of stars in open clusters as functions of age (e.g. Skumanich, 1972) lead to a similar dependence (Verbunt and Zwaan, 1981). Such estimates are of order

$$-\dot{M}_2 = \dot{M}_{\mathrm{MB}} \simeq 6 \times 10^{-10} \left(\frac{P}{3 \text{ hr}} \right)^{1+\epsilon} \quad M_\odot \text{ yr}^{-1} \tag{64}$$

where $0.2 \lesssim \epsilon \lesssim 0.7$. While such rates seem reasonable for CVs with $P \gtrsim 3$ hr, they do not explain the rates $\sim 10^{-8}$ $M_\odot$ yr^{-1} required to drive Eddington-limited LMXBs.

13. Secular Evolution of CVs

The lack of a good theory of orbital braking for $P \gtrsim 3$ hr (or even clinching identification of the mechanism responsible) means that there is as yet no standard theory of orbital (or secular) evolution of short–period binaries. However observational constraints on the evolution of CVs are sufficiently tight that theory has little room for manoeuvre, and this case is the best understood.

We note first that the mass transfer rates quoted in Sect. 12 imply mass transfer timescales $t_M = M_2/(-\dot{M}_2) \lesssim$ a few times 10^9 yr, significantly less than the age of the Galaxy. As the binary period must change on a timescale $\sim t_M$, this means that the present distribution in orbital period results from secular evolution, and is not simply a reflection of the initial orbital periods. Also it is not easy to think of observational selection effects that are strongly dependent on period. This explains the central importance in this field of the CV *period histogram* (Fig. 3).

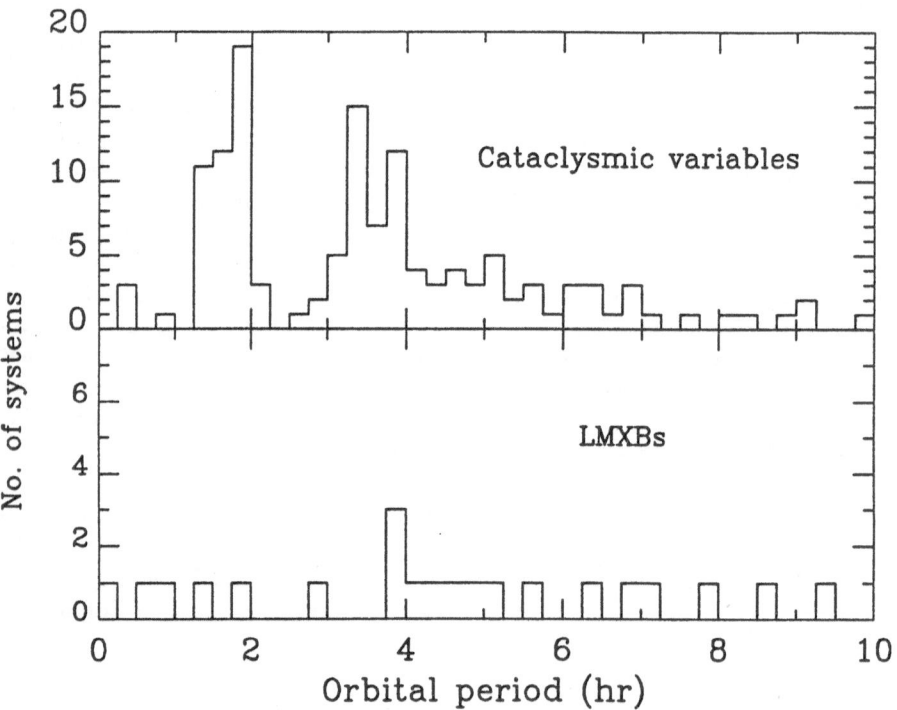

Fig. 3. Period histograms for CVs (top) and LMXBs (bottom) (data from Ritter, 1990). Two CVs with periods > 12 hr are not shown. The two periods between 1 and 2 hr in the LMXB distribution are unconfirmed

The CV period histogram reveals three main features which are all statistically significant:

(a) an apparent long–period cutoff $P \sim 10$ hr

(b) a period "gap" 2 hr $\lesssim P \lesssim 3$ hr

(c) a minimum period $P \sim 80$ min.

Explanations of all three features now exist, although that for the period gap is still controversial. The easiest feature to explain is the maximum period (a).

13.1. The Maximum Period

From (22) we have a relation between the period and the mass of a lobe–filling main-sequence secondary star. Also, mass transfer stability (Sect. 5.3) requires that the secondary mass not exceed a value of order the white dwarf mass M_1. But the latter in turn must not exceed the Chandrasekhar limit $1.44 M_\odot$. Combining these constraints we get

$$P \simeq m_2 \times (9 \text{ hr}) \lesssim m_1 \times (9 \text{ hr}) < 1.44 \times 9 \text{ hr} \simeq 13 \text{ hr}. \qquad (65)$$

In fact there is also a clear selection effect against observing periods $\gtrsim 8$ hr, as it is difficult to obtain periods in a few nights' observation.

13.2. The Minimum Period

The explanation of the minimum period (c) is now generally accepted, and introduces ideas which are important in discussion of the period gap. In secular evolution M_2 decreases, and from (22) this means that P decreases also, at least for a main–sequence secondary. But we can write (20) as

$$P \propto \left(\frac{R_2^3}{M_2} \right)^{1/2} \qquad (66)$$

From this form it is obvious that while CVs generally evolve towards shorter periods, this is entirely dependent on the simultaneous shrinking of the stellar radius R_2. Hence the minimum period is associated with a breakdown of the simple main–sequence mass–radius relation $R_2 \propto M_2$. There are two reasons for this to occur near 80 minutes. First, as M_2 decreases, the star's mass will get too low for it to burn hydrogen in its core, and it will try to become degenerate. From (24) we see that this will eventually lead to a period increase; as (22) and (24) intersect at a period ~ 40 minutes, CVs could not have periods less than this however they evolved. But in practice the minimum period will be longer than this because of the second type of deviation from the main–sequence mass–radius relation: the effect of mass loss is to make the low–mass secondary star oversized for its mass. A simple way of seeing this is to realise that if mass is lost from a star with a deep convective envelope, the surface of the star is "refilled" on a dynamical timescale with material from deeper within the star, which has slightly higher specific entropy $S \propto \log(p/\rho^\gamma)$, with $\gamma \simeq 5/3$ (in a convective zone the entropy decreases outwards slightly). This material retains its original entropy, as its rise to the surface is rapid and the process is nearly adiabatic. But the pressure p of this material must equilibrate to that of its surroundings on a dynamical timescale: thus its higher entropy requires the new surface material to have slightly lower density than the original surface material, which in turn means that the star is slightly larger than before the mass loss. Of course, the new stellar configuration is now slightly out of thermal equilibrium, in that the surface material has too much specific entropy, i.e. is too hot. This will be rectified in a fraction of the thermal (Kelvin–Helmholtz) timescale

$$t_{\rm KH} = \frac{GM_2^2}{R_2 L_2} \qquad (67)$$

where L_2 is the luminosity. The hotter surface layers will radiate more heat into space than they receive from below, ultimately from the nuclear reactions in the star's centre. If there is no further disturbance, the star will achieve thermal equilibrium once more, and adopt the main–sequence radius appropriate to its new smaller mass. Evidently slow mass loss will not greatly affect this conclusion. But if mass is removed from the star on a timescale comparable with $t_{\rm KH}$, the star will be unable to shrink back to its new main–sequence radius and will gradually become oversized for its mass. From (66) this will cause the period to stop decreasing.

Near $P = 80$ minutes the secondary has $M_2 \simeq 0.1 \rm M_\odot$, implying $t_{\rm KH} \gtrsim 2 \times 10^9$ yr, while GR alone makes $t_M \lesssim 2 \times 10^9$ yr, so this effect clearly operates. The main uncertainty remaining is the precise value of the minimum period: this depends sensitively on the secondary's structure, and ultimately on a knowledge of the low–temperature opacities which determine its surface boundary condition.

13.3. The Period Gap

While the CV maximum and minimum periods are reasonably well understood, explanations of the period gap still await confirmation. This is a direct consequence of our poor understanding of the nature of orbital braking at periods $\gtrsim 3$ hr. Note first that the gap is a *statistical* phenomenon: there are CVs in the gap, their numbers consistent with being born near these periods. Thus the gap results from something happening to CVs as they evolve into this period range. If we discard the rather unlikely possibility of selection effects discriminating strongly against finding periods in the 2 - 3 hour range, we have essentially two possibilities. We can either assume that braking continues uninterruptedly when a CV's period decreases to ~ 3 hr, or not. In the first case, $P \simeq 3$ hr must be a new minimum period for CVs born at longer periods, i.e. the braking must be strong enough to cause a period *increase* to set in at this point, just as it does near $P = 80$ min. In fact (66) and (67) (with $M_2 \simeq 0.3 \rm M_\odot$) show that

$$t_M \simeq t_{\rm KH} \simeq 5 \times 10^8 \ {\rm yr} \qquad (68)$$

at such periods, so this is not implausible, particularly given the wide dispersion in mass transfer rates inferred from observation. However this would only prevent CVs evolving below 3 hr from longer periods: we would still need a further reason why CVs apparently are born with much greater frequency just below ~ 2 hr rather than above. If instead we assume that orbital braking is severely reduced (e.g. to the GR rate) near 3 hr, we get an elegant explanation for all the observed features of the period gap. The result (68) means that the secondary star must be significantly out of thermal equilibrium as a result of the rapid mass loss, and in particular oversized for its mass. Reducing orbital braking to the GR rate will make $t_M > t_{KH}$; thus the star will now be able to shrink back towards its main–sequence radius, thus losing contact with the Roche lobe and switching off mass transfer. However, the star cannot shrink inside its main sequence radius, and orbital braking through GR alone will certainly reduce the Roche lobe size enough to restart mass transfer within the lifetime of the Galaxy.

It is clear that this type of behaviour is qualitatively what is needed to explain the gap, provided that we can arrange the cessation of mass transfer to occur between ~ 2 and 3 hr in most cases. It is clear that the sharp decrease in braking must occur near $P \simeq 3$ hr, where the secondary has $M_2 \simeq 0.3 M_\odot$. It was noted quite early on that this is about the mass at which lower–main–sequence stars lose their radiative cores and become fully convective. The idea that the period gap is somehow associated with this changeover is attractive, as it supplies a reason why the gap is "universal", i.e. occurs between essentially the same periods for almost all CVs. But we have no good theory of the orbital braking, let alone a reason why a fully convective secondary should sharply reduce it, and we have advanced little beyond plausible–sounding qualitative ideas. For example, within the magnetic braking picture one can argue that the lack of a radiative core robs the stellar dynamo process of a stable layer to anchor the footpoints of magnetic loops, which would then be strengthened by convective motions. But there is no compelling reason to assume this. Indeed Tout and Pringle (1992) for example make the opposite hypothesis, and assume that fully convective stars do undergo magnetic braking. Their derived angular momentum loss rate is high enough to drive mass transfer rates typical of systems above the gap, but predicts that systems below the period gap should have similar mass transfer rates also, contrary to observation.

The period gap explanation detailed above, independently advanced by Rappaport et al. (1983) and Spruit and Ritter (1983), thus has no firm theoretical foundation. However there is considerable circumstantial evidence supporting it. The position and width of the gap, as well as the sharpness of its sides, are very well reproduced by calculations in which the extra orbital braking (64) is simply switched off once the secondary becomes fully convective. If we consider only the strongly magnetic AM Her systems among CVs, no less than 6 out of the current total of 17 lie in the extremely small period interval $113.6 < P < 115.6$ min. This period "spike" has been explained by Hameury et al. (1988) as marking the resumption of mass transfer after passage through the period gap. Systems are much easier to find at this point because the fully convective secondary initially expands on mass loss and the orbital period increases slightly before decreasing. Not only are they somewhat brighter here, they spend longer near this period. Because the precise period at which mass transfer resumes is sensitive to the white dwarf mass, this explanation forces extremely tight constraints on the distribution of this quantity in AM Her systems (e.g Hameury *et al.* , 1989; Ritter and Kolb, 1992) which are nevertheless so far compatible with observation. Presumably these constraints do not hold for non–magnetic CVs, which show no period spike.

14. Short–Period LMXB Evolution

As LMXBs differ from CVs only in substituting a neutron star or black hole for the white dwarf, it is reasonable to suppose that their orbital evolution might be similar: nothing in the preceding Section depends crucially on this difference. However, the LMXB period histogram does not support this view (Fig. 3): there are too few systems in the 80 min. – 2 hr range below the period gap, particularly as the two systems shown there have unconfirmed periods. This lack of systems is probably not caused by selection effects, as many of the LMXB periods were determined by EXOSAT, which had a 4 day orbit, and thus no reason to discriminate against short periods (unlike low Earth–orbit satellites whose periods are themselves close to 100 min). The CV and LMXB distribution thus appear to be fundamentally different. The presence of a system with a well–confirmed period of 2.8 hr is unpromising for explanations involving stronger orbital braking and thus a wider period gap in LMXBs. Some physical effect associated with the compact nature of the primary apparently makes LMXBs effectively disappear around 3 hr. Current attempts to explain this centre around the possibility that X–ray heating might bloat up the companion star by changing its internal structure from convective to radiative (Podsiadlowski, 1991; Rappaport *et al.* 1991; Frank *et al.* , 1992) or that pulsar–type emission could evaporate the companion at short periods (van den Heuvel and van Paradijs, 1988).

References

Blondin, J.M., Kallman, T.R., Fryxell, B.A. & Taam, R.E., 1990, ApJ, 335, 862

Frank, J., King, A.R. & Lasota, J.P., 1992, ApJ, 385, L45

Frank, J., King, A.R. & Raine, D.J., 1992, *Accretion Power in Astrophysics*, 2nd Ed., Cambridge University Press.

Hameury, J.M., King, A.R. & Lasota, J.P., 1988, A&A, 195, L12

Hameury, J.M., King, A.R., Lasota, J.P. & Ritter, H., 1988, MNRAS, 231, 535

Horne, K., 1991, in *Structure and Emission Properties of Accretion Discs*, Proceedings of IAU Colloq. 129, eds. C. Bertout *et al.* , Editions Frontières, Paris.

Harpaz, A. & Rappaport, S.A., 1991, ApJ, 383, 739

Illarionov, A.F. & Sunyaev, R.A., 1975, A&A, 39, 185

King, A.R., 1988, QJRAS, 29, 1

King, A.R., 1991 in *Neutron Stars: Theory and Observation*, Proceedings of the NATO ASI, Agia Pelagia, Crete, p.493, Kluwer Academic Publishers, Dordrecht.

King, A.R., Regev, O. & Wynn, G.A., 1991, MNRAS, 251, 30P

Landau, L.D. & Lifschitz, E.M., 1958, *Classical Theory of Fields*, Pergamon, Oxford.

Livio, M., 1992, in *Evolutionary Processes in Interacting Binaries*, Proceedings of IAU Symp. 151, eds. Y. Kondo *et al.* , Kluwer Academic Publishers, Dordrecht.

Mestel, L. & Spruit, H.C., 1987, MNRAS, 226, 57

Morfill, G., Spruit, H.C. & Levy, E.H., 1991, in *Protostars and Planets III*, eds. E.H. Levy *et al.* , University of Arizona Press, Tucson.

Podsiadlowski, P., 1991, Nat, 350, 136

Rappaport, S.A., Verbunt, F., & Joss, P.C., 1983, ApJ, 275, 713

Ritter, H., 1988,, A&A, 202, 93

Ritter, H., 1990,, A&AS, 85, 1179

Ritter, H. & Kolb, U., 1992, A&A, in press

Shakura, N.I. & Sunyaev, R.A., 1973, A&A, 24, 337

Shapiro, S.L. & Lightman, A.P., 1976, ApJ, 204, 555

Skumanich, A., 1972, ApJ, 171, 565

Spruit, H.C. & Ritter, H., 1983, A&A, 124, 267

Tout, C.A. & Pringle, J.E., 1992, MNRAS, in press

van den Heuvel, E.P.J. & van Paradijs, 1988, Nat, 334, 227

Verbunt, F. & Zwaan, C., 1981, A&A, 100, L7

Waters, L.B.F.M. & van Kerkwijk, M.H., 1989, A&A, 223, 196

Watson, M.G. & King, A.R., 1991, in *Structure and Emission Properties of Accretion Discs*, Proceedings of IAU Colloq. 129, eds. C. Bertout *et al.* , Editions Frontières, Paris

White, N.E., 1989, in *Theory of Accretion Disks*, eds. F. Meyer *et al.* , Kluwer Academic Publishers, Dordrecht.

Part II

High Accuracy Timing and Positional Astronomy

Pulsars – The New Celestial Clocks

D. C. Backer [1]

[1] Astronomy Department, University of California, Berkeley, CA
94720, USA

Abstract: Pulsating radio sources, or pulsars, are rapidly rotating, highly magnetized neutron stars. A pulsar emits beams of radio emission along its dipole magnetic field axes. The star's rotation sweeps one of the beams past the Earth to create the observed pulse if the axis is tilted with respect to the rotation axis. New classes of pulsars – including ones with millisecond rotation periods – have been discovered in the last decade. The rotations of pulsars are extremely stable. This stability allows pulsars to be used as probes of the interstellar plasma, for tests of General Relativity in the case of binary pulsars, and for detection of a cosmic background of gravitational radiation from the early universe.

1 Pulsars

1.1 A Brief History of Neutron Stars

Pulsars are firmly associated with highly magnetized, rapidly rotating neutron stars. Neutron stars were first "invented" as a theoretical entity shortly after the discovery of the neutron. Calculations in the late 1930's indicated that a one solar mass neutron star would have a 10-km radius and a central density around 10^{15} gm cm^{-3}. Only theoretical progress was made on understanding the nature of neutron stars until the late 1960's. They were not expected to be observable owing to their small size. One of the most prescient remarks about neutron stars came in the year prior to the discovery of the first pulsar when Pacini (1967) suggested that a rapidly rotating, highly magnetized neutron star with a misaligned dipole moment would radiate sufficient magnetic dipole radiation to explain the energetics of the Crab Nebula.

Pulsars were discovered accidentally during an investigation of extragalactic sources by Hewish, Bell and colleagues (1967). Shortly thereafter other pulsating radio sources were found. The most important other object was the pulsar in the Crab Nebula because it provided the observational evidence for the direct link, proposed many decades earlier, between neutron stars and supernovae. Pulsars in general are confined to the galactic plane. The spatial distribution on the sky of 450 of the nearly 600 presently known pulsars shown in Fig. 1 does not overwhelmingly show a concentration in the galactic plane owing to the fact that the known objects are rather near the solar system. Pulsars are discussed in books by Manchester & Taylor (1977) and Michel (1991) and a review by Taylor & Stinebring (1986).

Following the pulsar discovery galactic, point, Xray sources were found that were soon identified with binary systems in which mass is being transferred onto the neutron star from the companion star. In these systems the source of energy is gravitational potential energy of the accretted matter. In pulsars the energy source is rotational kinetic energy. In the past decade or so a new class of pulsars – those with binary companions, and those associated with globular clusters, have been discovered (Backer & Kulkarni, 1990). As the field has progressed over the past two decades the study of accretion powered Xray sources and rotation powered pulsars has merged to become comprehensive study of the death of stars (see review by Bhattacharya & van den Heuvel 1991).

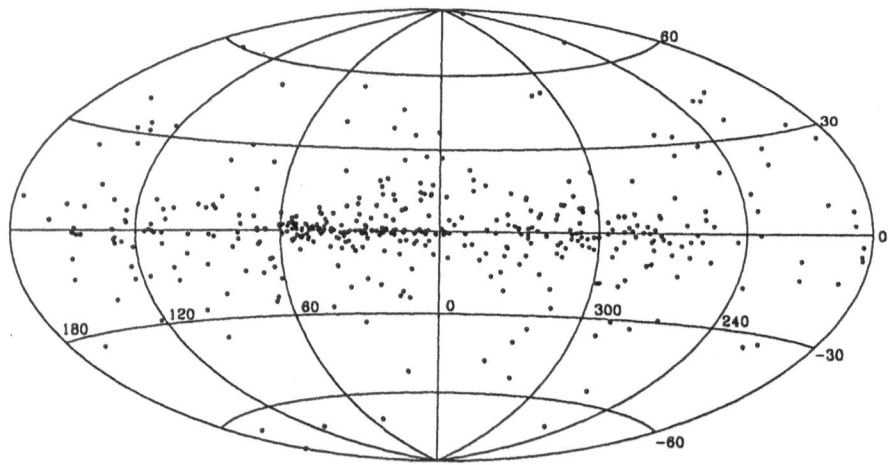

Fig. 1. Distribution of 450 pulsars in galactic coordinates: longitude l and latitude b. The concentration of objects near $b = 0°$ and $0° < l < 60°$ is the result of high sensitivity surveys. (From Backer 1988.)

1.2. Standard Model of Pulsars

Observations of pulsars have led to a number of highly developed theories about the structure and dynamics of pulsars. The extremely high matter density, the ultra strong magnetic field, and the rapid rotation of pulsars has pushed astrophysics into many extreme directions. I will discuss the main ideas about pulsars working from the inside out. The internal structure of a neutron star has a number of uncertainties owing to the extreme density of matter that is beyond the limits of laboratory measurements. Condensed matter physicists divide the neutron star into a core, which is a supercon-ducting, superfluid of $^3P_2$ neutrons and $^1S_0$ protons, an inner crust of $^1S_0$ neutrons, and an outer crust of heavy nuclei. The core contains 70% of the stellar radius and has 90% of the mass. The inner and outer crusts cover 25% and 5% of the radius, respectively, and the inner crust contains the most of the remaining 10% of the mass. The values presented here are for a stiff equation of state that would give a 1.4 $M_\odot$ star a 16-km radius.

In the rotating superfluid neutron star the rotation is carried by quan-tized vortices with a density of $6300/P$ cm^{-2} (P is the period in seconds). As the star slows down, which is the consequence of torques applied via the magnetic field to the outer crust, the density of vortices must fall. The vor-tices move outward and disappear. The inner crust is differentially rotating owing to the pinning of vortices to its lattice structure, a phenomenon seen in the laboratory in He^3. This leaves a piece of the star unconnected to the rotating body that we observe. In §4.6 I discuss instabilities in the smooth rotation and spin down of pulsars that are attributed to sudden or erratic events in the pinning and unpinning of vortices in the inner crust. For fur-ther information on neutron star interiors read the tutorial lectures by Sauls and by Alpar & Pines in the NATO ASI volume edited by Ögelman & van den Heuvel (1989) and the textbook by Shapiro & Teukolsky (1983).

Many observable phenomena in pulsars and in accreting Xray binaries lead us to conclude that neutron stars have intense magnetic fields that are dominated at large radii by a dipole component. Furthermore the dipole axis is required to be misaligned with the rotation axis. The lifetimes of pulsars require lifetimes for the currents that sustain the fields in excess of 10^6 years, and perhaps much longer. The crustal conductivity may or may not allow such stability. The properties of the pulsed radio emission lend the strongest support to the off-axis, dipole model. The radio emission is probably beamed along the dipole axis, and we observe a pulse as the star swings its beam past our line of sight (Fig. 2). Perhaps there are symmetric beams emanating from both poles that can be seen if the axes are orthogonal.

All isolated pulsars have period derivatives that indicate the presence of a spindown torque. A few objects, which I will discuss in §5.1, display ap-parent spinups, but these are almost certainly the result of a small intrinsic spindown rate coupled with an acceleration in the gravitational potential of a globular cluster star system. A number of young pulsars, with ages determined by their spindown rate, are surrounded by a nebula of energetic particles that one easily concludes are the result of energy emitted by the pulsar. The presence of a torque to spindown the pulsar, and the nebulae around other very young and rapidly slowing pulsars are consistent with Pacini's original idea that magnetic dipole radiation is responsible for both phenomena. If pulsars spindown in this manner, then we can estimate their energy loss rate with:

$$\dot{E} = I\Omega\dot{\Omega} = 4 \times 10^{31} \text{ erg s}^{-1} I_{45} \dot{P}_{-15}/P^3, \qquad (1)$$

where I_{45} is the moment of inertia in units of 10^{45} g cm^2, $\dot{P}_{-15}$ is the period derivative in units of 10^{-15} s s^{-1}, and $\Omega \equiv 2\pi/P$. Magnetic dipole radiation leads to the relation $\dot{E} \propto \mu_{d,\perp}^2 \Omega^4/c^3$, where $\mu_{d,\perp}$ is the equatorial component of the magnetic dipole moment. An estimate of the surface magnetic field is then

$$B_{d,\perp} = 10^{12} \text{ G } (P\dot{P}_{-15})^{0.5}, \qquad (2)$$

for conventional neutron star parameters. The actual surface field strength can be significantly larger as a result of the aligned component of the dipole and higher multipole moments.

Articles about the structure of the magnetosphere surrounding pulsars and the nature of the resultant torque make a clear case for the insufficiency of the vaccuum dipole model (Goldreich and Julian 1969; Gunn and Ostriker 1969; Arons 1979; Michel 1991). Nevertheless the energy loss rate and magnetic field calculated above seem to give values that are supported by other observations: the power required to sustain the Crab Nebula and the surface magnetic fields inferred from cyclotron lines in the spectra of accreting Xray binaries.

Goldreich and Julian discuss the formation of a charge separated magnetosphere driven by rotationally induced electric fields. A portion of this magnetosphere will corotate with the star as long as the field lines loop out to axial radii less than c/Ω, the light cylinder radius (Fig. 2). The remaining portion of the magnetosphere must reach out into the surrounding medium with field lines that sweep back to form a transverse, strong field wave far from the star. The sweep back requires a current that is driven by the rotation. This current can provide an additional braking torque on the star. While this description is certainly more complex than the vacuum dipole model, estimates of the resultant braking torque and magnetic field are comparable in magnitude and have similar scaling with neutron star parameters. Some models conclude that the current related torque is largely independent of dipole inclination angle. There are no exact solutions for a charged, rotating magnetosphere that are relevant to the pulsar case (see Michel 1991, Arons 1992).

The above discussion summarizes the main properties of a rotating, magnetized neutron star. But how does it emit beamed radio signals? Good

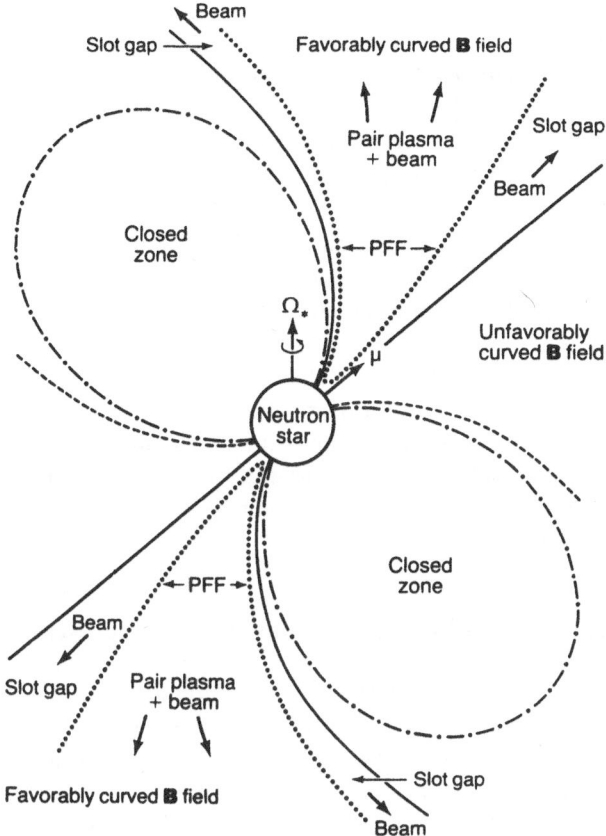

Fig. 2. Pulsars are modeled as rotating neutron stars with an off-axis dipole magnetic field. The closed zone rotates with the star and extends to a cylindrical radius of c/Ω where corotation speeds approach c. An e^+e^- plasma is created by a particle beam accelerated off the surface above a pair formation front (PFF) and along favorably curved field lines. Beamed radio emission is generated at low altitudes in the pair plasma. (From Arons 1983.)

question. The single idea that has held the attention of most investigators is that the emission arises in the relativistic current along the dipole axis. The large and rapid sweep of linear polarization that is observed as the pulse swings by our line of sight led early investigators to this conclusion (Radhakrishnan and Cooke, 1969). Nothing has swayed us from this in the intervening years. The qualitative idea is that the relativistic current is ducted along the intense magnetic field lines in the narrow zone of open field lines. Curvature of the field lines leads to an acceleration whose vector is in the plane of curvature. Curvatures ρ between that of the stellar surface and that of the light cylinder, when coupled with the expected radio emission frequency $\gamma^3 c/\rho$, suggest relativistic factors of $\gamma \simeq 10^2$. The voltage drop along these polar field lines would allow much larger factors, and indeed other models for the formation of radio emission exist.

The intensity of the radio emission, when coupled with the small size of the emitting region, which we determine either from this emission model or from the temporal fluctuations, indicate brightness temperatures of the emission as high as 10^{30} K. The extreme thermodynamic condition of high brightness temperature and low energy (very low) photons require the collective motion of many charges – sometimes called coherent radiation. Many charges moving in unison lead to the coherent addition of emitted voltages rather than the usual addition of incoherent powers. In summary, we typically view the radio emission as the result of turbulence convected relativistically along polar field lines, and carefully arranged to prevent any degrees of freedom for upconverting the copious number of low energy photons to a more balanced thermodynamic state.

1.3. Origin and Evolution of Isolated Neutron Stars

Once a class of objects is isolated in astronomical studies attention turns to understanding their origin and evolution. In a snapshot of the universe, one must make sense out of the different groups of related objects by connecting one class as the progenitors, for example, of another class. Sometimes there are even missing links. If a snapshot is taken of a steady process, then the number of objects in various evolutionary stages will be proportional to the lifetime of the objects in that stage. Students in school are a trivial example. Consider 'classes' as the groups. There are nearly equal numbers in each since the lifetime of all classes is the same, one year. Attrition of students from school will form an independent group and cause the class size to diminish with school year. On the other hand grouping according to primary school, junior high school, and high school, leads to relative lifetimes in the ratio of 6:2:4, for a typical US town. The relative numbers in each group will follow this ratio, apart from the attrition factor.

One of the principal large goals of the study of neutron star evolution is connecting the pulsars we see with the entire population of neutron stars, with the supernova progenitors, and with the massive star population which is continually replenished by star formation in the interstellar medium. This is a tall order. The more restricted task of understanding the life cycle of the pulsar phase of the isolated neutron star is where some progress has been made.

We are fortunate with pulsars in that their signals provide an evolutionary clock in their period derivatives. The ratio of the period to the period derivative gives an age, an age for doubling the period. Furthermore we can take the model presented above and create a more specific age after one or more assumptions are made. If the original period of the pulsar is P_0, and if the braking mechanism evolves with constant perpendicular magnetic moment, then the true age of the star is

$$t_c = \frac{P}{2\dot{P}} \left[1 - \left(\frac{P_0}{P} \right)^2 \right] . \tag{3}$$

Isochrones based on (3) are drawn on the $P\dot{P}$ plot of pulsars in Fig. 3; the logarithm of the age in years of each isochrone is noted to the right of each. The typical age of pulsars by this means is a few million years. The boundaries of the isochrones delineate high and low magnetic field extremes. The standard model allows us to interpret the coefficient in terms of the magnetic moment, and with further assumptions, the dipole component of the surface magnetic field (2). Another age estimate comes from the location of the pulsar with respect to the galactic plane, which contains in a narrow layer the massive stellar progenitors. In many cases we even have the transverse motion of pulsars. Distance estimates required for these parameters are fortunately available from the column density of electrons that disperse pulsar signals (§2.4).

What emerges from this abundance of parameters is that most pulsars are born in the layer that is associated with present day massive star formation. They live as pulsars for an interval of about 10 million years. Their "death" as radio emitting pulsars is the result of a combination of a slow decrease of luminosity combined with an abrupt turnoff (the e^+e^- limit in Fig. 3). During this time they travel away from the galactic plane to form a layer many times thicker than their progenitor stars. Also the torque on the pulsars seems to decay faster than expected based on the simple dipole spindown theory discussed earlier. This rapid decay has been attributed to a reduction of the dipole magnetic field by 1 or 2 orders of magnitude over 10 million years. This is shown in Fig. 3 by bends in the evolutionary paths of low and high field objects. These ideas are consistent with the observations, but there is a range of opinion at present. The data are not overwhelmingly in support of a single synthesis of ideas on the evolution of pulsars during their first 10 million years.

On the origin side, the association of ten young pulsars with supernova remnants is conclusive about the birth of neutron stars in the cataclysmic event of the death of a massive star. We know that a supernova remnant indicates a massive star's final act by totaling the mass and the kinetic energy of the present day remains. Significant questions such as, "Are all neutron stars formed this way?", remain unanswered. Larger numbers of associations with supernova remnants and more proper motions may be available in the future. In the interim, large monte carlo simulations are being conducted to find a consistent model for the pulsar population origin and evolution.

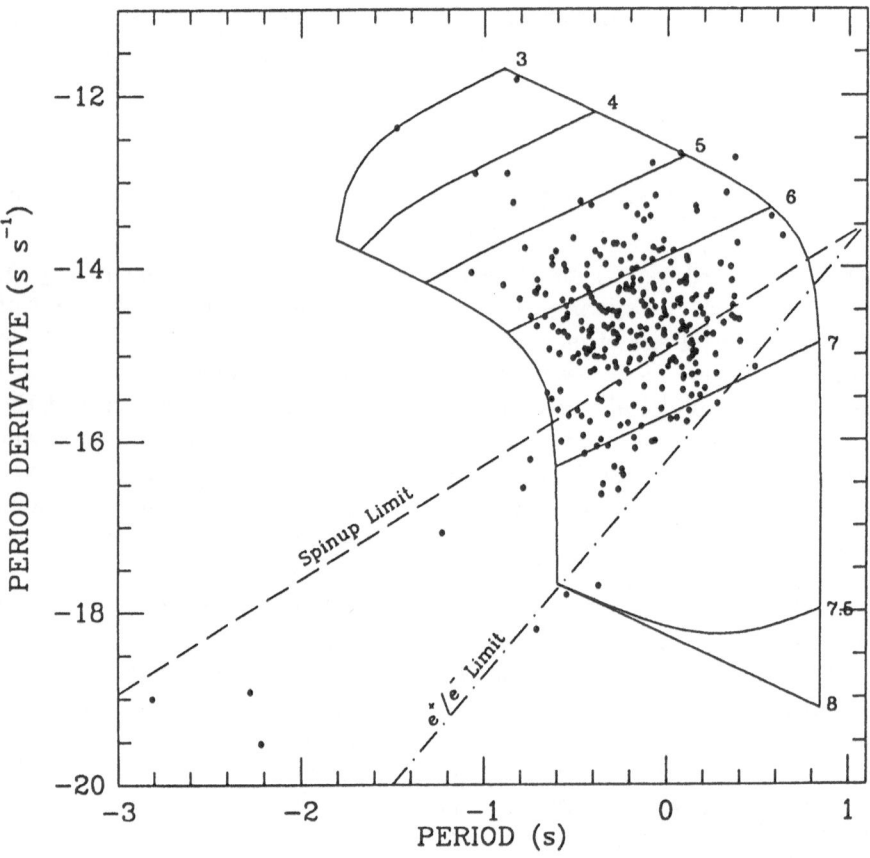

Fig. 3. Distribution of rotation periods and period derivatives of pulsars. The standard model discussed in the text leads to the various evolutionary tracks, or limits, in the diagram. (From Backer 1988.)

2 Radio Astronomy Fundamentals

In this section I provide a brief introduction to radio astronomy signals and pulsar signal processing. The fundamentals of radio astronomy and the properties of radiation can be explored further in the introductory radio astronomy and astrophysics texts by Kraus, Rohlfs, Rybicki & Lightman, and Shu.

2.1 Radiation Properties

We measure properties of the received electromagnetic radiation: the 2D intensity distribution, polarization, spectrum, and time variability. One then must face the inverse problem of constructing a model of the source emission: the emission mechanism, the distribution of emitters and propagation of emission both within the source and along the intervening path, the dynamics of emitting and associated matter, and, finally, the origin and evolution of the system. I will focus here on the former – radio astronomical measurements – and forego much discussion of the latter – astrophysics.

The radiation comes from an assemblage of many independent radiators – accelerated charges and charge bunches and quantum systems. As such we are interested in intensive quantities, and not in the electric field. These are presented below in a hierarchical manner associated with a chain of integrations over independent variables (beware, always, of exact meanings as associated jargon exists, at least in English). Order of magnitude values for two objects, the Sun and a typical pulsar, are given in columns 3 and 4.

Table 1. Radiation Definitions

Quantity	Units	Sun	Pulsar
I, Intensity, or B, Surface Brightness	$Wm^{-2}Hz^{-1}sr^{-1}$	10^{-7}	10^2 (large!)
F, Flux, or S, Flux Density	$Wm^{-2}Hz^{-1}$ 10^{-26} = 1 Jansky (Jy)	10^{-11}	10^{-28} @ 1 GHz
L, Luminosity	WHz^{-1}	10^{16}	--
P, Power	W	10^{31}=$L_\odot$	10^{29}
E or U, Energy	J	10^{49}	10^{45} (not much!)

Two further examples: the man on the moon imagery is a result of the variation of the surface brightness, or intensity (which comes from the reflected solar radiation); and the solar constant, 10^3 W m^{-2}, is the solar flux integrated over frequency.

An important property of intensity is that it does not change along the path of the radiation in the absence of absorbing material, and in Euclidean spacetime. This is shown by an energy conservation argument. Pick two cross sectional areas transverse to the path. The energy transmitted from 1 and passing through A_2 is equal to the energy received through 2 and coming from A_1:

$$I_1 dA_1 d\Omega_1 d\nu dt = I_2 dA_2 d\Omega_2 d\nu dt. \tag{4}$$

But the solid angle $d\Omega_1 = dA_2/r^2$, and, likewise, $d\Omega_2 = dA_1/r^2$. Hence $I_1 = I_2$. This also contains a reciprocity relation: the conclusion can be reached for rays going either direction, so that problems of reception can be treated as problems of emission.

An isolated box of particles at temperature T is bathed in isotropic electromagnetic radiation with an intensity spectrum given by the Planck Law:

$$I d\nu = \frac{2h\nu^3}{c^2} \frac{1}{[\exp(h\nu/kT) - 1]} d\nu. \tag{5}$$

No such perfect box exists, but some situations come close. The surface brightness of the Sun has a spectrum that approximates a Planck distribution with $T = 5850K$. At long wavelengths the Jeans approximation is used since $h\nu << kT$.

$$I d\nu = \frac{2kT\nu^2}{c^2} d\nu = \frac{2kT}{\lambda^2} d\nu \tag{6}$$

You can remember the factor of 2 as being associated with the 2 degrees of freedom of the photon spin. In this approximation you can see how surface brightness may be discussed as brightness temperature at radio wavelengths.

The 2 photon spins lead to 4 polarization intensities, commonly described as Stokes parameters. These are the total intensity (I), two linear polarizations (Q and U), and a circular polarization (V). The total intensity would be measured by a pair of dipole antennae oriented at 90 degrees ($X^2 + Y^2$). Q is measured as the difference ($X^2 - Y^2$), while U is measured by a similar difference with the dipoles rotated by 45 degrees ($X'^2 - Y'^2$). In this discussion, X and Y are the electric fields sampled by the dipoles. The right circular polarization could be measured by a helical antenna, or formed from the linears as $R = X + jY$; the left is $L = X - jY$. The V Stokes parameter (intensity) is then $|R|^2 - |L|^2 = 2XY$.

Recall the qualification about constancy of intensity: "in the absence of absorption". There is no vacuum, only approximations to a vacuum. Matter absorbs electromagnetic radiation in an attempt to come into equilibrium, either giving off more than it absorbs or v.v. Of course it may also absorb the radiation that you are interested in and emit at a different wavelength. The simplest form of this is the one dimensional propagation of intensity through a uniform slab of absorbing matter.

$$I_{out} = I_{in} \exp(-\tau) + I_{slab}[1 - \exp(-\tau)], \quad \text{or} \tag{7a}$$

$$T_{out} = T_{in} \exp(-\tau) + T_{slab}[1 - \exp(-\tau)], \qquad (7b)$$

where τ is the optical depth of the slab which is the product of an absorption coefficient and the physical thickness. Thermodynamics can be applied here: if $T_{in} = T_{slab}$, then $T_{out} = T_{in}$; i.e., everybody is at the same temperature.

The topic of radiative transfer, both within the source of interest and along the way in the intergalactic, interstellar, solar system and earth's atmosphere is extremely important in astronomy. Absorption is caused by the microphysics of the photon interaction with matter. There are other macroscopic effects caused by these interactions: reflection, refraction, diffraction, and, in the presence of a magnetic field, birefringence. The latter three effects are associated with thermal plasma interactions and are important in radio astronomy, particular with pulsars (§2.4, §3.0).

2.2 Radio Telescopes

The dipole antenna discussed above has an effective area for intercepting the electromagnetic flux density of λ^2. A parabolic reflector is used in astronomy to focus the E field. The focused field is then sampled with a dipole, or its equivalent, to convert the free space field into a current driven through an amplifier. A Jansky of flux density produces one fW for a 10^4 m^2 collecting area and a bandwidth of 10^7 Hz. A tremendous amplification is required to bring this feeble signal up to the level of a mW which is a reasonable signal strength to manipulate.

The radio telescopes in use are typically mounted in alt-az, or polar (ha-dec), configuration. Some of the largest antennas only allow transit observations with mechanical, or electronic, steering in declination. The largest reflector is in Arecibo, PR, and is a spherical surface with an aberration-correcting waveguide "feed"; n.b., the "feed" feeds the signal to the dish, and via reciprocity, is equivalent to accepting signal from the dish. Some telescopes use a surface that deviates slightly from a parabola to improve their performance; e.g., the elements of the VLA. The new Green Bank Telescope will have a radical design which is an offset piece of a parabola that will eliminate any blockage of the incoming signal on its way to the "feed". Many high frequency telescopes have a secondary reflector near the primary focus to bring the signals to a focus near the vertex of the paraboloid; a hyperboloid for the Cassegrain design where the secondary is between the primary focus and the primary reflector, and an ellipsoid for the Gregorian design where the secondary is behind the primary focus. Typical f/D ratios are near 0.43 at prime focus, and near 10 at Cassegrain.

The gain of a telescope is the ratio of the intensity at the focus to the incoming intensity. The efficiency of a telescope is the ratio of the effective area, which is what determines the intensity at the focus, to the geometric area. Efficiencies are less than unity owing to such things as blockage, which I just mentioned regarding the GBT, but more importantly owing to properties of the feed, which I come to next.

First let's take a simple case and introduce the use of the Fourier transform. Also I will consider the case of transmission of electromagnetic waves. If I transmit from two locations, then in the far field these signals will interfere with successive maxima spaced in angle by $\lambda/b_\perp$ radians, where $b_\perp$ is the antenna spacing projected along the direction considered. The intensity pattern, which is the absolute square of this interference, also has "fringes" which on reception would be called light and dark. This configuration is that of a Fourier pair: the E-field radiation distribution and the far-field interference pattern. We call the far-field pattern an angular spectrum. Spectrum implies Fourier transform, and it is the Fourier transform of the E-field radiation distribution converted via $\lambda/b_\perp$ to angle. Let's move on. For a uniform field over a telescope diameter d, we would get a pattern that is $\mathrm{sinc}(\pi\theta d/\lambda)$ $= \sin(\pi\theta d/\lambda)/(\pi\theta d/\lambda)$, and an intensity pattern of $\mathrm{sinc}(x\pi\theta d/\lambda)^2$. This intensity, or gain, pattern is the "beam" of the telescope.

2.3 Radio Astronomy Receivers

We have to convert the focused intensity into a current that will be amplified by, $e.g.$, a transistor. This task is not obvious. First we have to "coax" the converging radiation down into a waveguide; $i.e.$, we have to match the field patterns of the waveguide modes to those of the freely converging field just before the feed. Then we have to disturb the waveguide propagation by sticking a paper clip in there, a dipole antenna or its equivalent, in order to generate a voltage that will drive a current down a wire with its characteristic impedence (R). The power received at this point is the primary quantity that we need to calibrate. This power can be expressed in terms of the temperature of a resistor that would generate an equivalent power across its terminals (V^2/R) via the thermodynamic oscillations of charges. And the voltage signal characteristics are just that – random. We call this temperature the antenna temperature.

Typically a controlled source of electrical noise, which can be used as a secondary intensity standard at any time, is injected at this point. The signal then enters the first amplifier stage (see sketch in Fig. 4). In modern receivers from 400 MHz to 400 GHz the early stages of amplification, and often much of the waveguide, are cooled to 20 K, or below, to minimize losses. Losses, by the radiative transfer equation given above, show up as extra power; $L \equiv [1 - \exp(-\tau)]$ is the loss factor, and T_{slab} is here T_{amb}, the ambient temperature. The typical radiometer equation is then:

$$T_{\mathrm{total}} = (((T_{\mathrm{source}} + T_{\mathrm{back}} + T_{\mathrm{cal}})(1 - L_1) + T_{\mathrm{amb}}L_1)G_1 + T_2)G_2... \quad (8)$$

where T_{source} is that which we wish to measure, T_{back} is the excess noise entering the receiver (which includes the 2.7 K cosmic background, emission

from the galaxy and the earth's atmosphere, and scattered ground emission),
$T_1 \equiv L_1 T_{\mathrm{amb}}$ and G_1 are the excess noise and gain of the first amplifier,
and so on.

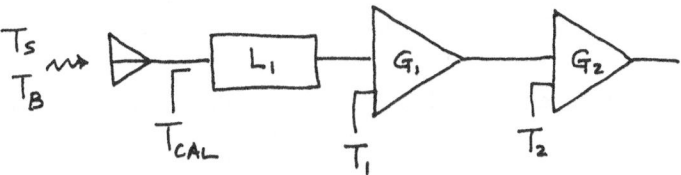

Fig. 4. Radio astronomy receiver schematic

The power that we measure, represented above as temperature, is proportional to the square of the voltage across a resistance, or load. This
voltage is, in turn, proportional to the electric field incident on the antenna.
The uncertainty in our measurement of power is ultimately limited by the
statistics of the independent samples of the received electric field after detection, and averaging over both bandwidth (b) and integration time (τ,
now a time interval and not opacity). The process is illustrated in Fig. 5.
First we represent the narrow band voltage as a slowly varying (on time
scale of $1/b$) amplitude modulating a rapidly varying (on time scale of $1/\omega$)
sinusoid at the center frequency (ω)

$$v_{\mathrm{nb}}(t) = c(t)\cos(\omega t) + s(t)\sin(\omega t) \tag{9a}$$

Independent samples are spaced by the crossing time of the modulation
$1/2b$. As a result power (temperature) fluctuations are given by

$$\Delta T = \frac{T}{\sqrt{2b\tau}}. \tag{9b}$$

One must detect a signal by ON-OFF comparison against a background of
the uncertainty ΔT which is proportional to the total system temperture,
T. The source portion of T is proportional to the telescope gain, or effective
area. Great care is taken then to minimize T while maximizing $b\tau$.

The narrow band representation is important for understanding the het-
erodyne, or frequency conversion, process in radio astronomy systems. The
signal is multiplied by a pure tone, or local oscillator, and filtered to shift
the center frequency from the high radio frequencies of reception to lower
frequencies more convenient for transmission and further amplification. The
translation starts with signals at Radio Frequencies and moves them first to

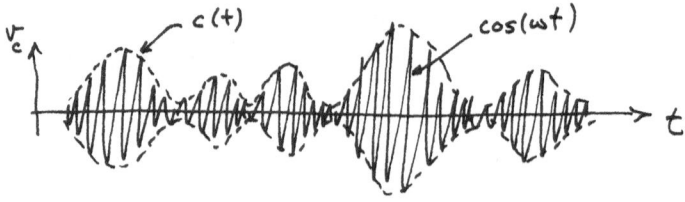

Fig. 5. Narrow-band electrical noise

Intermediate Frequencies and often to final Video Frequencies for detection, spectral analysis or other signal processing.

2.4 Propagation in the Interstellar Medium

Pulsar signals traverse interstellar paths with column densities of 3×10^{21} cm^{-2} and 10^{20} cm^{-2} for atomic hydrogen and thermal electrons over a kpc path, respectively. The paths are threaded by magnetic fields whose strength is a few μGauss. The ionization that produces the electrons comes from the ultraviolet flux of hot stars, cosmic rays, and shocks generated by star formation and star death events. The magnetic fields are generated by large-scale currents of the interstellar plasma. The plasma is turbulent on scale lengths stretching all the way from 10^{10} cm, which diffracts and refracts radio signals, up to 10^{15} cm, which leads to secular variation of the propagation parameters, and beyond.

2.4.1 Phase and Group Velocities

The phase velocity of an electromagnetic wave is $v_p \equiv \frac{\omega(k)}{k}$ which becomes c in vacuo. The index of refraction for weakly ionized, unmagnetized plasma is

$$n \equiv \frac{c}{v_p} = \frac{ck}{\omega} = \sqrt{1 - \frac{\omega_p^2}{\omega^2}}. \tag{10}$$

where $\nu_p = \omega_p/2\pi = 9$ kHz $\sqrt{n_e(\mathrm{cm}^{-3})}$. The group velocity v_g is determined by the derivative of the dispersion relation which is derived from consideration of the propagation of a wave packet, or disturbance, that has finite bandwidth.

$$v_g \equiv \frac{\partial \omega}{\partial k} = cn, \tag{11}$$

where the evaluation is given for the weakly ionized plasma.

If the medium has a variable index of refraction, then the radiation propagates along rays that bend (refract) according to

$$\frac{d(n\mathbf{k})}{dl} = \nabla n, \tag{12}$$

where $n, \mathbf{k}$, and l are the index of refraction, the ray direction, and the ray path length, respectively. The gradient is defined over a transverse dimension that exceeds the Fresnel zone scale, $R_F = \sqrt{x(1-x)\lambda d}$, where d is the source-observer distance and x is the disance to the plane at which one is evaluating R_F. You can see from this that the Fresnel zone defines a tube whose waist diameter grows to its largest dimension midway between the source and the observer. Gradients of the index of refraction across the tube make it snake its way along the path. There can be more than one refractive region that can send the signal to the observer. The *gain* of the refractive region is determined by the fraction of a Fresnel zone that is stationary (constant) phase, but this jumps ahead of our story to topics of propagation.

2.4.2 Dispersion

The dilute plasma along the line of sight delays low radio frequencies relative to high radio frequencies. If this medium were fixed, then we could measure a dispersion constant once, and apply this to all data recorded at various radio frequencies. However, all of the media mentioned are fully turbulent with a broad spectrum of length scales. Precise pulsar timing requires continuous determination of the dispersive effects.

The dispersive delay Δt_d is calculated from the path integral of the inverse of the group velocity which is determined using the cold, unmagnetized plasma index of refraction, n.

$$\Delta t_d = \int_P^E (\frac{1}{v_g} - \frac{1}{c})ds = 0.00415 \text{ s DM(pc cm}^{-3}) \ \nu(\text{GHz})^{-2}, \tag{13}$$

where DM is the column density of electrons, or dispersion measure. The integral is taken from the pulsar at P to the Earth at E. A DM of 25 leads to an excess delay of 1 second for signals at meter wavelengths. The interstellar density is around 0.03 cm^{-3}. A DM of 25 then indicates a distance of some 750 pc. The dispersion within a narrow band of signals places a limit on the resolution that can be obtained:

$$\tau_d = 8.3 \ \mu\text{s DM(pc cm}^{-3}) \ b(\text{MHz}) \ \nu^{-3}(\text{GHz}) \ < \ (2b(\text{MHz}))^{-1}. \tag{14}$$

Measurement of dispersion requires timing of the pulse arrival at two or more radio frequencies. This is straightforward until one reaches the precision where one must deal with the fact that the pulse shape changes with radio frequency. Then some fiducial point will scale with the inverse of the frequency squared, but other parts may disappear owing to spectral differences, or may be offset early or late. One approach to dealing with this complexity is to make a frequency dependent template of the pulse shape to compare with the data.

Figure 6 displays the temporal variations of DM for two pulsars. The PSR 1937+21 data are particularly sensitive and display an *event* in the record in early 1992 – a brief increase in the DM that lasted 2-3 months. Structures such as this may occur when the line of sight crosses a thin sheet of ionized gas seen edge on. Larger amplitude structures like this have been inferred from observations of refractive focussing (Fiedler *et al.* 1987) where the source, medium and observer are such as to bring a caustic past the observer in an interval of a few months.

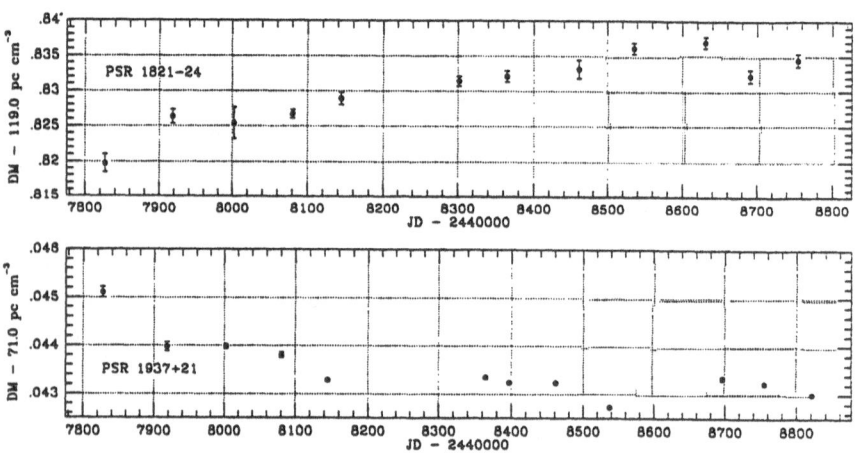

Fig. 6. Dispersion measure variations for PSR 1821-24 and PSR 1937+21

2.5 Search Techniques
2.5.1 Sampling

Pulsar intensity signals are characterized by their period, duty cycle, and dispersion. The period and duty cycle lead to requirement of sampling the intensity at a rate many times the pulse period, 10-50. The Nyquist theorem says we gather all the information in bandwidth limited noise if we sample at twice the bandwidth rate. In an analog system the bandwidth is defined by the post-detection filter. One desires a rectangular passband gain, but this is often difficult to achieve. A simple RC circuit produces an exponential gain curve which if sampled at $\delta t = RC/2$ will cause a significant amount of aliasing of higher frequency signals. Let's take a closer look at this. The RC filter gain is $G(f) = \exp(-fRC)$. We sample at intervals of $RC/2$. The spectrum of this signal is formed via a Fourier transform of the real data samples into a complex spectrum. The amplitude falls as $\exp(-fRC)$ initially, and would continue to do so if we sample the signal much faster than $RC/2$. However at the end of our spectrum, the Nyquist frequency of $f_N = 1/RC$, the higher frequency noise is aliased into the spectrum. Now noise "powers" add, so the amplitude will become,

$$A(f) = \sqrt{\exp(-fRC)^2 + \exp(-(2RC - f))^2}. \tag{15}$$

In fact even further aliasing, from the next folds of the spectrum, take place, but the contribution is small. The additive nature of noise power leads us to use the amplitude squared, or fluctuation power. Filters with sharper cutoffs are desireable to prevent this aliasing which reduces the signal to noise ratio by a factor of

$$R(f) = \frac{\exp(-fRC)}{A(f)}. \tag{16}$$

On the other hand some well-known pulsar discoveries, 1937+21 in particular, have made use of the extended range of frequency sensitivity that is possible with the aliased spectrum.

Another common low pass filter is created by integrating the detected signal for a time δt. Each sample is then independent and the effective low pass filter is given by the sinc function. This is seen by describing the process as a convolution by a top hat function of width δt, and then sampling on the uniform grid spaced by δt. The Fourier transform is then the product of the transforms of two first two convolved functions. The aliasing is now

$$p(f) = \text{sinc}(\pi f \delta t)^2 + \text{sinc}(\pi(1 - f\delta t))^2 + \text{higher folds.} \tag{17}$$

The data is taken over an interval $T = N\delta t$ and the Fourier transform assumes the data is replicated on a grid of T. In the power spectrum domain the samples are spaced by T^{-1}, the resolution, and the response varies as $\text{sinc}^2(\pi\delta fT)$ in amplitude and the corresponding phase response ramps linearly with frequency by π between adjacent Fourier bins.

2.5.2 Standard Approach

Most searches proceed by forming a set of 1D time series for a complete set of dispersions that can be formed from the output of a bank of narrow band filters. In recent Arecibo searches, this filter bank is effected by use of a digital correlator (*e.g.*, Backer *et al.* 1990). Single dispersion data can be formed from the observations by summing along staircases of decreasing slope in the 2D frequency-time plane of data. In the case of digital correlator sampling, a conversion to the frequency domain is effected by first symmetrizing the sampled correlation function, and then transforming. The effective filters thus have a sinc($2\pi\nu\tau$) gain response, where ν is the radio frequency. The time series for a complete set of staircases can be formed by a fast folding algorithm first discussed by Staelin (1969) shortly after pulsars were discovered. This eliminates redundant sums just as the FFT algorithm eliminates redundant sum/products.

The next step is to take the Fourier transform of the time series of length T that were formed for each independent DM channel, or staircase. This is effected with the efficient FFT algorithm which can be computed efficiently on modern vector processors. FFTs of lengths ranging from 32K to 32M are common.

The Fourier transformation results in the narrow duty cycle, pulsed signal being concentracted in a series of Fourier harmonics whose amplitudes are comparable out to a harmonic number given by the inverse of duty cylce, the ratio of the pulse width to the pulse period. These are delta functions with a sinc$^2(\pi\delta fT)$ shape that results from the sampling window; the harmonics are of course not necessarily centered on the FFT frequencies. In addition, scintillation of the signals (§3.3) can cause a noticable broadening of the responses. A harmonic summing algorithm is required to provide maximum sensitivity to the entire pulsed signal. One searches a range of possible harmonics required for every possible frequency. This allows for a range of possible duty cycles.

Dewey *et al.* (1985) succinctly stated the minimum detectable flux density of a pulsar search as

$$S_{\min} = \frac{10(T_{\text{sky}} + T_{back})}{G\sqrt{N_pB\tau}} \sqrt{\frac{w}{(P-w)}}, \qquad (18)$$

where the factor of 10 is the signal to noise ratio required for firm detection amidst the many search parameters, N_p is the number of polarizations summed (1 or 2), G is here the telescope gain in K/Jy, B is the total bandwidth, τ is the integration time, w is the apparent pulse width from the combined effects of intrinsic width and dispersion and scattering, and P is the pulse period.

There are many searches recently completed, or under way. These use the large telescopes at Arecibo, Parkes, and Jodrell Bank. Previous searches have mined the galactic plane for weaker and weaker pulsars with ever shorter periods. The discovery by Wolszczan of two high latitude pulsars with short periods (during a period when the Arecibo telescope was otherwise idled by mechanical failure) has led to a flurry of all-sky searches. The next few years will see a continually growing list of these new pulsars. The searches are particularly exciting because the millisecond period pulsars are a poorly understood distinct population, and the timing of these objects, as I will discuss, allows a variety of fundamental measurements to be performed.

2.5.3 Other Aspects

Hamilton *et al.* (1973) presented a novel two-dimensional approach to pulsar searching. The radio frequency-time matrix of samples is transformed in both coordinates to delay-fluctuation frequency. A sequence of diagonal lines in the rf-t domain, which represent the dispersed pulsar signal, are transformed into a set of Fourier harmonics along a diagonal path in the d-ff domain. A two-dimensional search algorithm is then required. The Fast Pulsar Search Machine (FPSM) that we constructed at UCBerkeley is based on the Hamilton *et al.* algorithm (Backer *et al.* 1990).

Current pulsar searches will have difficulty finding the shortest rotation period pulsars if they are in a very short orbital period binary system. The Doppler spread of the signal during the integration will reduce the amplitude of the harmonics as noted by Backer (1987) and as recently studied by Johnston & Kulkarni (1991). Searches for pulsars in globular clusters now regularly explore 'acceleration space'. Anderson *et al.* (1990) reported the first pulsar that was found with their acceleration search algorithm, but not with their fixed Doppler algorithm. Middleditch has developed an algorithm that does a full Keplerian orbit search. He earlier developed an acceleration search that he applied to an investigation of extragalactic optical supernovae (Middleditch and Kristian 1984).

2.6 Pulsar Timing Systems

The simplest pulsar timing system is a power detector that is followed by a sampler and a signal averager which accumulates data into "phase bins" synchronous with the apparent pulse period. The structure within a pulse, its profile, requires sampling at intervals between 1/100 and 1/1000 of a period. The bandwidth must be limited to keep the dispersion smearing τ_d (14) below the sample time. More sensitivity can be obtained by defining many adjacent channels each with bandwidth b. There are several signal processing techniques that provide many adjacent channels. Local oscillators and filters are used in the analog filter bank. Digital correlators were introduced for pulsar work, first in VLBI systems by Erickson *et al.* (1972), and later by our group at Berkeley for single dish work. Navarro and Kulkarni at Caltech are currently completing the latest version of an correlator based pulsar signal processor. In recent years a broad band FFT processor has been developed and employed for pulsar signal processing at the NRAO in Green Bank, WV, by Fisher and LaCasse. All of these systems create a two dimensional array of intensity measurements. The signals from individual channels can be summed after removal of the relative dispersion.

In all cases the summing must be synchronous with the pulse period that is constantly changing owing to the Doppler motion of the Earth, and, in the case of pulsars with companions, the motion of the neutron star itself. Two techniques are commonly employed. Either the time base of the sampler and signal processor is changed to maintain synchronism with the apparent period, or the time base is held constant, and the phase of each sample calculated to allow storing in the correct "phase bin". In the latter case, the number of entries in each bin must also be maintained. When the sampling is driven by the channel bandwidth, an odd number of phase bins per period is required to match the pulse period.

The detection techniques discussed above have a severe limitation in the case of very short period pulsars with modest DMs and for observations at meter and decimeter wavelengths. The equation above produces a limiting bandwidth, and therefore time resolution, when the dispersion smearing τ_d equals the sample interval $(2b)^{-1}$. If further resolution is required one must remove the dispersion in the voltage, or electric field, prior to detection. This is called coherent disperion removal in contrast to the post detection removal, or incoherent technique.

Several approaches have been taken to this. The very first work by Hankins (1971) used voltage data sampled on magnetic tape. The dispersion was removed in a computer, and then the signal was detected and processed further. The scientific goals of this computer intensive process were not pulsar timing. A surface acoustic wave technique was developed, but did not see intensive use. Hankins later developed with Stinebring and colleagues a hybrid analog/digital convolution device based on a Reticon device designed for chirp-z transforms (Stinebring *et al.* 1992). Biraud and colleagues at Meudon developed a swept oscillator approach to coherent dispersion removal. This clever scheme, which was first used by Staelin to look at the Crab pulsar, turns time, or pulse phase, into frequency so that an ordinary digital correlation spectrometer can be used for a signal averager. Coherent dispersion removal from within a narrow signal band can be most clearly stated in terms of the phase function with which one must multiply the spectrum to remove the dispersion. The convolution technique already mentioned employs the Fourier transform theorem that a multiplication in frequency is equivalent to a convolution in time. Ables at CSIRO has developed an FFT-based coherent dispersion removal pulsar timing processor.

3 Further Topics on Radio Wave Propagation

3.1 Absorption

In pulsar timing we are not too interested in the absorption of pulsar radiation in the column of neutral hydrogen atoms along the path. We do use this absorption along with a model of the galactic rotation to infer distances of the pulsars (see Frail & Weisberg 1990). For some of the nearest millisecond pulsars we can determine a trignometric parallax (§4.7), but for these there is typically no hydrogen absorption for comparison.

3.2 Birefringence

The group delay for two circularly polarized modes differs if the integral of the product of the parallel magnetic field $B_\perp$ with the electron density is nonzero. This difference is typically seven orders of magnitude less than the dispersive delay, and is therefore neglected in arrival time measurements. The phase delay difference does produce observable effects for linearly polarized signals. A linearly polarized signal can be decomposed into opposing circular modes whose relative electrical phase determines the PA of the linear. The PA will rotate with frequency by an angle χ equal to the phase delay between the modes $\omega(t_O - t_X)$. This rotation is called Faraday rotation. The Faraday rotation angle is derived from a path integral of the difference between the inverses of the phase velocities for the two circular modes, $c/n_{O,X}$.

$$\chi = \omega(t_O - t_X) = \frac{\pi}{c\nu^2} \int_0^L dz\, \nu_p^2\, \nu_B, \quad \nu_B = 2.8\ \mathrm{MHz}\ B(\mathrm{G}), \qquad (19a)$$

$$\chi = \mathrm{RM}(\mathrm{rad\ m^{-2}})\ \lambda(\mathrm{m})^2, \quad \mathrm{where} \qquad (19b)$$

$$\mathrm{RM} = 0.81\ \mathrm{rad\ m^{-2}} \int_0^L dz(\mathrm{pc})\ N_e(\mathrm{cm^{-3}})\ B_\perp(\mu\mathrm{Gauss}). \qquad (19c)$$

The parameter RM is the Rotation Measure, and ranges from 0.1 to 1000 rad m^{-2} for most objects in the galaxy.

The ratio of the rotation and dispersion measures gives an estimate of the galactic magnetic field strength (component parallel to signal path, and averaged along the path with weighting by the electron density):

$$< B_\perp > = 1.232\ \mu\mathrm{Gauss} \frac{\mathrm{RM}(\mathrm{rad\ m^{-2}})}{\mathrm{DM}(\mathrm{cm^{-3}pc})}. \qquad (20)$$

These magnetic field estimates along with pulsar distance estimates are used to create a galactic model of the magnetic field in the vicinity of the solar system (Rand & Kulkarni 1989; Clegg et al. 1992).

3.3 Scattering

Turbulence in the interstellar plasma is driven by injection of energy into a gas on a characteristic length scale, and the spread of turbulent cells to other scales through viscous interaction and dissipation. Turbulence in the ionized interstellar medium means that the radio emission wavefront is crinkled in proportion to the path integral of the corresponding index of refraction fluctuations along the unperturbed signal paths. If the phase of the wavefront is perturbed by more than a radian within a Fresnel zone, then diffraction develops. That is, across a plane perpendicular to the path at the location of the observer there will be bright intensity maxima and minima where the signal goes to zero. The Fresnel zone size for a kpc path (3×10^{21} cm) at decimeter wavelengths (30 cm) is 3×10^{11} cm, 4 solar radii for comparison. The instantaneous diffraction pattern will be very frequency dependent if there are many random turns of phase within the Fresnel zone owing to the quadratic dependence of the index of refraction on frequency. Typical transverse scales for a radian of phase change are around 10^9 cm. Eventually, at some high frequency, the phase perturbation within the Fresnel zone will drop below a radian and the diffractive effects will subside. Topics dealing with interstellar scattering are discussed in recent conferences edited by Cordes, Rickett & Backer (1988) and Baldwin and Wang (1991).

The Earth observer, the pulsar, and even the medium are all moving with respect to each other. This produces a time variable diffraction pattern. The signal intensity goes up and down with wide swings of intensity when the bandwidth is small enough. Astronomers refer to this as scintillation. At meter wavelengths and for DMs around 25 the bandwidth of scintillation is a fraction of 1 MHz and the scintillation time scale is minutes. Larger radio sources, quasars and radio galaxies, show no such effects owing to their size which blurs the diffraction pattern – stars twinkle and planets do not in the Earth's atmosphere. In summary the scattering observables are: the time scale for scintillation; the frequency scale for "spectral scintillation"; the radio frequency scaling law of these two parameters; and the size scale of the diffraction pattern (when it is smaller than the Earth so that one can observe it). From these parameters one can assess the strength of the turbulence along a given path, and the transverse velocity of the diffraction pattern.

Diffraction requires ray paths that, while formerly diverging from the source, converge on the observer. The path lengths of the various rays that reach the observer differ. Thus, an impulse spit out by the pulsar will now be spread out by a geometric factor,

$$\tau_b = \frac{d\theta_s^2}{2c}, \tag{21}$$

where d is the distance to the turbulent "screen", and θ_s is the "scattering" angle. Let me make a semantic point here. The microphysical phenomenon that does the damage to the wavefront is scattering of the electrical field by electron interactions. So the phenomenon is called *interstellar scattering*, and not interstellar scintillation which is just one of the effects of interstellar scattering. Equation 21 is for a source at infinity. An excellent exercise is to develop the scale factor for a source at a finite distance with the phase perturbing screen at an arbitrary fraction of the source-observer distance; simple point to point ray path geometry will do. Note that I have repeatedly referred to a collapse of the phase perturbing medium into a single screen. This thin-screen approximation is crucial to a simple analysis of the effects, a Born approximation approach. The scintillation bandwidth is the conjugate parameter to this broadening time since it is the differential delays between paths that cause the constructive and destructive interference as a function of frequency.

Now one might ask if there isn't also turbulence on scales larger than the Fresnel zone? Yes, and this produces refraction and the DM variations shown in Fig. 6. The source position is shifted by gradients of phase, and quadratic phase terms lead to amplification or diminuation of the intensity. Even more severe perturbations can lead to caustics, and double images which interfere coherently just like an interferometer. These are fascinating areas of current research.

3.4 Solar Wind and Ionosphere

Pulsar signals also traverse the solar wind flowing out from the Sun and the ionosphere that surrounds the Earth. At a radio frequency of 1 GHz typical delays are around 0.1 microsecond for both of these media with the solar elongation of 90°. These plasma delays also scale with the inverse square of the radio frequency. Both media are highly variable on a day to day and season to season basis. The solar wind delay is very dependent on the solar elongation owing to the sharp gradient in the density of the outflowing plasma (Krisher *et al.* 1991). The Earth's troposphere adds a rather small additional nondispersive delay

3.5 Relativistic Delay in Solar System Potential

Pulsar signals are delayed relative to those in a vacuum by the curvature of space that is synonomous with the presence of massive bodies. Photons from the pulsar follow null trajectories of the geometry. To first order, the line element is

$$d\sigma^2 = (1 - 2U)c^2 dt^2 - (1 + 2\gamma U)(dx^2 + dy^2 + dz^2) = 0. \qquad (22)$$

where U is the gravitational potential from all solar system bodies, and γ is one parameter in the Parametrized Post-Newtonian coordinate system of

Will & Nordvedt (1972). An expression for the time of flight extra delay owing to this effect is then

$$\Delta t_S = (1 + \gamma)\Sigma_p \frac{Gm_p}{c^3} \ln \left| \frac{\hat{\mathbf{n}} \cdot \mathbf{r}_p + r_p}{\hat{\mathbf{n}} \cdot \mathbf{R} + R} \right|, \tag{23}$$

where $\mathbf{r}_p$ is the position of the observatory relative to the pth solar system body, and $\mathbf{R}$ is the pulsar's position relative to the body p. This delay, which was first pointed out by Shapiro (1964), is the *longitudinal* counterpart of the more familiar *transverse* effect that was identified by Einstein as one of his three tests of General Relativity, the bending of light by the Sun. The factor $2Gm/c^2$ is the gravitational radius of an object, and is about 3 km for the Sun. The corresponding delay is then 10 microseconds times the logarithmic factor. This delay is 135 microseconds at the limb of the Sun for a distant pulsar, and is very difficult to measure in the presence of very large and variable plasma delays. The most precise measurements of γ come from dual frequency radio interferometry. The best current measurement is $\gamma = 1.0002 \pm 0.0020$ by Robertson & Carter (1991). There will also be large contributions to the relativistic delay as the signal traverses the interstellar gravitational potential. However, these are so nearly constant that their effect on pulsar timing will not be measureable. This is particularly true when one considers the stellar parameter estimation process that is the subject of the next section.

4 Pulsar Timing

The standard model of pulsars presented above in §1.2 has one or more beams of radiation continuously emitted in a radial direction. The space-time locus of this radiation is a spiral winding away from the neutron star like the stream of water from a lawn sprinkler. Pulses are detected at the observatory as radiation from successive stellar rotations cross the observer's path. The observer records the time of these crossing events as registered on a comoving clock. More precise definitions are given below. The first stage of pulsar timing analysis is a transformation of the space-time reception event of the pulse front crossing the observatory to the corresponding transmission event of the launching of this pulse at the neutron star. The transmission event is fictitious and covers our ignorance of the emission process. I will not deal with issues of aberration and retardation of the radiators and the radiation in a pulsar magnetosphere. I am also not interested in the exact orientation of the neutron star, but only its relative orientation between successsive rotations. The second stage of pulsar timing is an analysis of events in the reference frame of the neutron star: do the transmission events observed follow a simple model for the rotation that involves an initial epoch, period, and period derivative to within the noise of the measurement (and precision of the transformation)? Precise pulsar timing has been discussed recently by Backer & Hellings (1986), Hellings (1986), Backer (1989), Taylor (1989) and Doroshenko & Kopejkin (1990). References for timing binary pulsars will be given in §5.

4.1 Arrival Time Measurement

In the ideal case one time tags the arrival of individual pulses. In practice individual pulses vary considerably within a well-defined window of rotational longitude, and may be too weak to detect individually. Pulsar timing then requires averaging over many pulses, often thousands, to obtain a stable emission pattern, or profile of flux density, with sufficient signal to noise ratio, snr, for pulse detection. One of the remarkable features of the pulsar phenomenon is that stable profiles are obtained in spite of the dramatic pulse to pulse fluctuations. Helfand $et\ al.$ (1975) present a quantitative study of the stabilization of profiles as a function of integration time. Similar studies are needed for a wider set of objects to validate this assumed property. Experimentally the samples of flux density are often taken with a time base that is synchronous with the apparent pulse period as estimated from an existing model. This allows the summing of successive blocks of 1024 samples, which is a typical number per period, into an averaging array. The space-time event recorded is the time of the first sample of the first block. All successive samples are taken at deterministic times relative to the first sample. Examples of several pulse profile integrations are shown in Fig. 7.

The sensitivity of radio astronomical observations – the ability to detect signal power against a background of unwanted noise coming from the amplifier and other unavoidable processes – is determined by the inverse of the square root of the product of the bandwidth and the integration time. In the present example the snr for each sample of the profile is proportional to $\sqrt{bNP/1024}$, where N is the number of pulses averaged. Dispersion limits b, but we can increase our sensitivity by using many adjacent channels of width b. Signal processors that define many channels and then detect and average the results, are a continuing engineering challenge.

Pulsar signals are highly polarized. Most timing observations accumulate averages of orthogonally polarized signals. These are calibrated to a common flux density scale and added together to form the total intensity profile. Careful calibration is required to prevent distortion of the profile from polarization effects.

The next step in pulsar timing is the comparison of a pulse average with a template pulse. The template pulse typically comes from a long term average of the data itself. Alternatively an analytic model can be derived from the data and then used as a template. See example in Figure 7. The observed pulse is convolved with the template pulse to obtain the relative pulse arrival time from the peak of the correlation function. Other

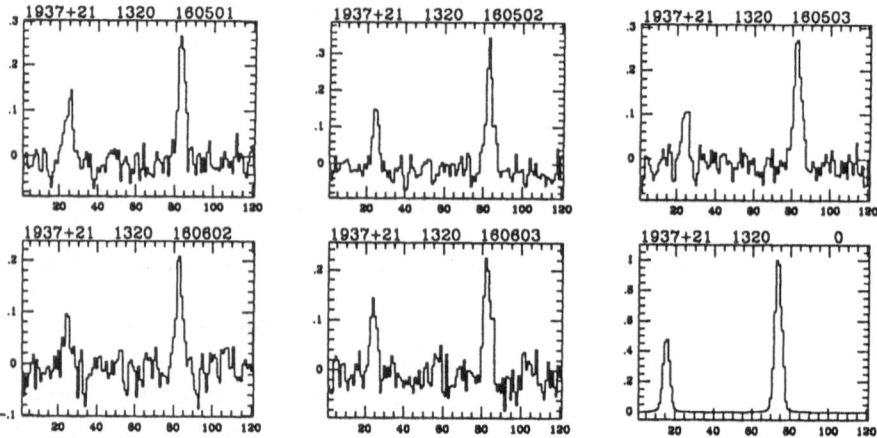

Fig. 7. Pulsar timing data profiles. The first five plots contain successive 2-min integrations of PSR 1937+21 at the NRAO 140ft telescope using 20 MHz of bandwidth at 1320 MHz; scale is flux density (Jy) *vs* bin number (1-121). In the lower right is a standard template which is convolved with the data to obtain relative arrival times.

approaches such as use of Fourier techniques have also been developed. The relative pulse arrival time is added to the first sample start time to obtain the observatory, or topocentric, time of arrival or TOA. The precision of determination of a TOA will depend on the source average flux density, the width of the pulse, and the instrumental factor discussed above. Short period pulsars can be timed more precisely than longer period objects with equal flux density owing to the narrower pulse width in the shorter period pulsars.

In standard practice the constant time interval in the template from its first sample to a selected fiducial point of the pulse, such as the peak, is added to the TOA. Also, the time interval in a many period integration from the first sample of the first pulse down to the first sample of a pulse in the middle of the integration is added to the TOA. This mid-integration correction reduces the sensitivity of the TOA to systematic drifts that will occur if there is an error in the assumed model. An additional correction of the TOA for propagation from the sampling hardware back to the fiducial point that corresponds to the observatory coordinates in the Earth reference frame should be done, but is not done in present work. This latter effect is essential only if absolute TOAs are compared between observatories.

4.2 Time Correction

Anyone who has some familiarity with astronomical observations knows that there are many, too many, time scales – sidereal time, universal time, atomic time, Julian Date, Gregorian Date, *et al.* See Jesperson & Hanson (1991) for a general discussion of time and frequency issues. We need to record our space-time pulse arrival event on a uniformly running time scale in the observer's reference frame. This time scale, generically, is atomic time. The standard atomic time scale, International Atomic Time - TAI, is based on an assumed value for the frequency of a transition in the Cesium atom, and a particular, but arbitrary epoch for the origin of the scale. The principal calendar scale is Julian Date which increments the day number after 24 hours of TAI and has no year; the day boundary is at noon. 1991 August 20 noon (in Graz) corresponds to JD 2448489.00.

Observatory clocks keep universal time, a 24-hour scale that is also known as Greenwich Mean Time. Coordinated Universal Time, UTC, is the most accurate universal time scale which is coordinated by Bureau International des Poids et Mesures in Paris (BIPM). UTC is deciminated by various means so that the observatories can keep approximate UTC. The UTC scale now runs at a rate identical to that of TAI. However, the goal of UTC is to assist in predicting the meridian transit of the Sun at any observatory. This means that it is a solar, or Earth rotation, time scale, and must be locked to the somewhat erratic rotation of the Earth. There is an offset in phase, or epoch, between UTC and TAI. This offset is adjusted by leap seconds that are inserted at midnight on December 31, or more often if needed. Observations of the orientation of the Earth are required to determine when leap seconds are needed. As an aside, the Earth's rotation was the basis of civilization's time scale for centuries. After electronic crystal clocks were developed in the 1930's the erratic behavior of the Earth's rotation was observed, and laboratory clocks took over the job of precise time keeping. Sidereal time gives the orientation of the Earth with respect to the stars, and is derived from UTC by a step in epoch called (UT1-UTC), and then a change in rate that reflects rotation with respect to the stars rather than the sun.

There are several means of transfering the epoch of a time scale from a standards laboratory to a remote astronomical observatory. The crudest is the long wavelength radio signals of WWV at 5, 10, 15, 20, and 25 MHz that are broadcast from Fort Collins, CO, and travel large distances around the globe by reflection off the ionosphere. Similar signals are broadcast by the time services of various nations. The accuracy is around 1 ms for distances of a few thousand kilometers. The US maintains a system of 100-kHz transmitters for radio navigation called Loran C. These signals travel as a ground wave and provide microsecond level time transfer for distances up to 1500 km. The most accurate time transfer is now done using the Global Positioning System of satellites. Signals are transmitted in bands between 1300 and 1700 MHz. The Cesium atomic clocks on board on these satellites are monitored from the ground, and a prediction ephemeris is sent back to them to allow real time transfer at the level of 100 ns. Even better transfer can be obtained by real-time tracking of the same satellites at the observatory and at a national standards laboratory – the common view technique. A transfer accuracy of a few ns has been obtained in an experiment where time was transferred between laboratories and satellites in three hops around the globe (Ashby & Allan 1984). Careful applications of relativity theory are required to conduct this experiment in the rotating, non-inertial reference frame of the laboratories.

The time transfer discussed above makes the best atomic time scale available to the remote observatory. This time scale currently is called Terrestrial Dynamical Time, TDT. TDT is computed at the BIPM by averaging over the International Atomic Time scales, TAI(x), as realized at various standards laboratories identified by x. TDT is proper time for an observer at sea level on the Earth, and is not a uniform time scale in the reference frame of the pulsar. In the pulsar frame the Earth clock is affected by the variable gravitational redshift that is the result of the influence of the Sun and the planetary masses in the solar system, and by the variable time dilation along the more or less elliptical orbit of the Earth. The ratio of the rate of a TDT clock (t_i) to the rate of a clock in the solar system barycenter (τ) in the absence of any solar system bodies is given by

$$\frac{dt_i}{d\tau} = \Sigma_p \frac{Gm_p}{r_{ip}c^2} + (\frac{v_i}{c})^2, \tag{5}$$

where m_p is the mass of solar system body p, r_{ip} is the distance from the clock to the body, and v_i is the velocity of the clock. These two effects, which are the result of the equivalence principle and special relativity, respectively, are inseparable in their time signature. The principle periodic term in the difference between the two clocks can be approximated by an analysis of a clock in a pure elliptical orbit around the Sun,

$$\Delta t_R \equiv t - \tau = 1.66 \text{ ms } [(1 - \frac{1}{8}e^2)\sin M + \frac{1}{2}e\sin 2M + \frac{3}{8}e^2\sin 3M], \tag{6}$$

where e is the eccentricity and M is the mean anomaly (roughly phase within year) of the Earth's orbit (Clemence & Szebehely 1967; Blandford & Teukolsky 1976).

If one conducts pulsar timing observations with submicrosecond precision, then this relativistic clock correction must be done much better than is possible by simple analytic expressions. Several groups who develop detailed models of the solar system dynamics have provided algorithms and ephemerides that have sufficient precision. The barycentric time scale τ is not what is obtained by these means. By international agreement the corrected time scale is declared to have only the periodic terms in $dt/d\tau$ removed. This modified scale is called Barycentric Dynamical Time, TDB.

In summary UTC(observatory) is converted to UTC(standards laboratory) by a satellite time transfer technique, and then to UTC(BIPM) according to an algorithm. UTC(BIPM) is, apart from a constant that is incremented slowly by integer seconds, atomic time TAI(BIPM). TAI(BIPM) is converted to proper time on the geoid TDT by a second algorithm, and then to coordinate time for solar system dynamics TDB through use of a relativistic clock transformation ephemeris.

4.3 Space Correction

The space correction relocates the arrival time from the observatory to the origin of a local inertial reference frame, the solar system barycenter. The Earth moves in a nearly circular orbit of radius 1.5×10^{13} cm, or 500 light seconds. The accuracy of the transfer of arrival that is required for the highest precision pulsar timing is 0.1 μs which is 2×10^{-10} smaller than the orbit size.

The space correction is typically effected in two steps (Fig. 8). First the arrival time is transferred from the observatory to the Earth center using a simple model for the Earth's orientation in an inertial frame and the star's celestial coordinates. This step is similar to the calculations that are required to perform Very Long Baseline Interferometry (VLBI) at radio wavelengths, although the accuracy requirements for VLBI are much more stringent.

The transformation from the Earth center to the solar system barycenter requires an accurate ephemeris, or prediction table, of the Earth's position as a function of date. There are two principal groups involved in solar system dynamics research and ephemeris production. They assemble all measurements of solar system bodies – optical eclipses and occultations, radar range measurements, planetary probe results – and feed this data into a comprehensive model fitting program. Masses and orbital elements are adjusted to obtain the best fit to the data. The resulting model of the solar system can then be extrapolated into the future for uses such as spacecraft tracking, observations of planets and other bodies, and pulsar timing. The ephemeris computer file typically consists of values evaluated every four days. Intermediate results are obtained by interpolation.

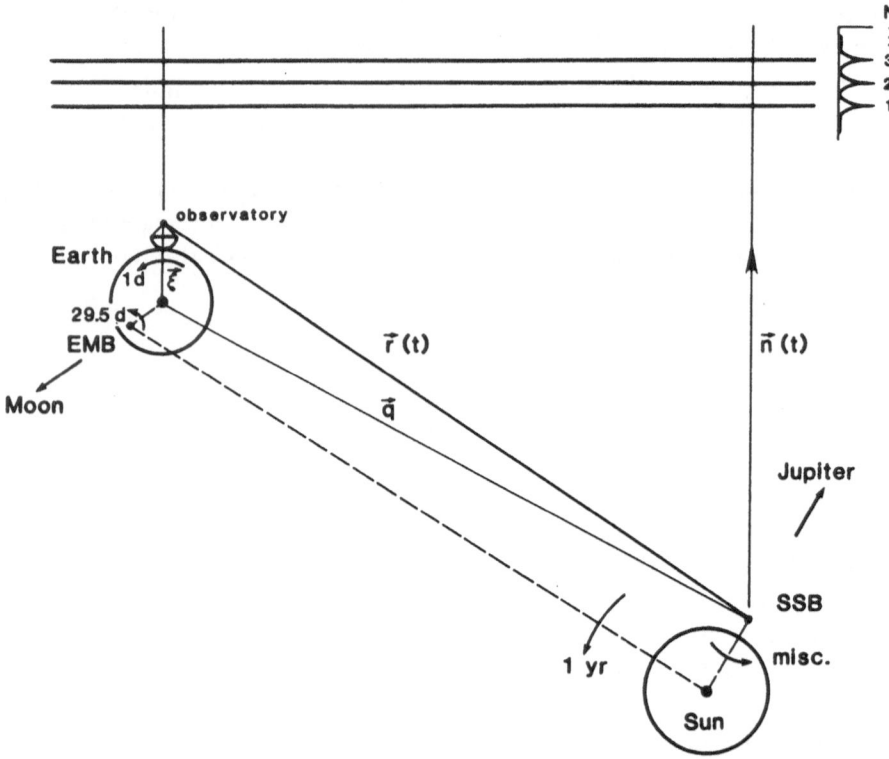

Fig. 8. Schematic of space transformation from the observatory to the solar system barycenter. While the relative sizes and locations of objects are inaccurate, the locations of the Earth-Moon mass center with respect to the Earth size, and the solar system barycenter with respect to the Sun size are nearly correct. Pulsar signals are represented as "pulse" fronts numbered by pulse number. (From Backer and Hellings 1986).

4.4 Pulsar Parameter Estimation

In the above sections I have stepped through the various terms that are removed in making the space-time transformation from the observer's reference frame to that of the neutron star. This provides a set of times of emission of pulses that can be matched to a model of the rotation of the star. The process, however, is not as linear as just implied. The celestial coordinates of the pulsar are required to remove the effects of the Earth's motion, and these, at the required level of precision, must be derived from the data.

The DM must also be derived from the data, and the multi-frequency observations that allow a DM determination may not be simultaneous. What must be done is to iterate from starting parameters that allow connection of data without rotation number ambiguities over limited intervals of time, and then work toward model parameters that are without ambiguity over longer intervals.

The rotation of the star, if it acts as a simple clock, can be described by a polynomial in pulse number,

$$t_N = t_0 + N\ P_0 + N^2\ P_0\ \dot{P}_0/2 + \ ...,\tag{13}$$

where $t_0, P_0,$ and $\dot{P}_0$ are the initial epoch, period and period derivative, and t_N is the epoch of the Nth pulse; epoch and arrival time are synonymous. Higher derivatives may also be present. However, if the star's spindown, which is described by $\dot{P}_0$, is the result of magnetic dipole radiation, then the next derivative can be estimated as

$$\ddot{P}_M = -\frac{\dot{P}^2}{P}.\tag{14}$$

In most stars this higher derivative produces unmeasureable effects. The rotation parameters are the strongest effects since they arise on the shortest time scales. After obtaining parameters that are suitable first for hours then for days, one can begin to include the astrometric parameters that have a time scale of one year. Observations of binary pulsars are further complicated by the required estimation of Keplerian parameters of the binary orbit as well (§5.2).

Figure 9 illustrates some of the steps that I have just discussed. In all cases the data displayed are timing residuals R – the difference between observations and a preliminary model. Note that in all cases the residuals are much less than the star's rotation period. This is extremely important since all we know is pulse phase, modulo one period. If the residuals are not small with respect to one period, then the model will have the pulse number, N, in error, and the residuals will not provide the correct data for new parameter estimation. At first data with imperfect estimates of the spin parameters are shown. Next data that extends beyond the region of a previous fit are shown. This illustrates the poor extrapolation capability of a fit over a restricted range; this is expected and is no methodological shortcoming. Finally data for an entire year are displayed that has a small position error, and for several years that shows the effects of proper motion.

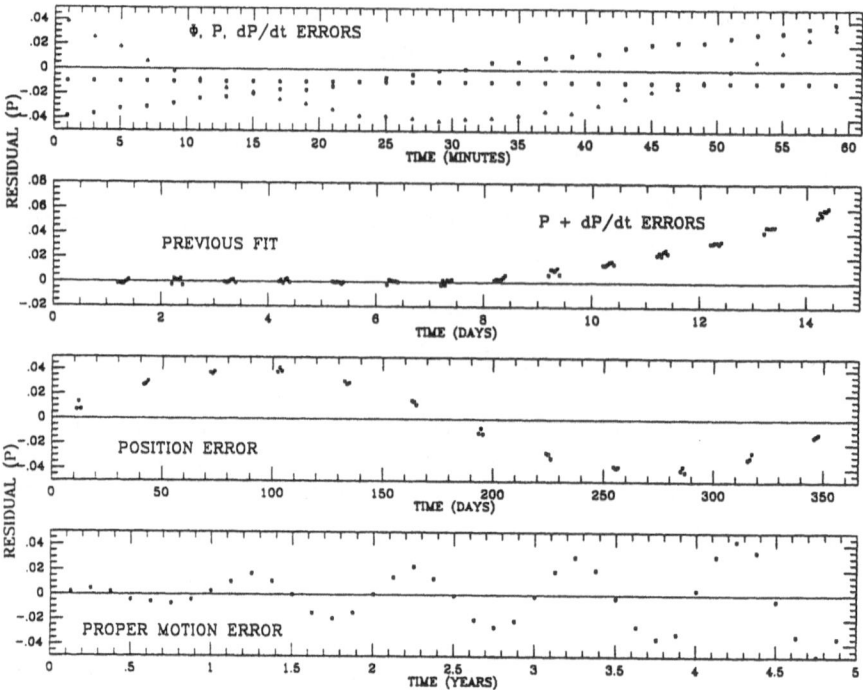

Fig. 9. Pulsar parameter estimation steps. (a) Illustration of residuals from model for errors in phase (filled squares), period (open squares) and period derivative (filled triangles) with data sampled every two minutes. (b) Demonstration of imperfect extrapolation of model that was fit over 7 days (4.5 hours each day) to next 7 days. (c) Sinusoidal residuals with one year period from position error. (d) Growing residuals from error in proper motion.

4.5 Rotation Noise

When proper account is taken of all the effects mentioned above, then one might expect the residuals to display the character of white noise which would result from uncorrelated errors. However, in many stars, particularly those younger than 10^5 y, the residuals are distinctly not white. Terms like "white" are used to indicate a frequency indendent spectrum of fluctuation, and "red" to indicate a spectrum where the low frequencies (*i.e.*, red) have larger amplitude variations then those at high frequencies. Qualitatively, a red spectrum of variations will appear as a slow meandering of arrival time, or phase, with an amplitude of a fraction of the pulsar period, 1-10%, and occasional sudden events, called glitches.

The glitches are most dramatic. In a matter of seconds to minutes the pulse period may jump by parts per million. The period event is accompanied by a similarly sudden increase in the spindown rate. The net effect of the subsequent evolution is that the excess spin is nearly removed over an interval of weeks to months, and the spindown rate returns to its preglitch magnitude. There is an extensive literature built up around these glitches that discusses the evolution of the internal structure of neutron stars from a condensed matter viewpoint. Briefly, the standard model here is that a thin, inner crust is normally decoupled from the rotation we observe. This inner crust is differential rotating with a specific angular momentum in excess of that in the rest of the star. Some catastrophe occurs that leads to an abrupt coupling of this inner crust to the rest of the star. Its extra momentum leads to the spinup. However, the star now carries an extra moment of inertia, and will spindown more rapidly. Eventually the inner crust will decouple again. The mechanism of coupling and decoupling has to do with the dynamics of the quantized rotation vortices which carry the angular momentum in neutron star interiors and the lattice structure of the crust (*e.g.*, see Sauls 1989).

The random noise behavior in pulsar timing residuals is also connected to the evolution of the neutron star figure. The neutron star crust has sufficient strength so that the crust can carry the equilibium figure of the star in a prior state of rotation – in other words, the equatorial bulge of the crust will be larger than one would calculate for an equilibrium, self-gravitating fluid. Eventually gravitational and centrifugal forces will need to balance at the appropriate radius. The transition from the old figure to the new may be marked by a glitch, or by slow meandering.

Investigators have quantified the degree of "red" rotation noise by a logarithmic activity index which is the variance of the timing residuals measured at a selected time interval relative to the variance in the somewhat noisy Crab pulsar. A background level of white noise, which is assumed to result from instrument sensitivity and not the pulsar, is removed. Activity indices vary from +1 for a few pulsars with variances ten times larger than the Crab's, down to upper limits of −4 for some some of the millisecond pulsars. The millisecond pulsars thus have the exciting combination of both short periods, which means short pulse widths and therefore high timing precision capability, and low activity which means they are very stable clocks.

4.6 Astrometry

In §4.3 the correction from the observatory arrival time to the arrival time at the solar system barycenter was described. This requires knowledge of the celestial coordinates of the pulsar in the reference frame of the ephemeris. Microsecond timing precision requires angular precision corresponding to the ratio of a microsecond to the light travel time for 1 AU, or 2 nanoradians which is $0.0004''$. At this level of precision the orientation of the reference frame of the ephemerides, which is based on ecliptic coordinates, and that of the extragalactic sky, which is based on VLBI observations and therefore is associated with the orientation of the Earth, is uncertain. Pulsar observations are playing a role in the tying together of these two reference frames.

A pulsar at 1 kpc moving with a transverse velocity of 1 km s^{-1} travels one nanoradian in one year. Pulsar velocities are much larger. This means that proper motion measurements are readily made by analysis of pulsar timing data. Unfortunately there is no way that we can measure the radial velocity component. The irregularities found in some pulsar timing data wreak havoc with these measurements, and have limited the determination of proper motions to the quieter pulsars. Fomalont *et al.* (1992) have made a detailed comparison between proper motions determined by timing and by radio interferometry.

The timing precision obtained with millisecond pulsars is so good that determination of a trigonometric parallax is possible. The pulses from a source at finite distance pass through the solar system along curved fronts, while those at infinite distance arrive on flat fronts. The curvature can be detected as early arrival times twice per year. Ryba & Taylor (1991) present a parallax of 1.2 mas for PSR 1855+09.

5 Binary, Millisecond and Globular Cluster Pulsars

5.1 Origin and Evolution

Most stars one sees in the night sky are members of a binary system. If supernova events do not unbind binary systems, then a significant fraction of neutron stars will also be members of binary systems. The Xray binaries provide firm evidence that at least the first supernova event in a binary, *i.e.*, the first neutron star production event, does not always unbind binaries. Hulse and Taylor (1974) discovered the first binary pulsar, PSR 1913+16. In §5.2 I will show that pulse arrival time measurements of a binary pulsar give us the mass function that allows an estimate of the binary companion mass if one assumes the pulsar mass is 1.4 M$_\odot$. For PSR 1913+16 and a few others the companion mass is a similar 1.4 M$_\odot$. These are the high mass binary pulsars which have progenitors first in the high mass Xray binaries, and earlier in high mass OB binary star systems. They evidently have survived two violent neutron star formation events. However their numbers are small, and the conclusion is that survival is improbable. The spindown lifetimes of these pulsars are an order of magnitude larger than those of most isolated pulsars. The production rate, which is obtained from the number divided by the lifetime, is then much lower than that of ordinary pulsars.

In 1982 an isolated pulsar PSR 1937+21, which is spinning at 642 Hz, was discovered (Backer *et al.* 1982). The evolution of this millisecond period pulsar is a great puzzle. However shortly after this discovery a second millisecond pulsar was found PSR 1953+29 (Boriakoff *et al.* 1983). This object was in orbit around an inferred low mass companion which presumably is a white dwarf. A swarm of models were generated at this time with the leading idea being that these millisecond pulsars had their progenitors in the low mass Xray binaries which, in turn, arose from main sequence systems with one high mass and one low mass star. The division between high mass and low mass is around 8 $M_\odot$, which is the nominal dividing line between stars that form neutron stars and those that evolve to white dwarfs. The models do not answer the question of how such a system could evolve to an isolated millisecond pulsar like PSR 1937+21. One distinguishing characteristic of these very short period pulsars is that their spindown rates are many orders of magnitude smaller than those of "normal" pulsars. Their small spindown rates suggest low magnetic fields (2) and low magnetic fields are required in most models of the low-mass Xray binaries. The age of these systems places a firm limit on how much decay of pulsar magnetic fields can take place. Certainly the fields cannot decay exponentially with a time scale of 10^{6-7} years as many authors had suggested in the 1970's. The role of the magnetic field is critical to the binary formation models. A low magnetic field on an old neutron star is required to allow mass accretion into the neutron star magnetosphere at small enough radii so that an accretion torque can spin up the old, and presumably slowly rotating, neutron star to a millisecond period. The process is limited along a line shown in Fig. 3.

The next episode in pulsar discovery in the 1980's was the detection of pulsars in globular clusters. The clusters are self gravitating systems of 10^{5-6} stars. They form a spherical population hovering around the center of our galaxy and were created very early in the evolution of the galaxy. As a self gravitating system, they have negative heat capacity and will collapse endlessly unless there is a source of heat to stir up the stars. One source of heat is the formation of binary systems in the core where the density is a million times higher than that in the solar neighborhood. Imagine what the night sky would look like there! One type of binary will be a neutron star and a low mass companion. These binary systems are made harder by further collisions, and eventually will evolve first into Xray binaries and then into millisecond pulsar binaries. The tree of possibilities is much richer than just outlined. In the cluster 47 Tucanae there are no less than 10 millisecond pulsars.

Fruchter *et al.* (1988) reported the discovery of a millisecond binary pulsar PSR 1957+20 that is eclipsed once per orbit by a halo of material surrounding the companion. This system had the properties of a system that Ruderman *et al.* (1989) had considered theoretically – that of a close binary where a portion of the rotational energy lost by the pulsar is deposited on the companion and ablates material from the companion's surface. They were considering scenarios by which low mass binary pulsars could eventually become single. A second system undergoing eclipses was found in the globular cluster Terzan 5. In this second case the "weather" in the ablation and subsequent wind formation processes is such that occasionally eclipses last for several orbits. These objects are very exciting. They have whittled their companion down to just 0.02 $M_\odot$, and provide new constraints on the form of energy and momentum loss by the pulsar as a consequence of its rotation. These systems seem to fill the gap between binary millisecond pulsars (and low mass Xray binaries) and solitary millisecond pulsars. The gap is further filled by the recent (post-school) firm detection of a pulsar with planetary companions (§5.5).

Scientists are often swayed into believing, and stretching, a strong, comprehensive idea or set of ideas. What you have just read may well be such a case. There is still no solution to the formation of an isolated, low field millisecond pulsar in the binary evolution model described. Now we have an object that, while not isolated, only has planetary mass companions. Other ideas may need strong consideration. One is that single millisecond period pulsars can be formed *ab initio* with low magnetic fields. Another is that single pulsars may be formed from the coalescense of either white dwarf or neutron star binaries. Neutron stars can also form in binaries where both systems were low mass as normal stars. The formation of one white dwarf can be followed by mass accretion that supercedes the limit for a white dwarf, 1.4 $M_\odot$, and leads to neutron star formation. The collapse to a neutron star may be accompanied by a residual disk within which planetary mass coagulation (§5.5) can occur in a manner similar to that proposed for the solar nebula.

In conclusion let me mention that a neutron star spinning 642 times per second has an equatorial speed of 0.15 c. At rotation rates only slightly larger, and here slightly is dependent quantitatively on the equation of state for the stellar interior, the star will be unstable to the growth of a triaxial figure that would emit gravitational radiation and tend to slow the star down. The fastest neutron stars may be very close to their rotational limit. Future pulsar surveys with good sensitivity for 1.0 millisecond rotation periods are required to support this contention.

Table 2. Binary, Millisecond and Globular Cluster Pulsars

BINARY, MILLISECOND AND GLOBULAR CLUSTER PULSARS - 92sep25

NAME	ROTATION PERIOD (ms)	PERIOD DERIVATIVE (10**-18 s/s)	DISPERSION MEASURE (pc/cm**-3)	ORBITAL PERIOD (d)	MASS FUNCTION (Msun)	ECCEN-TRICITY	ASSOCIATION	REFERENCE
0021-72C	5.757	-40+/-50	24.4	--	--	--	47 TUC (NGC 104)	89 IAU 4892
0021-72D	5.357	..	24.7	--	--	--	47 TUC (NGC 104)	90 workshop
0021-72E	3.536	..	24.2	-2	..	..	47 TUC (NGC 104)	90 NATO ASI
0021-72F	2.624	..	24.4	--	..	--	47 TUC (NGC 104)	90 NATO ASI
0021-72G	4.040	..	24.2	--	..	--	47 TUC (NGC 104)	90 NATO ASI
0021-72H	3.211	..	24.3	-1?	..	..	47 TUC (NGC 104)	90 NATO ASI
0021-72I	3.485*	..	23.7	-1?	..	..	47 TUC (NGC 104)	90 NATO ASI
0021-72J	2.101	..	24.6	0.12	..	..	47 TUC (NGC 104)	90 NATO ASI
0021-72K	1.786	..	24.9	..	..	..	47 TUC (NGC 104)	90 NATO ASI
0021-72L	4.346	..	24.5	..	..	..	47 TUC (NGC 104)	90 NATO ASI
0655+64	195.671	0.68	8.7	1.03	0.0712	<.00005	field	82 APJ 253 L57
0820+02	864.873	103.9	23.7	1232.47	0.0030	0.01187	field	80 APJ 236 L25
1257+12	6.218	..	10.16	--	--	--	field	90 IAU 5073
1310+18A	33.163	..	25.0	255.84	0.0098	<0.01	M53	89 IAU 4853
1516+02A	5.553	..	29.5	--	--	--	M5	89 IAU 4880
1516+02B	7.947	..	29.5	6.85	0.00065	0.126	M5	89 IAU 4880
1534+12	37.904	..	11.6	0.421	0.32	0.27	field	90 IAU 5073
1620-26A	11.076	0.82	62.9	191.44	0.0080	0.02532	M4 (NGC 6121)	88 NAT 332 45
1639+36A	10.378	<0.045	30.4	--	--	--	M13 (NGC 6205)	89 IAU 4819
1639+36B	3.5	..	30.4	1.26	0.008	--	M13 (NGC 6205)	
1745-20	288.60	..	210	--	--	--	?NGC6440	89 IAU 4905
1744-24A	11.563	-.05+/-0.09	242.2	.0708	0.00032	<0.003	TER 5	90 IAU 4974
1802-07A	23.10	..	187	2.62	0.0097	0.22	NGC 6539	90 IAU 5013
1820-11	279.828	1378.	428.4	357.76	0.068	0.79462	field	89 preprint NAT
1820-30A	5.440	..	86.0+/-0.3	--	--	--	NGC 6624	90 IAU 4988
1820-30B	378.59	..	86+/-3	--	--	--	NGC 6624	90 IAU 4988
1821-24A	3.054	1.62	120	--	--	--	M28 (NGC 6626)	87 IAU 4401
1831-00	520.947	14.3	88.3	1.81	1.2E-4	0.0001	field	87 NAT 328 399
1855+09	5.362	0.017	13.3	12.33	0.0056	0.00002	field	86 NAT 322 714
1908+00A	3.6	..	200	--	--	--	NGC 6760	90 IAU 5010
1913+16	59.030	8.64	171.6	0.32	0.1323	0.61713	field	75 APJ 201 L55
1937+21	1.558	0.11	71.0	--	--	--	4C 21.53	82 NAT
1953+29	6.133	0.030	104.6	117.35	0.0024	0.00033	2CG 095	83 NAT
1957+20	1.607	0.016	29.1	0.38	5.2E-6	0.000	field	88 NAT 333 237
2127+11A	110.665	-20.0	67.25	--	--	--	M15 (NGC 7078)	89 NAT 337 531
2127+11B	56.133	+8.8	67.25	--	--	--	M15	89 IAU 4762
2127+11C	30.529	+4.99	67.14+/-0.02	0.335	0.153	0.6814	M15	89 IAU 4772
2127+11D	4.803	-1.1	67.	--	--	--	M15	90 workshop
2127+11E	4.651	+0.18	67.	--	--	--	M15	90 workshop
2303+46	1066.371	569.3	61.	12.340	0.2455	0.65838	field	85 NAT 317 787

5.2 Keplerian Binary Pulsar Timing

If the neutron star under observation is a member of a binary system, then one must transform the arrival time from the barycenter of the binary system to the neutron star before proceeding with analysis of the spin parameters of the star. A classical description of this transformation is based on Kepler's laws of motion. Figure 10 is an aid to understanding the geometry of a distant binary system. The center of the sphere is both the center of mass of the binary, and one focus of the pulsar's elliptical orbit. The plane of the orbit is inclined by an angle i with respect to the plane of the sky. The intersection of the two planes defines the line of nodes, which has a position angle on the sky χ. The periastron point in the orbit, the point of closest approach of the two bodies, is at an angle ω measured from the line of nodes. The masses of the pulsar and its companion are m_1 and m_2, respectively; $m \equiv m_1 + m_2$.

The location of the pulsar in the orbit with respect to the line through the binary barycenter and the periastron point is given by the true anomaly ν. This line, which is along the major axis of the ellipse, is the line of apsides. The mean anomaly is a uniformly increasing phase angle $M = \Omega_b(T - T_o)$, where T_o is the epoch of a periastron passage, $\Omega_b = 2\pi/P_b$, and P_b is the binary orbital period. The eccentric anomaly E is an angle defined on a circle that just encompasses the orbital ellipse. The origin of E is the line of apsides, and the projection of the radius of the circle at angle E onto the line of apsides is identical to the similar projection of the pulsar location on the ellipse. Kepler's equation relates the mean and eccentric anomalies:

$$M = \Omega_b(T - T_o) = E - e \sin E, \qquad (28)$$

where e is the eccentricity of the orbit. Further trigonometry relates the eccentric anomaly E to the true anomaly ν, so that the plane polar coordinates (R_1, ν) of the pulsar relative to the binary barycenter are determined by

$$R_1 = a_1(1 - e \cos E), \quad \text{and} \qquad (29)$$

$$\sin \nu = \frac{(1 - e^2)^{0.5} \sin E}{1 - e \cos E}, \quad \text{and} \quad \cos \nu = \frac{\cos E - e}{1 - e \cos E}. \qquad (30)$$

The component of this motion along the line of sight is the one that affects the arrival time,

$$\delta t_b = a_1 \sin i(1 - e \cos E) \sin(\nu + \omega). \qquad (31)$$

For an approximation one can substitute M into the above equation for both ν and E to obtain an idea about the variations of δt_b. There are five Keplerian parameters that must be determined for an elliptical orbit:

$P_b, a_1 \sin i, e, \omega$, and T_o. These provide a constraint on the mass of the companion which is known as the mass function:

$$f(m_1) = \Omega_b^2 (a_1 \sin i)^3 / G = (m_2 \sin i)^3 / m^2. \tag{32}$$

The timing of a binary must proceed from an initial determination of its orbit using Doppler data alone, apparent periods, to a more precise orbit which is obtained by phase connecting the data which uses the above formulation for δt_b.

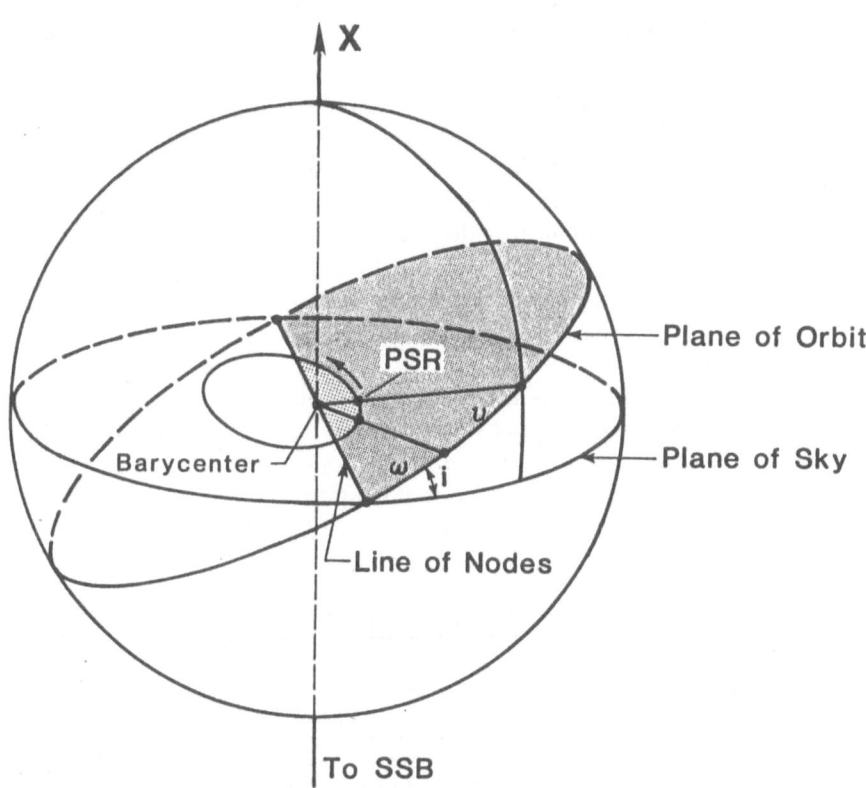

Fig. 10. Schematic of orbital motion of binary pulsar about the binary barycenter relative to the line of sight and plane of sky. (From Backer and Hellings 1986.)

5.3 Relativistic Binary Pulsars

A pulsar in a short orbital period binary with a massive companion will require post-Keplerian corrections to its timing model. The largest effect is the secular advance of the line of apsides in an inertial reference frame. The corresponding effect in the solar system is the advance of perhelion of the planet Mercury. A second relativistic effect is the counterpart of the relativistic clock correction for an Earth clock which was presented in §4.2. The neutron star rotation is a clock, and we must transform from coordinate time in the binary barycenter to proper time in the orbital frame of the neutron star. These relativistic parameters depend on orbital phase with independent signatures so that the masses and the inclination of the system can be determined. A fundamental development in pulsar research is the detection of the orbital period decay in PSR 1913+16 that is most likely the result of a back reaction to gravitational radiation by the binary mass system (Taylor *et al.* 1979; Damour & Taylor 1991). The over determination of the elements of the system that is provided by the three effects mentioned above allow one to actually predict the gravitational wave orbital decay and hence compare the observations to theory. Binary pulsars provide the only experimental evidence for the existence of gravitational radiation.

The light travel time of signals from the pulsar through the gravitational potential of its companion provides a further effect, just as it does for propagation past our Sun, which is detectable in several binary pulsar systems. Damour & Taylor (1992) have developed a post-Keplerian parameterization of the weak and strong field effects that are, or may be, detected in binary pulsar signals.

The most prominent relativistic effect is apsidal motion of the orbit. This is so large, and the measurements of PSR 1913+16 by Taylor and his colleagues have been so precise, that proper modeling of the data must take into account that the motion is not uniform along the elliptical orbit. The time averaged rate of apsidal motion can be calculated from the simple Einsteinian formula:

$$\dot{\omega} = \frac{3(Gm)^{2/3}(\Omega_b)^{5/3}}{(1-e^2)c^2}, \quad \text{or} \tag{33}$$

$$\dot{\omega} = 0.20° \ y^{-1} \left(\frac{P_b}{\text{days}}\right)^{-5/3} \left(\frac{m}{M_\odot}\right)^{2/3} (1-e^2)^{-1}. \tag{34}$$

Parameters for the three relativistic binaries, PSR 1534+12, PSR 1913+16, and PSR 2127+11C, are given, or can be estimated, from the data in Table 2.

The relativistic clock transformation leads to a periodic correction Δt_{Rb} which is similar to that discussed in §2.3 for the Earth clock.

$$\Delta t_{Rb} = \frac{G^{2/3}}{c^2} \frac{m_2(1+m_2/m)}{(m\Omega_b)^{1/3}} e \sin E, \quad \text{or} \tag{35}$$

$$\Delta t_{Rb} = 0.0069 \text{ s } \left(\frac{P_b}{\text{days}}\right)^{1/3} \left(\frac{m_2}{M_\odot}\right) \left(\frac{m}{M_\odot}\right)^{-1/3} \left(1 + \frac{m_2}{m}\right) e \sin E. \quad (36)$$

The fourth relativistic effect is the decay of the orbit that results from the back reaction to emission of gravitational radiation. The field equations of general relativity for a dynamic system reduce to a wave equation for the perturbation of the spacetime metric about its flat space form in the limit of small perturbations, or weak fields and slow speeds. While the perturbation of the spacetime metric in the immediate vicinity of a neutron star is not small (*i.e.*, does produce strong field effects), the binary system is much larger and weak field calculations are probably sufficient. The question of constraints on strong field parameters from binary pulsar timing observations is now under consideration (Damour & Taylor 1992; Taylor *et al.* 1991). The orbital decay rate $\dot{\Omega}_b$ and time scale $T_b \equiv \frac{\Omega_b}{\dot{\Omega}_b}$ for two point masses in General Relativity is given by

$$\dot{\Omega}_b = \frac{96}{5} \frac{G^{5/3}}{c^5} \frac{m_1 m_2}{m^{1/3}} \frac{(1 + \frac{73}{24}e^2 + \frac{37}{96}e^4)}{(1 - e^2)^{7/2}} \Omega_b^{11/3}. \quad (37)$$

$$T_b = 1.3 \times 10^{11} \text{ y } \left(\frac{m_1}{M_\odot}\right)^{-1} \left(\frac{m_2}{M_\odot}\right)^{-2/3} \left(1 + \frac{m_1}{m_2}\right)^{1/3} \left(\frac{P_b}{\text{days}}\right)^{8/3} f(e). \quad (38)$$

In (38) the two masses are taken to be 1.4 $M_\odot$ neutron stars and $f(e)$ is the ellipticity factor from (37). For the dual neutron star binaries with orbital periods of 0.3 days and modest eccentricities gravitational radiation will lead to a coalescence, or explosion, of the two stars on a time scale of 100 million years. This means that there are many events like this in the lifetime of our galaxy, and that these are moderately common events in the manifold of galaxies.

5.4 Globular Cluster Pulsars

One fascinating result from the cluster pulsars is their use as probes of the gravitational field in cluster cores. The apparent period of a pulsar is affected by its space motion. Motions in clusters have small velocities of a few tens of km s^{-1}, and these Doppler shifts are inseparable from the period measurement. The acceleration of a pulsar would similarly not be detectable as it contributes to the period derivative. However for several pulsars now the period derivative is negative, unlike any noncluster pulsars. A direct inference is that acceleration by the cluster potential which results from the enclosed mass, and perhaps a contribution from nearest neighbors, has reversed the sign of the apparent effect. For the M15A pulsar this requires an acceleration of 6×10^{-5} cm s^{-2} – roughly akin to accelerating uniformly from our school into downtown Graz in 24 hours!

During the past year the uncertainties in the timing parameters of the 11-ms pulsar in M4, PSR 1620-26, have been resolved – it now is clear that there is a large period second derivative. The probable interpretation of this effect is that this pulsar, which is in a 191 orbit with a low mass companion, is encountering another object as it moves through the cluster. While this is a straightforward explanation, one can predict the rate of encounters of a given accelaration, or acceleration derivative, from the presumably known quantities of cluster density and velocity dispersion. In this case the observed effect is many orders of magnitude larger than expected – a riddle to be solved in future studies. An alternative explanation is the presence of a third bound object in the system. Its orbital period would need to be sufficiently large so that all we detect is the cubic phase residual which is what a period second derivative produces. Furthermore its orbital period would need to be large to maintain stability – three body systems with commensurate orbital periods are unstable. However a weakly bound third object can be stripped easily from the system as it moves around in the cluster encountering other stars. The riddle remains.

5.5 Planets around Pulsars

At the time of these lectures Bailes *et al.* had just announced the detection of a perturbation of the arrival times of PSR 1829-10 that indicated the presence of a 6×10^{28} g companion in a 6-month orbit. The key word here is "6-month orbit". By 1992 January the authors had retracted this stunning result. They had made a serious error in data analysis which is worth reviewing for its lesson (and not to make fun of the authors). Pulsar model parameter fitting is an iterative process as outlined in §4. Improved model parameters, if they are valid, will always predict arrival times better than former ones. The method of determining model parameter *updates* is often an approximate one where intrinsically nonlinear functions are linearized by differentiation, and perhaps simplified, to form the equation of condition. The fit for position components is an example. The motion of the Earth is complicated, but is approximately an annual sinusoid in ecliptic longitude. This approximation is used in the determination of improved coordinates of a pulsar once there is no period ambiguity in the pulsar model over intervals of one year, or more. The approximation is satisfactory, *provided* that one iterates until the model fitting no longer updates the parameters. At that time, it does not matter at what level the fit model approximates the correct equation of condition – the data agree with the model applied exactly. Bailes *et al.* had the misfortune of starting with a large position error. As I understand their process, they used an approximate model for adding the position correction they determined into further analysis runs. This left them with a 6-month sinusoidal residual that was easily fit to perturbation of the distant pulsar by a planetary companion.

The story ends on an upbeat note. Wolszczan & Frail (1992) reported the detection of a *pair* of planets around the 6-ms pulsar 1257+12. The amplitude of the perturbation is around 1 ms for each and the periods are nearly in 3:2 resonance, 66.6 and 98.2 days. Comparison of their model with our data confirms the presence of these periodic residuals from a simple spin model (Backer *et al.* 1992). While various authors have put forward alternative models involving precession of the pulsar neutron star, the proof of the planetary hypothesis will come in a few years when the expected Newtonian orbital perturbations of the two orbits are adequately observed (Rasio *et al.* 1992). Various parameters of the two orbits are expected to "osculate", with a small amplitude and a period of 6 years, owing to the near resonance of the orbital periods.

6 Pulsar Timing Array

In §4.0 the methods of pulsar timing and the parameters required for investigations of isolated and binary pulsars were discussed. The data for each pulsar were treated independently using a model of the Earth's space-time coordinate as a function of epoch that is assumed to be more accurate than the precision of the pulsar data. In this section the limits of this model are discussed, and the influence of a stochastic background of gravitational radiation on pulsar arrival times is presented. The section concludes with a description of the use of an array of precisely timed pulsars to solve for *global* parameters that affect all objects in a correlated manner. The ideas presented here were the subject of a recent workshop – *The Impact of Pulsar Timing on Relativity and Cosmology*, 1990 June, Berkeley (workshop notes are available). Ideas concerning a Pulsar Timing Array are developed by Foster & Backer (1990) and Romani (1989).

6.1 Time Coordinate

In §4.2 I described the transfer of *time* from a standards laboratory to the observatory. We are interested, of course, not in some fuzzy, metaphysical concept like *time*, but in the very concrete concept of *time interval*, or its inverse, frequency. The time coordinate for us is just the numbered sequence of cycles of a frequency. We are comparing rotations of our star to the fundamental oscillations of an atomic quantum transition. There are two concepts used in evalation of the quality of a standards laboratory time scale: accuracy and stability.

The atomic time scale is determined by defining the second, and by identifying an epoch. The second of Terrestrial Time is 9,192,631,770 cycles of the frequency corresponding to a fine structure transition in ground state of the Cesium 133 atom at sea level. (An accurate definition of what we mean by sea level is currently being negotiated.) The epoch when atomic time reads some value, such as zero or 2440000.0 days 0 seconds, is not of particular concern here. What does concern us is the accurate realization of this definition. How do we make a device that produces this countable frequency that is stable from now until forever? The answer is that it is not easy. The measurements of transitions of Cesium atoms are affected by temperature, magnetic fields, vibrations of the room, interactions with container walls, and so on. Many of these effects can shift the frequency. The task of achieving high accuracy is to identify all such systematic influences and remove them.

If the systematic shift in an atomic clock is constant, then the clock can be very stable, but not accurate. If the shift varies with time, then the clock is both unstable and inaccurate. Time averaging can improve the stability of the clock if the shifts average to zero sufficiently rapidly with increased averaging time. A stability analysis always involves a time scale of measurement. Instability mechanisms have a characteristic time scale and amplitude. For example, if the clock is affected by temperature, then one would expect changes at various rates: (a) an air conditioner cycle; (b) daily; (c) yearly. Often the amplitude of instabilities increase with an increase of time scale. It is just tougher to stabilize a clock over long scales, than over short time scales. Allan at the National Institutes of Standards & Technology developed a useful algorithm to characterize oscillator stabilities as a function of measurement interval. Consider a series of time difference measurements, $\delta t_{ij} = t_{ij} - t_{0j}$ of clock i relative to clock 0 with measurement index j. Assume that clock 0 is perfect. Then form the Allan standard deviation:

$$\sigma_i(\tau) = \frac{< [(\delta t_{ij+1} - \delta t_{ij}) - (\delta t_{ij} - \delta t_{ij-1})]^2 >_j^{0.5}}{\sqrt{2}\tau}, \tag{39}$$

where $\tau = t_{j+1} - t_j$ is the interval between measurements. The double difference in this measure removes slopes between clocks since one is interested in stability not accuracy. The measure is also known as the fractional frequency stability since one can write the statistic in terms of fractional frequency differences, $(\delta t_{ij+1} - \delta t_{ij})/\tau$. If the numerator is independent of time, the stability is limited by measurement interval only. This is often true on short time scales where a signal is being measured against a background of white, frequency independent, noise. On longer time scales unwanted, and uncontrollable influences ultimately set a limit on σ. A stability floor is reached. On longer time scales the stability typically degrades even further. This characterization is true for every oscillator ever measured: electronic crystals, atomic clocks, the rotation of the Earth, and my Casio watch (Fig. 11).

The stability floor for the best Cesium atomic time scales is around 10^{-14} on time scales of six months to a year or so (Fig. 11). The PTB scale, from the German standards laboratory, is evaluated as the best in existence.

The BIPM in Paris is responsible for the collection and comparison of time scale data from around the globe. They form two time scales: International Atomic Time (TAI) and Terrestrial Time (TT-BIPM); the former is by agreement with all participants, and the latter is their formulation that they claim is the best scale for use in precise pulsar timing (Guinot and Petit 1991). A stability of 10^{-14} corresponds to a numerator of 0.4 μs in (39), or a typical timing residual of 0.3 μs for time intervals of one year. On longer time scales estimates of stability suggest a modest rise with interval. The atomic time community has made steady progress with accuracy and stability since time keeping began – starting with the Stonehenge era. Keep this limit in mind when we come to discuss the timing of an array of pulsars.

6.2 Space Coordinate

In §4.3 the transformation of the pulse arrival time event at the observatory to the corresponding arrival time event at the solar system barycenter, the local inertial reference frame was discussed. This transformation requires accurate estimation of the barycenter to observatory vector at the time, which is on an inertial clock time scale, of the observatory pulse arrival event. We use an ephemeris of the Earth's motion as a function of time to estimate this vector. The ephemeris is constructed from a detailed model of solar system dynamics that includes the Sun, all planetary systems, minor planets, and a band of matter representing the asteroid belt, and uses an appropriate level of relativistic effects (*e.g.*, Ash *et al.* 1967, Standish 1990). The parameters of this model are adjusted to fit data from *all* observations – from ancient eclipses to recent satellite planet flybys. The table below gives the current estimates of the errors in planetary masses, and the corresponding offset between the estimated barycenter location and the "true" location. These mass errors are large with respect to asteroid and cometary masses, but comparable to masses of the largest minor planets and larger moons. The barycenter will also be shifted by errors in the celestial coordinates of the planets, but a list of these errors is not readily available.

We can take these mass errors and simulate their effect on pulsar timing. The arrival time transformation is effected by a scalar product between the observatory vector, which is of order 500 s – the AU – in length, and a unit vector in the direction of the source. This means that the planetary timing error for pulsar j in direction $\hat{\mathbf{n}}_j$ will be

$$c\delta t_{pj}(t) = \Sigma_p \left(\frac{\Delta m_p}{\mathrm{M}_\odot} \right) \mathbf{r}_p(t) \cdot \hat{\mathbf{n}}_j(t), \qquad (40)$$

where Δm_p and $\mathbf{r}_p$ are the mass error and position, respectively, of planet p. Figure 12a gives a simulation of this error for one pulsar with the Δm_p's taken as those in Table 3.

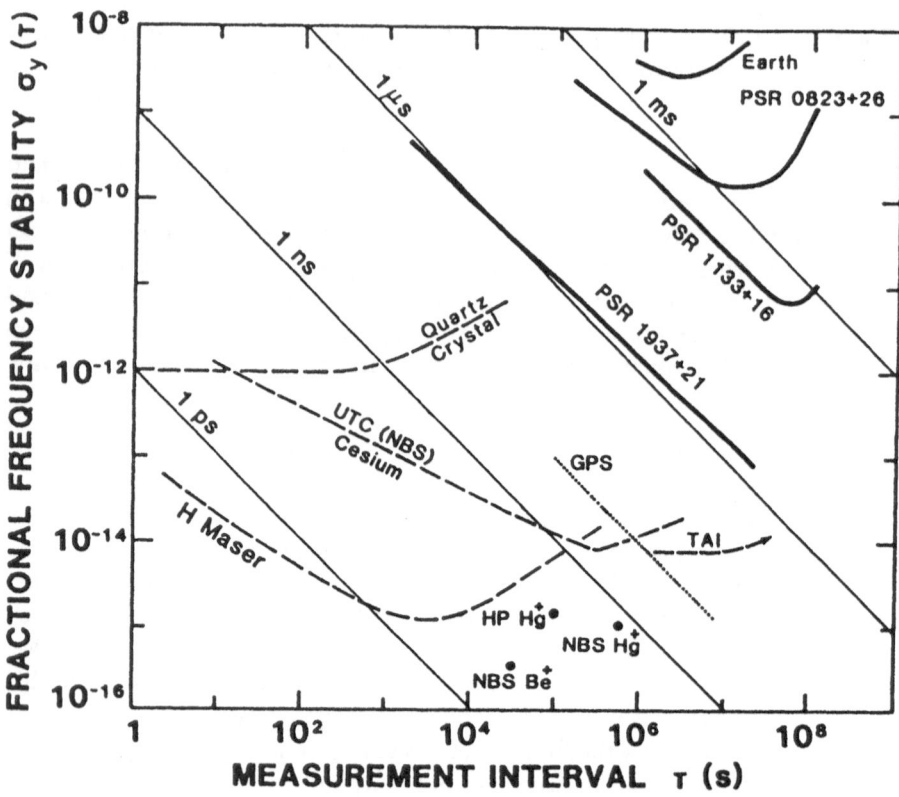

Fig. 11. Fractional frequency stability of both natural (solid) and laboratory oscillators (dashed). Dotted line gives satellite time transfer stability. White noise time errors correspond to lines identified with values and slope of -1. (From Backer & Hellings 1986.)

The simulation in Figure 12a is for 90 years. If we have only 10 years of data, then a large part of the variations, which are dominated by the orbits of Jupiter and Saturn, will be absorbed in fits for the pulsar epoch, spin, and spindown rate. Figure 12b shows such fits in 10-year segments. Figure 12c shows the final residuals that one would expect to see from the mass errors in the current ephemeris calculations. An amplitude of a fraction of one microsecond is typical. The residuals for different pulsars will appear different as a result of the dot product in (40). There will be a slow variation around the sky with a dipole signature in the residuals. The dominant temporal variation will be cubic.

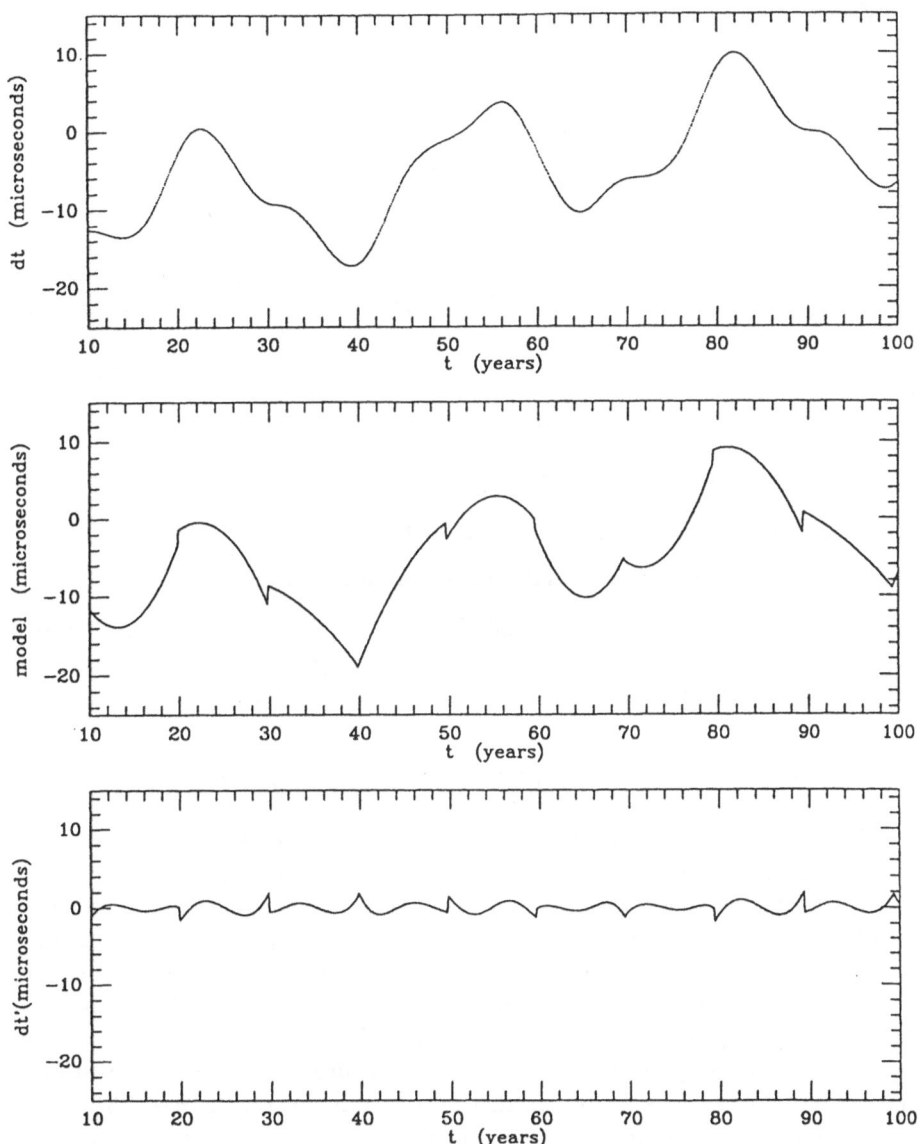

Fig. 12. Simulation of timing errors that result from errors in planetary masses for PSR 0021-72. (a) The errors. (b) The results of fits to errors for pulsar spin parameters in 10-y segments. (c) The final residuals that result from subtracting (b) from (a).

Table 3. Planetary Mass Errors

System	Δm_p $(10^{22}$ gm)	P_p (y)	r_p (AU)	$\Delta r/c$ (ns)	Reference
Mercury	1.4	0.2 5	0.39	1	Mariner
Venus	1.5	0.62	0.72	3	Mariner
Mars	0.2	1.88	1.52	1	Mariner
Jupiter	150	11.9	5.20	1950	Voyager
Saturn	300	29.5	9.52	7140	Voyager
Uranus	15	84.0	19.2	720	Voyager
Neptune	30	165	30.0	2250	Voyager

6.3 Gravitational Wave Background

6.3.1 Influence of Gravitational Radiation on Pulsar Timing

Following the review of Backer and Hellings (1986) and references therein I start with the metric of a spacetime interval as perturbed by gravitational radiation that is specified by its dimensionless strain amplitude, h:

$$ds^2 = c^2 dt^2 - dz^2 - (1 + h \cos 2\psi) dx^2 - (1 - h \cos 2\psi) dy^2 - h \sin 2\psi \, dx \, dy. \quad (41)$$

This expression results from a vacuum solution of Einstein's linearized field equations. The solution corresponds to a temporal spectrum of plane gravitational radiation propagating along the $+z$-axis with amplitude $h(t - z/c)$, where h is an arbitrary function of $(t - z/c)$, and ψ is the constant polarization angle between the principal polarization direction and the x-axis. The Earth and the pulsar are assumed to be at rest. We choose the origin of the coordinate system at the Earth, and place the pulsar in the xz-plane at an angle θ from the propagation direction of the gravitational radiation (the $+z$-axis) and at a distance l from the Earth (Fig. 13). A photon emitted at the pulsar and moving toward the Earth will follow a path in flat space that may be written as

$$x = (l - \sigma) \sin \theta, \qquad y = 0, \qquad z = (l - \sigma) \cos \theta, \qquad (42)$$

where s is the distance parameter. The null geodesic may be written as

$$c^2 dt^2 = (1 + h \cos 2\psi \sin^2 \theta) d\sigma^2. \qquad (43)$$

Inspection shows that the influence of the gravitational radiation appears as an index of refraction, albeit one that changes with space and time, and that the photon propagation will be advanced or retarded compared to

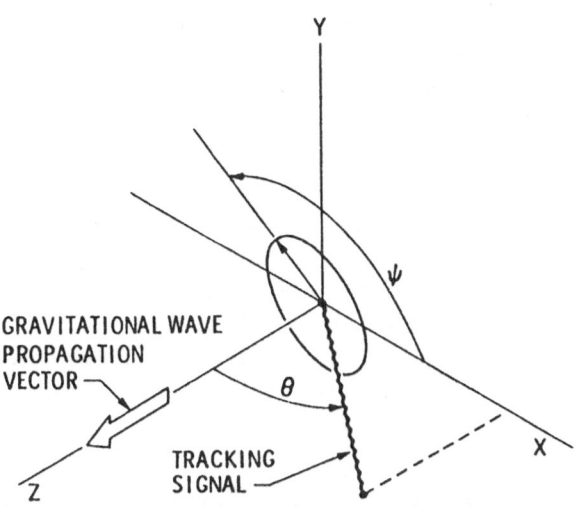

Fig. 13. Geometry of observer, pulsar, and gravitational wave propagation vector. (From Backer & Hellings 1986.)

its vacuum path in flat space. Integration of the above equation from the spacetime event of emission at the pulsar (T, l) to the spacetime event of reception at the telescope $(t, 0)$ gives

$$c(t - T) = l + \frac{1}{2}c(1 - \cos\theta)\cos 2\psi[H(t) - H(T - l\cos\theta/c)], \qquad (44)$$

where $H(u)$ is the indefinite integral of $h(u)$.

Consider the plane wave case where $h(t - z/c) = h_0 \sin(\omega t - kz)$, where $k = \omega/c$. Substitution leads to

$$c(t - T) = l - \frac{1}{2}\frac{h_0 c}{\omega}(1 - \cos\theta)\cos 2\psi[\cos(\omega t) - \cos(\omega t - (\omega l/c)(1 + \cos\theta))].$$
$$(45)$$

Photons from the pulsar arrive early or late according to the product of an amplitude h_0/ω, an angular factor in (θ, ψ), and the sum of two phase factors. The second phase factor is from the incomplete phase of the wave in the vicinity of the emission event, near the pulsar, and the first is from the incomplete phase in the vicinity of the reception event, near the Earth. The first term will be correlated for all pulsars observed, as discussed later, while the second term is uncorrelated amongst the objects for all but very nearby emitters – small $\omega l(\pi - \theta)^2/c$.

The angular factor winds through two ± cycles as the polarization angle moves around the sky transverse to the emission direction. This demonstrates the quadrupolar signature of the radiation. As θ goes to zero the residual delay goes to zero, where the photons travel in the reverse direction of the wave. This demonstrates the transverse nature of gravitational radiation. As θ goes to π the absolute value of the residual delay increases toward 2 from the angular factor in θ, but eventually goes to zero as the two phase factors cancel exactly at π where again the photon path is parallel to the wave. For a wave period of one year and a pulsar distance of 1 kpc, this cancellation will occur only for θ much less than $0.4°$.

Pulsar timing has reached the level of 1 μs for timing several pulsars over intervals of a year. This provides a sensitity to gravitational radiation with dimensionless strains of $h_\circ \simeq 3 \times 10^{-14}$.

6.3.2 The Spectrum of a Stochastic Background of Gravitational Radiation

In the preceding section the effects of a single wave on the pulsar arrival time were considered. In the more general case $h(t)$ will contain a full spectrum of waves. Furthermore there will be waves coming from arbitrary directions. Consider first the spectrum. An arbitrary function $h(t)$ has the Fourier spectrum

$$\tilde{\mathbf{h}}(\omega) = \int_{-\infty}^{\infty} h(t) \exp[jwt]dt. \tag{46}$$

For a stochastic variable one is interested in the power spectral density $P_G(\omega) \equiv |\tilde{\mathbf{h}}(\omega)|^2$. If the statistics of h(t) are stationary, then the integral over the power spectrum is equal to the integral of the variance over the data (Parseval's Theorem).

$$\int_{-\infty}^{\infty} P_G(\omega)d\omega = 2\pi \int_{-\infty}^{\infty} h(t)^2 dt. \tag{47}$$

The units of h are dimensionless, while those of $\tilde{\mathbf{h}}$ are t, and those of P_G are then t^2. This exercise in transforms and units is necessary if one wishes to take a segment of the spectrum and turn it into an effective wave amplitude. In the case of discretely sampled data, the corresponding relation is:

$$\Sigma_i P_{Gi} \frac{\delta\omega}{\delta t_i} = \Sigma_i h_i^2, \tag{48}$$

where δt is the sample spacing, $\delta\omega$ is the resolution of the spectrum, $1/T$, and T is the duration of the data sampled. $P_G/\delta t$ has "density" units of variance per frequency interval in the spectrum.

One assumption about the gravitational wave background is that it is scale invariant with equal energy density in each octave, or decade, of the spectrum. The energy density of a wave with amplitude h and frequency ω is given by

$$\rho_G = \frac{\dot{h}^2}{8\pi G} = \frac{h^2\omega^2}{8\pi G}. \tag{49}$$

A spectrum of waves then leads to a spectrum $\tilde{\rho}_G(\omega)$. If this spectrum contributes equal energy per octave, then $\rho_G(\omega) = \rho_o\omega_o/\omega$. The constant $\rho_o\omega_o$ is referred to as the logarithmic spectrum because one integrates it with $d\ln\omega$ which is dimensionless.

The results of §6.3.1 can now be combined with this spectrum discussion to convert an arrival time precision over a time interval T into a limit on the gravitational wave background logarithmic energy density spectrum. Consider timing residuals with an rms of 1 μs over ten years. If sampled ten times per year, then these would produce a discrete power spectral density with amplitude 2 μs^2 y, i.e., 2 μs^2 per frequency interval of a cycle per year. This corresponds to a power spectral density $P_G/\delta t$ of about 10^{-19} at one cycle per year frequency. The energy density per cycle per year of the spectrum is then 2.5×10^{-26} g cm^{-3}. And finally the energy density per logarithmic interval is 2×10^{-35} g cm^{-3}. This value scales with the amplitude of the residuals squared, which follows directly from (45,49). The value also scales with frequency to the fourth power: two factors from (49), one from the conversion of timing residual into wave amplitude (45) and one from expressing the results in terms of a logarithmic spectrum.

The energy density of background radiations are typically expressed in terms of the density that is required to close the universe, $\rho_c = 2 \times 10^{-29}$ g cm^{-3}:

$$\Omega_G = \frac{\rho_G(\omega)\omega}{\rho_c} \approx 10^{-6}\frac{R(\mu s)^2}{T(y)^4}, \tag{50}$$

where a Hubble constant of 100 km s^{-1} Mpc^{-1} is assumed in the computation of ρ_c, and R is the timing residual defined in §4.4. Present limits on Ω_G from pulsar timing are presented in §6.4.

6.3.3 Sources of the Stochastic Gravitational-Wave Background

In this discussion the focus is on sources from energetic phenomena in the early universe. The origin of the stochastic background can be discussed in a number of ways. One classification of sources is by their temporal character: short bursts from isolated, violent astrophysical events, quasi-periodic sources from rapidly spinning or orbiting bodies, and the broad-band, stochastic background. Carr (1980) summarizes the possible contributions according to whether the source is *primordial*, by which he means having its nature related to the structure of the early universe, or *generated*, by which he means some object forms and then radiates gravitational radiation according to the quadrupole formula. A related categorization simply

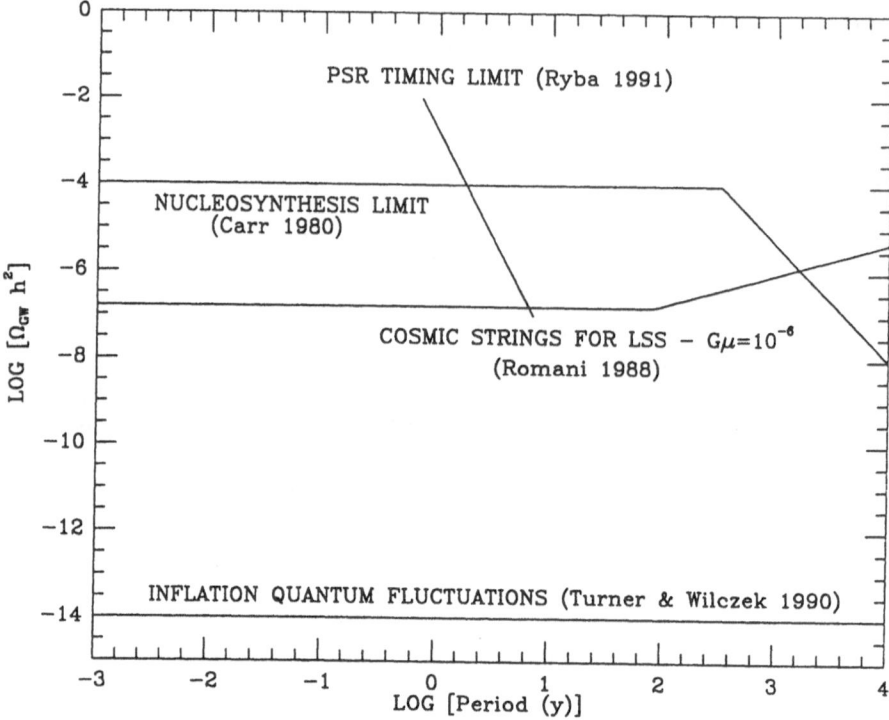

Fig. 14. The logarithmic gravitational wave background spectrum. Estimates of the background resulting from dissipation of cosmic strings that might be seeds for large scale structure and galaxy formation and from inflation quantum fluctuations are shown. Limits from nucleosynthesis and pulsar timing are given.

uses the epoch, or temperature, of the universe at the time the stochastic background is formed.

Let's start with the primordial sources and work forward in time through the evolution of the detectable universe. A thermal distribution of gravitons from the Planck era is possible. If this spectrum decouples and evolves into the present era, then it will be redshifted down below the microwave background spectrum to wavelengths of order 3 mm, or a temperature of 1 K. This is not of interest to the pulsar timing measurement. Grishchuk (1977) has given arguments for a nonthermal spectrum. The existence of a stochastic background of gravitational radiation at this early epoch creates an anisotropy which, if sufficiently large, would affect nucleosynthesis of

Helium and other light elements. Carr (1980) places a limit of $\Omega_g = 10^{-4}$ from this consideration for periods ranging from 1 s to 300 y (Fig. 14). Further limits on the spectrum resulting from early epoch anisotropy can be placed using the isotropy of the microwave background. Gravitational radiation with amplitude h_o, in the present era can induce a quadrupole term in the microwave background comparable to h_o, and radiation at the time of recombination can produce smaller scale anisotropies.

The next class of primordial sources occur during the QCD transition from the false-vacuum phase to the radiation-dominated phase, and are a mix between Carr's primordial and generated sources. Starobinskii (1979) considered the production of gravitational radiation by structural inhomogeneities during this transition interval, and came up with a broadband, scale invariant spectrum. Witten (1984) discusses processes that occur at the beginning of the radiation era where bubbles of low-temperature phase, quark nuggets, amidst the high-temperature phase will collide. He summarizes the principal sources of the stochastic background in an excellent appendix. Turner and Wilczek (1990) have developed the ideas of gravitational-wave production by nucleating and percolating bubbles during *extended inflation* era between phases. Their spectrum is not very promising for the pulsar timing experiment (Fig. 14). Other sources have been considered which involve nucleation of the low temperature phase by impurities such as magnetic monopoles, cosmic strings and texture. Furthermore black holes could be produced in this era whose collisions and evaporation could add to the stochastic background.

Vilenkin (1981) developed the spectrum that would be produced by a cosmic strings that come out of the phase transition era and survive to act as seeds for galaxy formation. These topological defects in spacetime form closed oscillating loops and decay be emission of gravitational radiation. Matter falls into the loop and remains after the decay. Vilenkin's spectrum is divided into three parts according to whether the loop forms and decays wholly in the radiation era, or forms in the radiation era and decays in the matter era, or forms and decays wholly in the matter era. The strings form on the scale of the horizon and then evolve with twists and connections to form loops. Various authors have made detailed models of the evolution of a network of strings with both galaxy formation as one result and observable consequences in the microwave and gravitational wave backgrounds as the constraint (Bennett and Bouchet, 1990).

Finally let me turn to a post-galactic source of generated radiation. The standard model for quasars, and active galactic nuclei in general, is mass accretion onto a massive black hole, $M = 10^{6-9} M_\odot$. Now galaxies and presumably quasars live in clusters where tidal interactions and collisions are likely during the lifetime of the system. There is considerable evidence for galaxy mergers, cosmic cannibalism. The existence of massive, binary black holes is a clear possiblity. Sillanpaa *et al.* (1989) have recently discussed the case for a binary black hole driving the nearly periodic outbursts of the radio source OJ287 with interval of 10 y. Begelman *et al.* (1980) have previously looked at the mechanism and time scales for binary black hole production, while Blandford (1979) considered the coalescence rate for massive black holes in the universe as a source of short wavelength gravitational radiation. The new lines of evidence in favor of energetic events in galactic nuclei suggest that the contribution of these objects to the stochastic background needs to be reconsidered. In particular, the accretion of smaller black holes into orbit around a principle black hole is a likely scenario given the observations. A single source, as calculated by Detweiler (1979), will produce a dimensionless strain of

$$h \simeq 2 \times 10^{-17} (\frac{1+z}{n})^{0.67} (\frac{T}{\text{y}})^{-0.67} (\frac{D}{\text{Gpc}})^{-1} (\frac{m_p}{10^8 \text{M}_\odot})^{1.67} f(\frac{m_s}{m_p}, n, e), \quad (51)$$

where z is the redshift, n is the harmonic number, T is the observed time scale which is connected back to the binary period using the redshift and harmonic number, D is the distance, m_p is the primary black hole mass, m_s is the secondary black hole mass, and f is a function of order unity. A hole of mass 10^9 would lead to a 30 ns effect over one year period. While this is currently unobservable, a contribution of 100 such objects emitting with random periods and phases would bring the effect up to observable levels.

6.3.4 Basis Function for Pulsar Timing Array Measurements

The program of timing an array of millisecond pulsars for the detection of a background of gravitational radiation requires a set of basis functions which can be used to fit to the data. There are five degrees of freedom in the perturbations of the space time metric that result from an arbitrary form of gravitational radiation. In §6.3.2 a wave along one direction with an arbitrary polarization was analyzed. The effects of three waves can then be superposed to obtain the net effect of an arbitrary signal.

The formulation of Detweiler (1979) is most useful for computation of the angular distribution of the perturbation that each degree of freedom has on the pulsar data. His formulation gives the Doppler shift for a wave incident along the z axis with orthogonal polarization components h_+, $h_\times$ as a function of the direction cosines of the pulsar α, β, γ.

$$z = \frac{(\alpha^2 - \beta^2)h_+ + 2\alpha\beta h_\times}{2(1+\gamma)}. \quad (52)$$

The Doppler shift is just the time derivative of the timing residuals, $R(t)$.

The perturbations of the space-time metric are then formed by permuting (51) for waves along the x, y, z axes:

$$
h_{\mu\nu} = \begin{pmatrix} h_{x+} & h_{x\times} & 0 \\ h_{x\times} & -h_{x+} & 0 \\ 0 & 0 & 0 \end{pmatrix} + \begin{pmatrix} h_{y+} & 0 & h_{y\times} \\ 0 & 0 & 0 \\ h_{y\times} & 0 & -h_{y+} \end{pmatrix} + \begin{pmatrix} 0 & 0 & 0 \\ 0 & h_{z+} & h_{z\times} \\ 0 & h_{z\times} & -h_{z+} \end{pmatrix},
$$
(53)

which sums to

$$
h_{\mu\nu} = \begin{pmatrix} h_{x+} + h_{y+} & h_{x\times} & h_{y\times} \\ h_{x\times} & h_{z+} - h_{x+} & h_{z\times} \\ h_{y\times} & h_{z\times} & -h_{y+} - h_{z+} \end{pmatrix}.
$$
(54)

This final matrix is symmetric and trace free. The five degrees of freedom are then h_{11}, h_{12}, h_{13}, h_{22}, and h_{23} with $h_{33} = -h_{11} - h_{22}$. The instantaneous Doppler shift patterns can then be computed for each degree of freedom from the constituent parts that are expressed by Detweiler's equation. Fig. 15 displays the h_{11} pattern. Note that there is considerable angular structure in this pattern beyond the quadrupole. The signature of gravitational radiation on an array of pulsars will be distinct from that of the other effects – the clock which has a monopole angular signature and the earth orbit which has a dipole angular signature.

This discussion has avoided mention of the temporal signature of the various effects. Only time variations of any effect will be detectable owing to the fact that we do not have an absolute measure of the pulsar distance. Furthermore linear and quadratic effects will also be absorbed into the determination of the pulsar rotation parameters which are also unknown *a priori*. Only cubic and higher order temporal modulations will be available for fitting to these global effects on the pulsar timing array (Foster & Backer 1990). In the case of the Earth orbital perturbations the angular and temporal basis function for various solar system parameters will be available because the errors arise in limited precision of known physical effects.

6.4 Pulsar Timing Array Experiments

The goal of timing an array of millisecond period pulsars is just coming within our reach with the experiments in progress. The most detailed study has involved timing the original millisecond pulsar PSR 1937+21 since the installation of improved data acquisition hardware in late 1983 (Stinebring, Ryba, & Taylor 1990; Ryba 1991). Since 1985 data was recorded at two radio frequencies so that dispersion measure variations could be removed. Data from a single pulsar can be used to place a limit on the gravitational wave background spectrum as outlined in §6.3.2. The importance of the results presented by Stinebring *et al.* is that they show a "signal" in that the residuals are not consistent with white noise. The residuals appear to be dominated by a cubic term in phase over the 5 years of dual frequency data. The amplitude of this cubic term is about 1 μs. The origin of this effect is unknown. There are a number of candidates. The atomic time scales

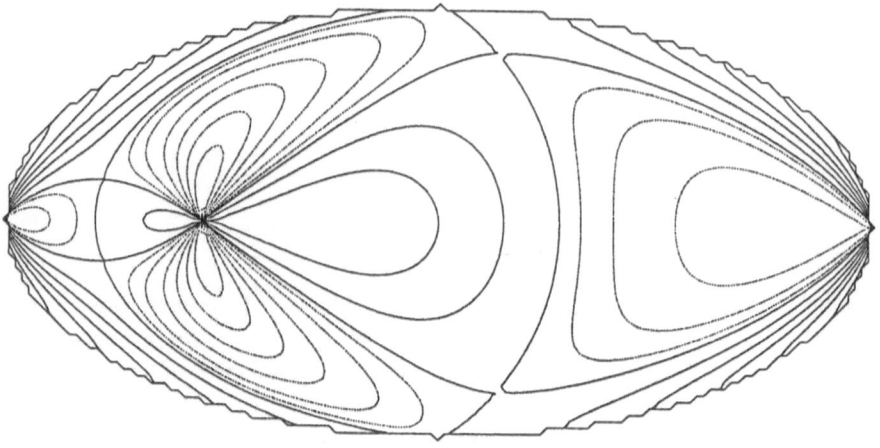

Fig. 15. The Doppler shift pattern that results from gravitational wave background component h_{11} in (44). Celestial coordinate system is arbitrary, but can be thought of as right ascension of 360° horizontally and declination of 180° vertically.

available are no more stable than 10^{-14} over durations of a few years (Fig. 11). This corresponds to a residual of 1 μs which is comparable to the observed timing residual. The ephemeris of the Earth's motion as presently provided does not include the latest values of the outer planetary masses. The resultant contribution of the ephemeris to timing residuals is then worse than that presented in Fig. 14 and Table 4; a microsecond of ephemeris error is also possible. Stinebring *et al.* attribute the systematic residual to a combination of intrinsic timing noise (§4.5) and interstellar propagation effects (§3.3). Most of their discussion is reserved for the important new limit that they can establish on the gravitational wave background,

$$\Omega_G < 9 \times 10^{-8} \text{ for frequencies near } 0.14 \text{ y}^{-1} \tag{55}$$

with 68% confidence. Ryba (1991) improves on these limits by a factor of four with somewhat extended data span and improved data analysis.

$$\Omega_G < 2.5 \times 10^{-8} \text{ for frequencies near } 0.14 \text{ y}^{-1} \tag{56}$$

with 68% confidence.

A number of programs are underway at several observatories to time the best millisecond pulsar "clocks". These are summarized in Table 4. Other observatories such as Parkes and Jodrell Bank are also involved in millisecond pulsar timing, but with less emphasis on pulsar timing array aspects. We started a pulsar timing array experiment in 1987 shortly after the 3-ms pulsar in the globular cluster M28 was discovered. Our first results are described in Foster & Backer (1990). We have made steady progress in sensitivity improvements, and have started dual frequency observations that allow removal of effects of dispersion measure variations. Timing residuals for three of the pulsars in our program are displayed in Figure 16. We are continuing to explore methods of improving the sensitivity of this experiment by improved receivers and new signal processing hardware. The goal is to have timing residuals at or below one microsecond for a modest number of pulsars distributed across the sky.

The present timing array is starved for objects distributed around the sky. The recent discovery of strong millisecond pulsars in the southern sky will help in this distribution once their timeability is established. The high latitude pulsar PSR 1257+12 seemed to be an ideal object owing to its location and its flux density. However observations now demonstrate that it has at least two planetary mass companions around it and possibly more. This circumpulsar "space junk" would seem to ruin it as a good clock, while making the astrophysics of the system a delight! Searches are underway to discover more objects throughout the sky. The future is reasonably bright that a significant number of new clocks will be found.

Table 4. Millisecond Pulsar Timing Array Programs

Observatory	Investigator	Pulsars	
Arecibo	Taylor	1855+09,	1937+21,
		1953+29,	1957+20
	Wolszczan	1257+12,	1534+12
Green Bank	Backer	1257+12,	1620-26,
		1713+07,	1821-24,
		1855+09,	1937+21
Nançay	Lestrade	1821-24,	1937+21
VLA	Taylor	1937+21,	others
Usada	Hirao	1937+21	
Kashima	Imae	1937+21	

Acknowledgements

This discussion of pulsar timing and related subjects is a summary of the work of many scientists over the past 20 years. I have given credit when specifically needed, but not for the general development of the field.

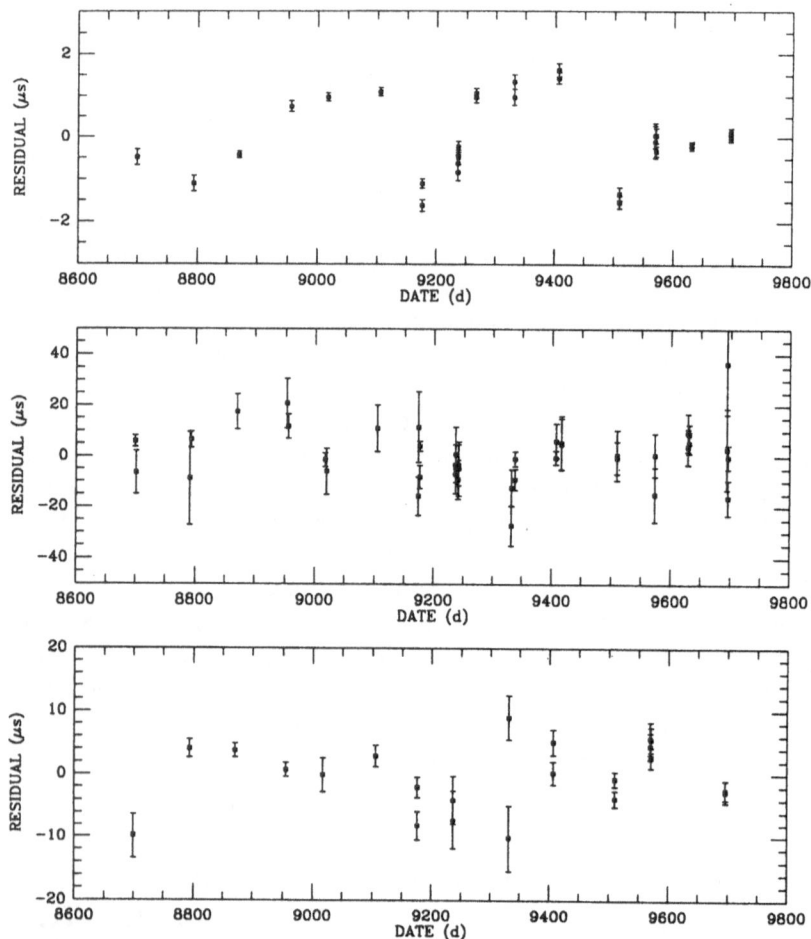

Fig. 16. Pulsar timing array data from experiment in progress at the NRAO 140ft telescope. The display contains timing residuals for PSRs 1821-24, 1855+09 and 1937+21. The 6-parameter model that is fit to the PSR 1821-24 data leave 3 degrees of freedom. The PSR 1855+09 data is compared to the model derived by Ryba (1991) with only a relative epoch removed. The PSR 1937+21 data is also compared to a model from Ryba (1991) with a relative epoch and a small period difference removed.

References

Anderson, S. B., Gorham, P. W., Kulkarni, S. R., Prince, T. A., and Wolszczan, A. (1990): "Nature" vol. 346, p. 42

Arons, J. (1979): "Space Science Rev.", vol. 24, p. 437

Arons, J. (1983): "Astrophys. J.", vol. 266, p. 213

Arons, J. (1992): in "The Magnetospheric Structure and Emission Mechanisms of Radio Pulsars", ed. T. H. Hankins, J. M. Rankin, and J. A. Gil, [Pedagogical University Press : Zielona Góra], p. 62

Ash, M. E., Shapiro, I. I., and Smith, W. E. (1967): "Astron. J.", vol. 72, p. 338

Ashby, N., and Allan, D. W. (1984): "Radio Sci.", vol. 14, p. 649

Backer, D. C., Kulkarni, S. R., Heiles, C., Davis, M. M., and Goss, W. M. (1982): "Nature", vol. 300, p. 615

Backer, D. C., and Hellings, R. W. (1986): "Ann. Rev. Astron. Astrophys.", vol. 24, p. 537

Backer, D. C. (1987): in "IAU Symposium No. 125", ed. D. J. Helfand and J.-H. Huang, [Reidel : Dordrecht], p. 13

Backer, D. C. (1988): in "Galactic and Extragalactic Radio Astronomy", ed. K. Kellermann and G.,Verschuur, [Springer-Verlag : Berlin] p. 480

Backer, D. C. (1989): in "NATO ASI C262", ed. Ögelman and van den Heuvel, p. 3

Backer, D. C., and Kulkarni, S. R. (1990): "Physics Today", vol. 43, p. 26

Backer, D. C., Clifton, T. R., Kulkarni, S. R., and Werthimer, D. J. (1990): "Astron. Astrophys.", vol. 232, p. 292

Backer, D., Sallmen, S., and Foster, R. (1992): "Nature", vol. 358, p. 24

Baldwin, J. E., and Wang, S., ed. (1991): "URSI/IAU Colloquium on Radio Astronomical Seeing", [Pergamon : Oxford], pp. 275

Begelman, M. C., Blandford, R. D., and Rees, M. J. (1980): "Nature", vol. 287, p. 307

Bennet, D. P., and Bouchet, F. R. (1990): "Phys. Rev. D", vol. 41, p. 720

Bhattacharya, D., and van den Heuvel, E. P. J. (1991): "Physics Reports", vol. 203, p. 1

Blandford, R. D., and Teukolsky, S. A. (1976): "Astrophys. J.", vol. 205, p. 580

Blandford, R. D. (1979): in "Sources of Gravitational Radiation", L. L. Smarr, ed., [Cambridge, Cambridge], p. 191

Boriakoff, V., Buccieri, R., and Fauci, F. (1983): "Nature", vol. 304, p. 417

Carr, B. J. (1980): "Astron. Astrophys.", vol. 89, p. 6

Clegg, A. W., Cordes, J. M., Simonetti, J. H., and Kulkarni, S. R. (1992): "Astrophys. J.", vol. 386, p. 143

Clemence, G. M. and Szebehely, V. (1967): "Astron. J.", vol. 72, p. 1324

Cordes, J. M., Rickett, B. J., and Backer, D. C. (1988): "AIP Conference Proc.", vol. 174, pp. 330 Damour, T., and Taylor, J. H. (1991): "Astrophys. J.", vol. 366, p. 501

Damour, T., and Taylor, J. H. (1992): "Phys. Rev. D", vol. 45, p. 1840

Detweiler, S. (1979): "Astrophys. J.", vol. 234, p. 1100

Dewey, R., Stokes, G., Segelstein, D., Taylor, J. H., and Weisberg, J. (1985): "Astrophys. J.", vol. 294, p. L25

Doroshenko, O. V., and Kopejkin, S. M. (1990): "Sov. Astron.", vol. 34, p. 496

Erickson, W. C., Kuiper, T. B. H., Clark, T. A., Knowles, S. H., and Borderick, J. J. (1972): "Astrophys. J.", vol. 177, p. 101

Fiedler, R., Dennison, B., Johnston, K., and Hewish, A. (1987): "Nature", vol. 326, p. 675

Fomalont, E. B. *et al.* (1992) preprint

Foster, R. S. (1990): Ph. D. thesis, University of California, Berkeley

Foster, R. S., and Backer, D. C. (1990): "Astrophys. J.", vol. 361, p. 300

Frail, D. A., and Weisberg, J. M. (1990): "Astron. J.", vol. 100, p. 743

252

Fruchter, A. S., Stinebring, D. R., and Taylor, J. H. (1988): "Nature", vol. 333, p. 237

Goldreich, P., and Julian, W. H. (1969): "Astrophys. J.", vol. 157, p. 869

Grishchuck, L. P. (1977): "Ann. NY Acad. Sci.", vol. 302, p. 439

Guinot, B., and Petit, G. (1991): "Astron. Astrophys.", vol 248, p. 292

Gunn, J.E., and Ostriker, J. P. (1970): "Astrophys. J.", vol. 160, p. 979

Hamilton, P. A., McCullough, P. M., Ables, J. G., and Komesaroff, M. M. (1973): "Astrophys. Lett.", vol. 15, p. 63

Hankins, T. H. (1971): "Astrophys. J.", vol. 169, p. 487

Helfand, D. J., Manchester, R. N., and Taylor, J. H. (1975): "Astrophys. J.", vol. 198, p. 661

Hellings, R. W. (1986): "Astron. J.", vol. 91, p. 650

Hewish, A., Bell, S.J., Pilkington, J. D. H., Scott, P. F., and Collins, R. A. (1967): "Nature", vol. 217, p. 709

Hulse, R. S., and Taylor, J. H. (1974): "Astrophys. J.", vol. 201, p. L55

Jesperson, J., and Hanson, D. W., ed. (1991): 'Proc. IEEE", vol. 79, pp. 891-1079

Johnston, H. M., and Kulkarni, S. R. (1991): "Astrophys. J."", vol. 368, p. 504

Krisher, T. P. et al. (1991): "Astrophys. J.", vol. 375, p. L57

Manchester, R. N., and Taylor, J. H. (1977): "Pulsars", [San Francisco : Freeman], pp. 281 (look for 2nd edition in 1993)

Michel, F. C. (1991): "Theory of Neutron Star Magnetospheres", [University of Chicago : Chicago], pp. 517

Middleditch, J, and Kristian, J. (1984): "Astrophys. J.", vol. 279, p. L57

Ögelman, H., and van den Heuvel, E. P. J. (1989): "NATO ASI C262" [Kluwer : Dordrecht], pp. 774

Pacini, F. (1967): "Nature", vol. 216, p. 567

Radhakrishnan, V., and Cooke, D. J. (1969): "Astrophys. Lett.", vol. 3, p. 225

Rand, R. J., and Kulkarni, S. R. (1989): "Astrophys. J.", vol. 343, p. 760

Rasio, F. A., Nicolson, P. D., Shapiro, S. L., and Teukolsky, S. A. (1992): "Nature", vol. 355, p. 325

Robertson, D. S., Carter, W. E., and Dillinger, W. H. (1991): "Nature", vol. 349, p. 768

Romani, R. W. (1989): in "NATO ASI C362", ed. Ögelman & van den Heuvel, p. 113

Ruderman, M., Shaham, J., and Tavani, M. (1988): "Astrophys. J.", vol. 336, p. 507

Ryba, M. F. (1991): Ph. D. thesis, Princeton University

Ryba, M. F., and Taylor, J. H. (1991) "Astrophys. J.", vol. 371, p. 739

Sauls, J. (1989): in "NATO ASI C262", ed. Ögelman and van den Heuvel, p. 457

Shapiro, I. I. (1964): "Phys. Rev. Lett.", vol. 13, p. 789

Shapiro, S. L., and Teukolsky, S. A. (1983): "Black Holes, White Dwarfs, and Neutron Stars", [Wiley : New York], pp. 645

Sillanpää, A., Haarala, S., Valtonen, M. J., Sundelius, B., and Byrd, G. G. (1988): "Astrophys. J.", vol. 325, p. 628;

Staelin, D. H. (1969): "Proc. IEEE", vol. 57, p. 724

Standish, E. M. (1990): å, vol. 233, p . 252

Starobinskii, A. A. (1979): "JETP Lett.", vol. 30, p. 682

Stinebring, D. R., *et al.* (1990): "Phys. Rev. Lett.", vol. 65, p. 285

Stinebring, D. R., Hankins, T. H., Ryba, M. R. (1992): "Rev. Sci. Instr.", vol. 63, p. 3551

Taylor, J. H., and Weisberg, J. M. (1979): "Astrophys. J.", vol. 345, p. 434

Taylor, J. H., and Stinebring, D. R. (1986): "Ann. Rev. Astro. Ap.", vol. 24, p. 285

Taylor, J. H. (1989): in "NATO ASI C262", ed. Ögelman and van den Heuvel, p. 17

Taylor, J. H., Wolszczan, A., Damour, T., and Weisberg, J. M., (1991): "Nature", vol. 355, p. 132

Turner, M. S., and Wilczek, F. (1990): "Phys. Rev. Lett.", vol. 65, p. 3080

Vilenkin, A. (1981): "Phys. Lett.", vol. 107B, p. 47

Will, C. M., and Nordvedt, K. (1972): "Astrophys. J.", vol. 177, p. 757

Witten, E. (1984): "Phys. Rev. D", vol. 30, p. 272

Wolszczan, A., and Frail, D. A. (1992): "Nature", vol. 355, p. 145

THE MISSION HIPPARCOS

J. Kovalevsky
CERGA - Observatoire de la Côte d'Azur
F - Grasse

1. What is stellar astrometry ?

This series of lectures is meant to present the most accurate and perfor-
mant astrometric instrument that has ever been put into operation. We
are going to describe this instrument, how it works and what results are
to be expected on the basis of what has already been achieved by this satellite. However, I believe that before entering into a detailed study of an
astrometric instrument, one should have some background on this domain of astronomy. I shall restrict it to stellar astrophysics since this is
the very first objective of HIPPARCOS even if a few 10^{-4} parts of its observing time is spared for minor planets and a couple of natural satellites.

So let us ask ourselves the following questions :
i) What is stellar astrometry?
ii) How was it done before HIPPARCOS?
iii) Why do we need astrometry
iv) What is the expected contribution of HIPPARCOS?

1.1. *What is stellar astrometry?*
Astrometry is the domain of astronomy that is assigned as an objective to
determine the positions of celestial objects. Since these positions generally vary with time, the objective is also to describe their motions as a
function of time. I shall not deal here with tasks that, by extension, belong also to astrometry, such as the determination of other geometrical
properties of celestial bodies such as their dimension and their shape.

In the case of stars, the primary objective of an astrometric program is to determine as accurately as possible positions for a given
epoch, and then, as time goes, to derive their variations that are used to
determine their proper motions and their parallaxes. Another objective
is to describe quantitatively double and multiple star systems. If these
three objectives have definitely an astrophysical impact, the study of positions and their time variations is also basically an astrometric objective
that permits to make better astrometry (see section 2.2) through the
construction of reference frames and fundamental catalogues.

1.2. *Methods of astrometry*
There are three different categories of astrometry depending upon the
portion of sky that can be tied into a single reduction of observation.

i) <u>Semi-global astrometry</u> - It is characterized by the possibility to link positions of stars that may be in different parts of the sky. The only limitation is that they should be accessible to the instrument. The main instruments of semi-global astrometry are the transit circle (or meridian instrument) and the astrolabe. They are the basic instruments for the construction of star position catalogues. Recently a visual Michelson interferometer in Mount Wilson Observatory achieved the same types of measurements (Shao et al., 1988). There is only one global astrometry instrument: it is HIPPARCOS, the only which can make measurements over the entire sky.

ii) <u>Small field astrometry</u> - Its objective is to determine the positions of stars in a limited sky field. Typical instruments are photographic astrographs or Schmidt cameras. Their contribution is to extend catalogue positions obtained by semi-global astrometry to many more - generally fainter - stars. In long focus telescopes, the objective is to measure parallaxes and study widely separated double or multiple stars. For such very small fields, CCD's advantageously replace photographic plates. Several other techniques exist; let us mention only the FGS (Fine Guidance Sensors) of the Hubble Space Telescope.

iii) <u>Star image analysis</u> - In these techniques, there is no reference to stars with known positions. The objective is to recognize and measure double or multiple star systems, or to measure stellar diameters. Interferometers, whether speckle or Michelson, are the main instruments for this type of astrometry.

A description of the presently existing astrometric instruments can be found in Kovalevsky (1990).

1.3. *Why stellar astrometry?*

There are many good reasons to determine apparent motions of stars. They fell essentially into two categories.

i) <u>To improve the knowledge about the star and its physical state</u> - The basic parameter is its parallax. Trigonometric parallaxes are the calibrated origin of all other methods to extend the distance scale by comparison of apparent magnitudes of objects having the same physical characteristics such as spectrum, variability or temperature and therefore the same luminosity. Trigonometric parallaxes allow to calibrate the bases of these extrapolations. All possible types of stars should have accurately measured parallaxes. If, in addition, orbital elements of double or multiple stars are determined from astrometric or photometric observations, the parallax permits to determine the masses of the components.

The knowledge of parallaxes is also necessary to transform apparent magnitudes into real luminosity, apparent diameters into diameters expressed in kilometers, and proper motions into velocities.

ii) <u>To study the structure, the kinematics and the dynamics of stellar groups</u> - The basic observable is, for these studies, the annual proper motion. It allows to study motions within clusters, to recognize stellar association membership, to analyze motions in the Galaxy, etc... They are a statistical indicator of stellar population types and are a major input in the studies of the relations between kinematic and chemical composition, age and/or spectra of stars (Galactic evolution).

From kinematical observational data, on can infer dynamical properties of the Galaxy and its evolution with time. As examples, one may mention the evolution of spiral arms, the search for matter distribution explaining the galactic gravitational field or the evolution of clusters.

It is to be highly encouraged to have parallaxes and radical velocities for as many stars as possible in the Galaxy since it would allow much more detailed analyses of the kinematic properties of the Galaxy.

1.4. *Present situation and expectations from HIPPARCOS*

The present situation of astrometry is summarized in table 1 which gives the order of magnitude of the precision of various astrometric techniques. This precision depends upon the magnitude of stars and upon the quality of the instrument.

Table 1

Type of instrument	Internal precision	Catalogue precision	Magnitude Limit
Schmidt cameras	0".10 - 0".25		18
Long focus astrographs	0".02 - 0".04	Parallaxes: 0".001 - 0".005	16
Photoelectric astrolabes	0".08 - 0".13	Positions: 0".04	8
Photoelectric meridian circles	0".10 - 0".15	Positions: 0".08	14
Optical interferometry	0".03 - 0".05	Positions: 0".01	7
VLBI	0".001 - 0".003	Positions: 0".0005	Radio

It is to be remarked that the observing programme of astrolabes and interferometers include a systematic reobservation of the same objets so that more precise catalogues may be derived. The same would be true for meridian circles, but at present they are essentially used to increase the number of astrometrically observed stars. Long focus instruments (e.g. US Naval Observatory 61 inch astrometric telescope in Flagstaff) are used essentially to determine parallaxes.

The results of positional observations are collected and published in star catalogues. Table 2 gives the characteristics of the main star catalogues. It is to be remarked that the nominal precision given does not represent the actual precision in 1990 because of the degradation due to the absence of proper motions or their bad quality. This is essentially the case of the astrographic catalogue and of the SAO catalogue.

Table 2

Catalogue name	Date	Number of stars	Nominal Precision
Catalogue astrographique	1910	15 000 000	0".7
SAO Catalogue	1962	259 000	0".5
FK5	1985	15 000	0".01
IRS	1985	40 000	0".2
Guide star catalogue	1986	15 000 000	1".3

The HIPPARCOS mission is expected to provide a catalogue of 118000 positions, annual proper motions and parallaxes with a mean accuracy for stars of magnitude 9 of the order of 0.0015 arc seconds. Presently, we know only about 5000 proper motions to better than 0.002 arc seconds and, for 100000 other stars the errors are of the order of 0.005 to 0.015 arc second per year. In addition there are probably systematic errors of the same order of magnitude. There are a few hundred parallaxes that are known to one millisecond of arc (mas) and may be another thousand have a precision of 2 or 3 mas. The other 7000 parallaxes are evaluated with errors of 5 mas or more. So one can say that it is expected that HIPPARCOS will improve by about one order of magnitudes the present quality of known astrometric parameters of stars and multiply by ten the number of measured parallaxes. Furthermore, the additional TYCHO experiment will provide the same astrometric parameters of 800000 stars with a precision of about 10 mas.

2. Principles and problems of stellar astrometry

The physical information that an astrometric instrument gets, is the direction from which the light emitted by a star arrives on the telescope or whatever light collector simple or complex is used.

This short statement immediately brings up two fondamental questions :
1. Has this direction any relation to the direction in which the star really is? Is'nt this direction time and observatory dependent?
2. When one speaks about directions, one has to mention in what coordinate system or, as astronomers call it, in what reference frame it is reckoned?

Let us elaborate somewhat on each of these two aspects.

2.1. *What direction is of interest to astronomy?*

A simple example suffices to give a positive answer to the first question. Atmospheric refraction depends upon the zenith distance of the star which varies with time during the night. In addition, zenith distances of a given star vary from one observatory to another. Clearly, astronomers want directions that are independent of the observing conditions and consequently of the presence of terrestrial and even solar system environment. Therefore, we shall define the direction of a star as the direction from which the light would arrive if the solar system would not exist. Let us note that this is different from the direction in which the star really is. It is actually the direction in which the star was when it emitted the light that is received, provided that between the star and the solar system rays had not been bent by interstellar matter refraction or relativistic light deflection. To know where the star is at present has not much astrophysical interest. To obtain it, would imply that correction for proper motion are made and that the distance is known so that the light time can be computed.

To correct the observed apparent positions in order to obtain positions fit for astronomical use is a rather complex procedure which requires a very accurate knowledge of some physical quantities concerning the environment or the kinematic conditions in the solar system. The corrections that have to be applied are :

i) Atmospheric refraction - In order to make this correction, one needs to have a model of the atmospheric refraction, to measure the parameters that enter into the model (temperature, humidity, atmospheric pressure at the telescope) and add some additional parameters determined in the process of observation reduction.
ii) Aberration - This effect is proportional to the velocity of the observer with respect to the only point that would be unchanged if the solar system was to be neglected, namely the barycenter of the solar system. This includes accurate knowledge of the motion of the Earth around the Sun and of the Earth rotation.
iii) Relativistic light deflection by the Sun and, if necessary, by other bodies of the solar system.

2.2. *Astronomical reference frames*

Because instruments have always been attached to the Earth, the definition of the system of reference celestial coordinates has been and still is strongly connected with the orientation of the Earth in space. Because it is an observable common to all geographie locations, the direction of the Earth axis of rotation has been chosen as the fundamental OZ axis. In the principal plane, the equator, an origin has to be chosen. It is the equinoxial point, at the intersection of the equator with the ecliptic. The problem is that both equator and ecliptic are moving in time, so that while observations are related to the instantaneous pole, the actual reference system must be fixed conventionally. This is done by defining a reference at the conventional epoch J.2000.0 together with an ecliptic and an equator in which all periodic motions would have been averaged out. In order, then, to express observations made in the instantaneous reference system to the conventional fixed system, the following corrections have to be applied.

i) **Earth rotation parameters** - They include the angular rotation of the Earth as refered to the instantaneous equinox, expressed generally in terms of UT1 and the polar motion. Both are unpredictable and are monitored by various techniques (VLBI, Laser ranging), the result being compiled and distributed the International Earth Rotation Service (IERS). These parameters are used to transform local celestial coordinates generally obtained by astrometric instruments into geocentric instantaneous celestial coordinates.

ii) **Nutation and precession** - They are respectively the periodic and the secular parts of the time dependent developments that describe the motion of the instantaneous axis of Earth rotation with respect to the conventional fixed reference system and therefore are to be used in transforming the instantaneous coordinates of a star into coordinates in the fixed reference frame. Theoretical developments are presently unsufficient for very accurate astrometry and whenever necessary, observed values published by IERS must be used.

iii) **Stellar parallaxes** - The apparent direction variations due to the motion of the Earth around the Sun have to be taken into account whenever known.

The coordinate grid system is obviously not directly observable and even the direction of the celestial pole is not accessible to many instruments. It is therefore necessary to have fiducial points all over the sky to represent the coordinate system. This is realized by fundamental star catalogues (presently FK5) in which star positions are given at the reference epoch together with their time variations due to proper motions. Other catalogues extend the FK5 to a larger number of stars (e.g. IRS: international reference stars). They allow to obtain star coordinates in local relative astrometry, for instance for photographic plate reductions. A detailed description of all the transformations involved in the reduction of astrometric observations of stars can be found in Green (1985).

2.3. *Application to space astrometry*
The two principles applied in the two preceding sections for Earth based observations do apply also to space astrometry.

i) Apparent to true position transformation - Evidently, refraction correction desappears. Concerning the aberration, one has to know accurately the orbital motion of the satellite about the Earth and add it to the motion of the Earth about the solar system barycentre. The speed of satellite rotation is a negligible contribution. Finally, the problem of relativistic light deflection remains the same with the addition of the effect of the Earth itself.

ii) Reduction to a fixed reference frame - When observations made by a satellite are devised to give star coordinates, and not simply relative local coordinates as in the astrometric mode of the Hubble Space Telescope, the transformation between the observed quantities referred to some coordinates in the satellite and the stellar reference frame implies an accurate knowledge of the position of one with respect to the other. In other terms, the orientation in space of the satellite must be know with an accuracy comparable to the expected precision of the observations. This orientation (called attitude by space engineers) is the equivalent of the ensemble Earth orientation parameters, nutation and precession in on-ground astrometry. The difference is that there is no independent networks to determine it, so that an astrometric satellite must be also a satellite that determines by itself its own orientation in space. This is not the simplest part of the satellite design and of the data reduction, even if, as is the case for HIPPARCOS, both functions (attitude determination and astrometric observations) are interrelated and are mutually supporting each other.

In any case, neither the celestial equator nor the ecliptic being observable by a satellite, fundamental stars must be observed so as to ensure the necessary link with the conventional reference frame.

3. The HIPPARCOS mission

The first astrometric satellite ever built, HIPPARCOS was constructed by the European Space Agency and launched on the 8th of August 1989. It has a very elongated orbit around the Earth with an apogee around 500km and apogee close to 36 500km. Its period is 10h 40m. The telecommunications with Earth are secured by three ground stations, Odenwald (Germany) which is the master station of ESOC (European Space Operation Center) controlling the other two stations: Perth in Australia and Goldstone in California. Well distributed in longitude, they cover about 95% of the orbit and almost 100% of the useful observing time.

The name of this satellite, apart from recalling Hipparchus, the author of the first star catalogue and discoverer of precession, is an acronym that stands for *"HIgh Precision PARallax COllecting Satellite"*. It is based upon an idea of P. Lacroute in 1966 but could be realized only much later when space techniques became sufficiently advanced and after many other scientists and engineers have come to a feasible and performing project that was approved in 1981 by ESA. Later, it was extended by the adjunction of the TYCHO experiment proposed by E. Hoeg. Nominally, the satellite should have been launched in a geostationnary orbit which would have simplified the operations and avoided some difficulties in the scientific data treatment, particularly for the attitude reconstitution. But due to the failure of the apogee motor, the mission had to be adapted to the new orbit. It is this actual HIPPARCOS mission that is described here and the reduction procedures take into account the real orbit.

3.1. *General principle of HIPPARCOS*

Astrometry reduces itself to measurements of angles. In global astrometry, one must measure large angles. This is done on-ground by measuring times between transits of stars through well defined circles (meridian or almicantarats). Using the relation between time and the angle of rotation of the Earth, time is transformed into angles. In the case of HIPPARCOS, this is also done through the satellite attitude, but this would be unsufficient. So the fundamental idea is to have a standard angular yard-stick materialized by the angle between two mirrors called the beam-combiner which are reflecting into the telescope two stellar fields separated by twice the angle between the mirrors. This basic angle γ, is equal to 58°. Within the field of view, one has to measure small angles.

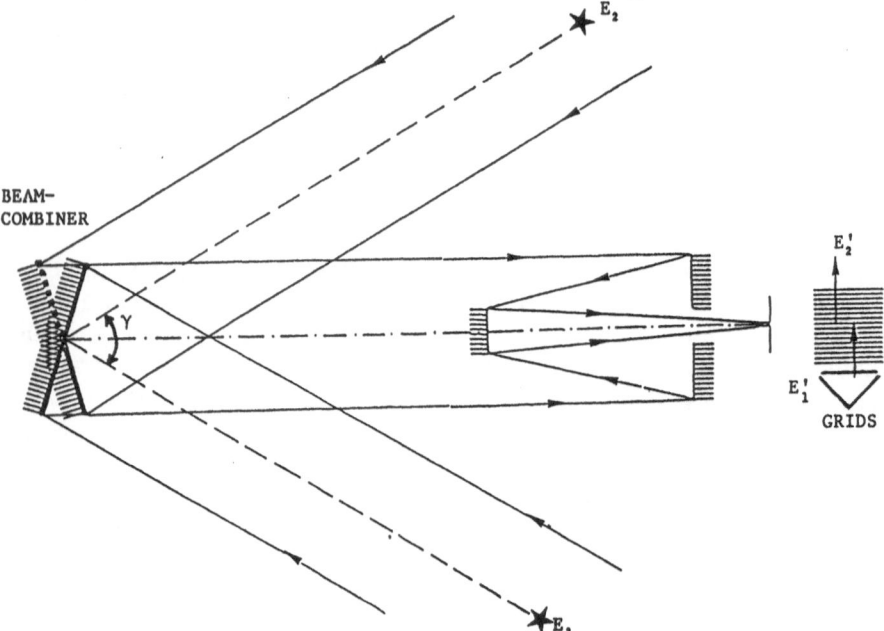

Figure 1 - Principle of HIPPARCOS. The images E'₁ and E'₂ of stars E₁ and E₂ in different fields of view combine on the focal grids.

This is done in the focal plane by letting the satellite slowly rotate around an axis parallel to the intersection of the mirrors of the beam-combiner. In the focal plane, a grid made of equidistant slits, modulates the light of the stars and the apparent angular distance between stars is determined by comparing the phases of the modulation curves (figure 1).

The combination of these three types of acces to angular determinations is the basis of the astrometric use of the satellite.

3.2. *Description of the satellite and of its payload*

The satellite has a general six faced parallelepipedic shape with a prismatic top and three large solar cell panels perpendicular to the main corpus. Light enters through two baffles and falls on the beam combiner which is a 29cm miror cut into two halves and glued at a 29° angle. This angle is very stable (see figure 2). It has no variation larger than 1 mas in 24 hours and its secular variation was 2 mas per month in the beginning of the mission and is getting smaller with time. From the beam combiner the light is reflected by a plane mirror towards a spherical mirror with a focal distance equal to 140cm. The optical configuration is actually a refracting Schmidt telescope and the fourth order corrections are carved in the beam combiner mirrors. The focus is somewhat behind the plane mirror which has a central hole. In the focal surface, set on a silicon substrat, are the grids. In the center is the main grid consisting of 2688 parallel slits with a period as measured in projection on the sky is 1.208 arcsec with 39% transparent width. This covers a field of 0°.9 x 0.9° on the sky. In both sides of the main grid are two *"star-mappers"* , one of which is redundant. It consists first of fourth *"vertical"* (that is parallel to the main grid) slits the width of which is 0".9 and the separation respectively a, 3a, 2a with a=5".625. A second system of four similarly separated slits but inclined by 45° in a chevron configuration completes the star-mapper (figure 3).

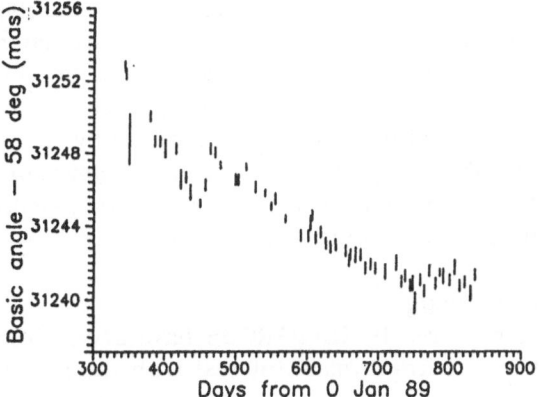

Figure 2 - Observed variations of the basic angle γ during the first one and a half year of mission.

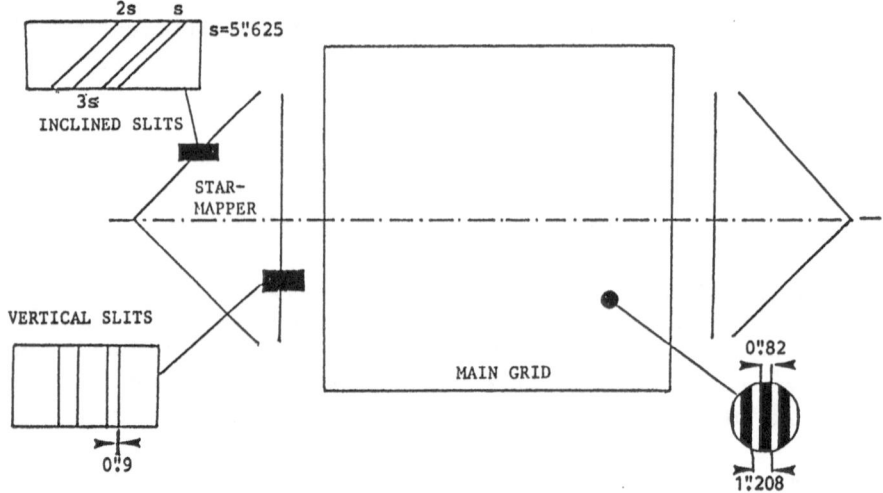

Figure 3 - Star-mapper and main grid structure.

The light received by a star-mapper is split into two beams by a dichroic plate so that each of the two photomultipliers receives different wavelengths corresponding roughly to B and V filters in the UBV system. These photomultipliers sample the received light intensity with a frequency of 600 Hz.

In the main grid, it was not possible to adopt the same system because it is necessary that each star be observed independently of others. This is made possible with the use of an image dissector instead of a plain photomultiplier. An image dissector can separate from the rest of the field of view a very small portion of it. In the case of HIPPARCOS, it is a circle of about 30" in diameter. The centre of the receiving area is continuously computed on-board using the star coordinates provided by a priori known approximate star positions and the on-board attitude. As the result the deflection circuits of the image dissector are controlled so as to make the recorded photons to be produced by the area centered on the tracked star. The sampling from the image dissector is registered at a 1200 Hz rate. A very detailed description of the satellite and the pauload is given in Perryman and Hassan (1989).

3.3. *Observing strategy*

The stars to be observed by HIPPARCOS have been carefully chosen for their astrometric or astrophysical interest with the additional constraint that they are evenly distributed over the entire sky. A little more than 118 000 stars have been selected and their positions, magnitudes and other already known features from ground-based observations, are collected in an *"Input Catalogue"* which is being distributed among the astronomical community. The sky is scanned with the satellite rotating in 2 hours and 8 minutes so that a star crosses the main grid in about 19 seconds. The rotation axis moves slowly, circling the Sun in 57 days keeping an angular distance of 43° from the Sun. In a $3^{1/2}$ year mission that is presently contemplated, each star will be observed between 30 and 100 times depending upon its position on the sky.

Clearly such a complex attitude behaviour cannot be achieved naturally. Whenever the attitude as monitored on-board departs from the nominal scanning law by more then 10 arc minutes, gas jet actuators are activated so as to reverse the natural attitude trend. Actually, this happens about 12 times per satellite rotation and more often near perigee. The on-board attitude determination is based upon three active gyroscopes and observations of star transits through the star-mapper in a manner that is analoguous to the way that data reduction consortia do it a posteriori (see section 5).

At a given time there are several program stars in the combined field of view. During an observation frame of 32/15th of a second, a maximum of ten stars are observed by a rigourous sequencing of individual observations. Unless a program star enters or leaves the field of view during the observation frame, the same stars in identical configurations are observed 16 times forming 16 interlacing periods. In each interlacing period, each star is observed an integer number N of 8 sampling periods of 1/200s; N is large for faint stars so as to compensate at least partly the small number of photons. The 20 groups of 8 sampling periods are distributed among stars proportionnaly to N in such a way that there is no unused time. When there are entering or leaving stars, there may be two or three different strategies adopted with and without these stars.

The satellite transmits to the ground all photon counts from the main grid and continuous photon counts from the star-mapper. In addition, it sends the values of the coil currents that have been used for the image dissector pointing, the on-board computed attitude and some engineering data on how the instrument behaves (temperature, tensions, etc...). This is elaborated on ground by the ESOC control centre and sent to the Data Reduction Consortia together with relevant data on gas jet actuations, orbital motion of the satellite, quality and availability of data, gyroscope readings, eclipse or penumbra situations for the satellite. Concerning the star-mapper data, all the readings are sent to TYCHO data reduction consortium. For the main mission reduction, only observations of program stars on the star-mapper are transmitted.

3.4. *The INPUT catalogue*

The principle of HIPPARCOS implies that one observes stars on a program. This means that a list of stars to be observed had to be made and the positions and magnitudes of these stars must be known in advance sufficiently well so that

1. they can be used for the on-board attitude determination: from an approximate knowledge of the positions and the crossing times observed by the star-mapper one can obtain a sufficiently accurate attitude (better than 1") permitting the IDT to be centered on the stars to be observed;
2. for most of the stars, the position error be significantly better than 0".5 to avoid grid-step errors due to the undiscernability between slits in a periodic grid;
3. magnitudes be known to better than 0.5 magnitude so that the observing time be correctly determined. Special treatment in almost real time is made for variable stars;
4. for the calibration of photometric determinations, some 15 000 stars have magnitudes known to better than 0.02 magnitude.

All these informations and some other have been gathered or obtained by observations by a large consortium of more than 50 astronomers of 9 countries to which many other observers joined. A total of more than 150 persons contributed to this work.

The result is a data base containing data for about 215 000 stars accessible at the Centre de Données Stellaires in Strasbourg. The data consists whenever available or pertinent, of
- Position
- Proper motion
- Parallax
- B and V magnitudes
- Spectral type
- Radial velocity
- Variability type
- Double or multiple star information

These 215 000 stars are all stars requested in 210 scientific proposals as an answer to an ESA announcement of opportunity.

Only 118 000 of these stars were finally selected for the actual program of observation. The same data, but generally better because of the observations performed in 1982-1989 will be published as the HIPPARCOS INPUT Catalogue and is used by the data reduction consortia (Perryman and Turon, 1989). The position accuracy as referred to 1990-1991 proved to have a r.m.s. of 0".25, although some stars may be wrong by 1".5 or 2" and a few misidentification have now been detected.

In the next sections, we shall discuss the methods used for the reduction of the data collected by the satellite. A much more detailed description can be found in Perryman et al. (1989).

4. Reduction of grid data

Whatever is the technique adopted to make astrometric measurements, one must determine the position of star images on the focal surface either directly or on a reproduced image like in photography or CCD detection. A star image is not point-like, but is a certain distribution of light over a finite area. This distribution is the diffraction figure in the best of cases as in HIPPARCOS, or blurred in addition by atmospheric agitation in ground-based astrometry.

It is well known that for circular apertures, the diffraction pattern is composed of a central circular speck and a number of fainter rings. The dimension of the central spot is of the order of $1.22\lambda/D$ where D is the diameter of the entrance aperture and λ the wavelength. For different shapes of the entrance pupil, the light distribution is more complicated, but as a general rule, the characteristic dimension in a given direction is of the order of λ/D where D is the diameter of the aperture in this direction.

In the case of HIPPARCOS, there are two entrance pupils, one per field of view, but they are equal and symmetric. The shape is half an ellipse, projection of the beam combiner mirror on the wave front with a central semi-circular obturation due to the hole of the plane mirror. The overall size is 29cm by 13cm. The larger dimension is along-track. The resulting diffraction pattern is elongated along the vertical direction and is thinner along-scan. For $\lambda=0.50\mu m$, the characteristic size is $2.1\mu m$, representing 0".31 on the sky.

The objective is to recognize some particular point of the diffraction pattern that will, by definition, represent the image of the position of the star. This is quite legitimate, provided that the algorithm would give the same result in different conditions or, in other terms, that the distances between such conventional points are correctly representative of the angular distances between stars independently of their magnitudes or colour. Let us present the situation in the case of single stars for which the characteristic point should be the maximum of the light distribution for a symmetrical diffraction pattern.

4.1. *Light modulation by a periodic grid*

Let us consider a grid with a period s and a transparent section whose width is s'. Let x_0 be the abscissa, perpendicularly to the grid system of a characteristic point of the star image and let $f(x-x_0)$ be the integrated intensity of light for any abscissa x. Thus, the light intensity collected by an elementary slit dx is

$$dI = f(x-x_0)\,dx \ .$$

If the abscissa of the centre of a transparent slit is u, the total illumination transmitted by the grid is

$$I(u) = \sum_{k=-\infty}^{+\infty} \int_{u=ks-\frac{s'}{2}}^{u=ks+\frac{s'}{2}} f(x-x_0)\,dx \ .$$

One can see that if $k \to k+1$, I remains unchanged. Therefore it is a periodic function with a period equal to s. Let us now assume that the image moves along the abscissa axis at a rate $-v$. This is equivalent to say that the slit moves with respect to the image with the rate $+v$. One has

$$I(t) = I_0 + I_1 \cos\left(\frac{2\pi v}{s}t + \phi_1\right) + I_2 \cos 2\left(\frac{2\pi v}{s}t + \phi_2\right) + \ldots$$

where $\phi_1 \ldots$ are phases and $I_1 \ldots$ are modulation coefficients. It has a spatial frequency equal to $1/s$ and a time frequency equal to s/v. This development is rapidly converging. If the characteristic width of $f(x-x_0)$ is w, the cut off spatial frequency is of the order of s/w. One may take for w the half intensity width of the diffracted image along the x axis. Note that it is wavelength dependent. In practice in most of the useful cases, it is sufficient to keep only the two first harmonics

$$I(t) = I_0 + I_1 \cos(\omega t + \phi) + I_2 \cos 2(\omega t + \phi') \ .$$

where $\omega = 2\pi\,v/s$ is the pulsation.

If the image is symmetric, $\phi = \phi'$. This may not always be the case, in particular for double stars or in case of coma. We shall therefore keep both phases in what follows.

4.2. *Photon counts produced by the modulation*

Illumination $I(t)$ is not directly accessible. The light is collected by the IDT during a fixed time Δt. Individual counts are produced by

$$F(t) = \int_{t-\frac{\Delta t}{2}}^{t+\frac{\Delta t}{2}} \left[I_0 + I_1 \cos(\omega t + \phi) + I_2 \cos 2(\omega t + \phi')\right] dt$$

$$= I_0 \Delta t + \frac{I_1}{\omega}\left[\sin(\omega t + \phi + \frac{\omega \Delta t}{2}) - \sin(\omega t + \phi - \frac{\omega \Delta t}{2})\right]$$

$$+ \frac{2I_2}{2\omega}\left[\sin2(\omega t+\phi'+\frac{\omega\Delta t}{2}) - \sin2(\omega t+\phi'-\frac{\omega\Delta t}{2})\right]$$

$$= I_0\Delta t + \frac{2I_1}{\omega}\cos(\omega t+\phi)\sin\frac{\omega\Delta t}{2} + \frac{I_2}{\omega}\cos2(\omega t+\phi)\sin(\omega\Delta t).$$

Δt is chosen to be small with respect to ω. We can develop the last equation to the second order of $\omega\Delta t$ replacing $\sin x/x$ by $1-x^2/6$. Then, $F(t)$ becomes

$$F'(t) = \left[I_0+I_1\left(1-\frac{\omega^2\Delta t^2}{24}\right)\cos(\omega t+\phi)+ I_2\left(1-\frac{\omega^2\Delta t^2}{6}\right)\cos2(\omega t+\phi')\right]\Delta t$$

Time integrated counts reduce the modulation coefficients which we shall call I'_1 and I'_2 but the phases remain unchanged.

Let us consider N counts $F(t_i)$ at times t_i ($1 \le i \le N$) and let us assume that the transit velocity of the star is constant so that ω is constant between t_1 and t_N. Each observation gives the following equation obtained after a division of (5.16) by Δt so as to express counts in Hertz

$$\frac{F'(t_i)}{\Delta t} = J(t_i) = I_0 + I'_1\cos(\omega t_i+\phi)+ I'_2\cos2(\omega t_i+\phi'),$$

Or, introducing new parameters

$$B_1 = I'_1\cos\phi \quad , \quad C_1 = - I'_1\sin\phi \quad ,$$

$$B_2 = I'_2\cos2\phi' \quad , \quad C_2 = - I'_2\sin2\phi' \quad ,$$

$$J(t_i) = I'_0 + B_1\cos\omega t_i + C_1\sin\omega t_i + B_2\cos2\omega t_i + C_2\sin2\omega t_i.$$

These equations are linear in the five unknown parameters and N is large compared to 5. It is therefore possible to solve this system by least squares and transform back the results to get I_0, I_1, I_2, ϕ and ϕ'.

Generally, the astrometric quantity required is the abscissa of the star at time $t=0$. All that can be obtained from the phase s is the distance from the centre of the transparent section, since if $\phi = \phi' = 0$, the intensity is maximum.

Assuming $\phi = \phi'$, the offset from the centre of the slit for $t=0$ is

$$\frac{s\phi}{2\pi} = \frac{s\phi'}{2\pi} = \frac{s}{2\pi}\left(\frac{C\phi + C'\phi'}{C + C'}\right)$$

where C and C' are coefficients that may be chosen proportional to the weight of the determination of ϕ and ϕ'. If $\phi'\ne\phi$, a choice between the above algorithms which are no more equivalent must be made.

But in what slit the image was at $t=0$ is not known and cannot be derived from this analysis. Other means, involving an a priori approximate knowledge of the positions of stars, must be used as it will be described in actual examples. If the slit number k in which is the star is known, then the abscissa of the image is

$$C = ks + \frac{s}{2\pi} \left(\frac{C\phi + C'\phi'}{C + C'} \right)$$

In the case of double stars, $\phi' \neq \phi$, but the form of the response is the same, since the function $I(u)$ is periodic too. The interpretation in function of the star centers is much more complicated (see section 8). However, during one RGC, when the stars present themselves in the same relative way, the same "pseudo centroid" is observed and determined on the main grid.

4.3. Light modulation by a single slit

The star-mapper consists of four slits whose separation (5".6) is such that a star illuminates only one slit at a time. Therefore, the time of crossing of the ensemble of the four slits is the mean of the time of crossing of each slit (giving the transit time through the barycenter of the four slits).

So let us consider one such slit (width s'). The instantaneous illumination is given by

$$I(u) = \int_{u=ks-\frac{s'}{2}}^{u=ks+\frac{s'}{2}} f(x-x_0)\, dx$$

where u is the abscissa of the centre of the slit. While the satellite is rotating, $u=vt$

$$I(t) = F(v(t-t_0))$$

where t_0 is the instant at which the photocenter of the image is at the centre of the grid. Photons are counted during a sample interval Δt. The corresponding illumination is

$$J(u) = \int_{-\Delta t/2}^{+\Delta t/2} f(x-x_0)\, dx\, dt$$

while the normalized illumination, called single slit response is

$$R(u) = \frac{J(u)}{\int_{-\Delta t/2}^{+\Delta t/2} \int_{-\infty}^{+\infty} f(x-x_0)\, dt\, dx}$$

$R(u)$ is calibrated from observations of a number of single star transits. Let $R^*(t)=R(u-u_0)$ centered at the maximum of the response. Let us assume that, for a given observation, one obtains a series of photon counts $\Phi_n(t+n\Delta t)$ and compute the correlation function

$$\sum (t) = \sum_{n_1}^{n_2} \Phi(t+n\,\Delta t)\,.R^*(\,\Delta t\text{-}\,\delta t)$$

This function is maximized for a certain δt equal to the correction to be applied to t in order to fit the best the observation to the single slit response.

The four times t_1 t_2 t_3 and t_4 are determined in this way for the four slits. A barycentric line of the four slits is defined as the line such that its abscissa α is

$$\alpha = \frac{\sum_{i=1}^{4} \alpha_i}{4} \quad ,$$

where α_i are the abscissae of the four slits. The time of crossing this line is given by

$$\tau = \frac{1}{4} \sum_{i=1}^{4} t_i$$

4.4. *Precision of the results*

The precision with which the time of crossing τ of the mean slit is determined, depends on the number of collected photons and therefore on the magnitude of the star. One may express this precision in milliseconds of arc on the sky, using the mean rotation rate of the satellite of 168".75 per second, so that one millisecond of time represents 6 mas on the sky.

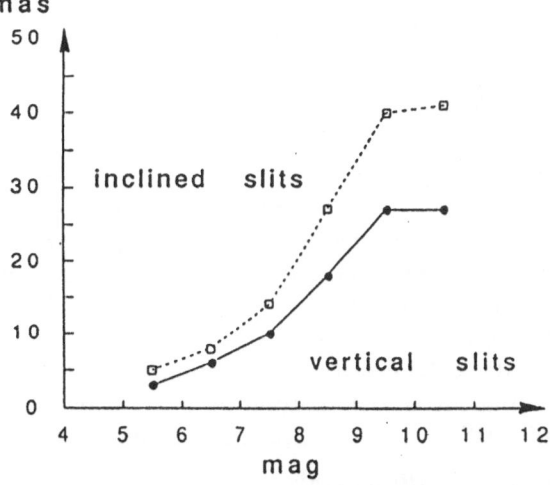

Figure 4 - **Typical average r.m.s. errors of the determination of the transit of a star through the star-mapper. One millisecond of time corresponds to 16 mas.**

Figure 4 gives the mean precision obtained from the reduction of actual data. The results are better for vertical slits than for inclined slits. Stars of magnitude larger than 10.5 are not observable by the star-mapper because the signal desappears in the background noise.

5. Attitude determination

This is the most involved part of the HIPPARCOS data reduction and also the one on which rests the final accuracy of the results as well as the rapidity of convergence towards the solution.

5.1. *Satellite to sky transformation*

Time or phase observations arc made on a grid fixed on the satellite, while the stars that are observed are referred to some celestial system of coordinates. A different celestial reference is defined for each data set covering an observing period included in one orbital period. The OZ axis is arbitrarily chosen as the mean position of the rotation axis of the satellite during the useful observation. The OXY plane is called "*Reference Great Circle*" (RGC), the origine OX being its ascending note on the ecliptic. It is essential to know what are the functional relations that exist between these two systems. Let us first define them. On the grids, the origin Ω will be the centre of the main grid. The ΩG coordinate will be taken along track so that G increases with time and ΩH is directly perpendicular being parallel to the main grid system. In this system of grid coordinates, the equations of the mean lines of the preceding star-mapper have ideally the form

$G = G'_0$ for the vertical slits

$G = G''_0 \ -|H|$ for the chevron slits

Or generally

$G = G_0 + \varepsilon H$

with

$$\varepsilon = -1, 0 \text{ or } +1 \tag{5.1}$$

For the satellite, we shall define a system of rectangular axes such that O being the centre of mass of the satellite, Oz axis is taken parallel to the edge of the beam combiner and Oxz is the bissecting plane of the two optical axes of the telescope. The intersection with the celestial sphere of the Oxy plane is the instantaneous great circle of rotation (IGC). It is refered to the RGC celestial coordinate system by three Eulerian angles α, β, δ so that the transformation from the body axes $Oxyz$ to the celestial RGC axes $OXYZ$ is defined by the rotation matrix (figure 5)

$$\mathcal{C} = \mathcal{R}_3(\alpha) \, \mathcal{R}_1(\beta) \, \mathcal{R}_3(\delta)$$

It could also be written under the form of

$$\mathcal{C} = \mathcal{R}_1(\alpha') \, \mathcal{R}_2(\beta') \, \mathcal{R}_3(\delta')$$

where α', β' and δ' are respectively the rotation angles around Ox, Oy and Oz.

It is somewhat advantageous to use a combined system that separates the along track component

$$\psi = \alpha + \delta$$

from small angles representing rotations around axes in the RGC or the IGC: $\theta = \beta'$ and $\phi = \alpha'$. So that

$$\mathcal{C} = \mathcal{F} (\psi, \theta, \phi) \qquad (5.2)$$

Here $\mathcal{F}$ cannot any more be expressed by matrix products, but still is a defined function.

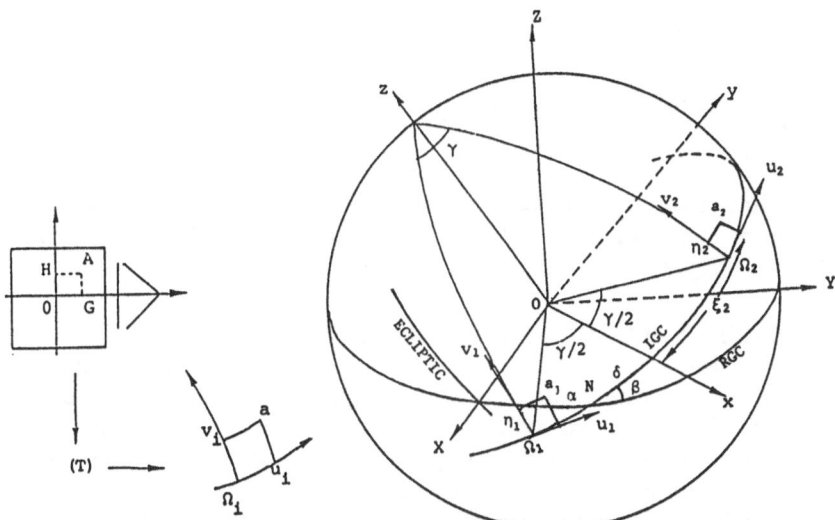

Figure 5 - Grid-to-field transformation, coordinate system with respect to the IGC and its relations with the RGC system.

Finally, let us define local celestial coordinates (figure 5). Let ω_1 and ω_2 be points for the preceding and following fields on the IGC separated by $\pm \gamma /2$ from Ox . These points are very close to the points whose images on the grid are both in Ω. The $\omega_1 u_1$ and $\omega_2 u_2$ axes are tangent to the RGC and $\omega_1 v_1$ and $\omega_2 v_2$ perpendicular in the same direction as OH. Let us note that if u_i and v_i (i=1 or 2) local coordinates of a star, the coordinates in the IGC system are $\xi_i = u_i = \gamma /2$ and $\eta_i = v_i$.

The transformation between local grid and local celestial coordinates is close to an identical transformation, provided that we use on the grid a grid angular unit called grid radian such that the grid period is 1".208 expressed in radians. It is however not an identity because of shifts and tilts of the grid with respect to the optical axis and of various aberrations of the optical system. The transformations

$(T_1):(GH) \rightarrow (u_1\ v_1)$

$(T_2):(GH) \rightarrow (u_2\ v_2)$

are called grid-to-field transformations for each field of view.

Since there is also some colour dependent aberration in the system, terms proportional to the colour index c are included in the transformation (i=1 and 2)

$u_i = G + U_i\ (G,H) + cU'_i\ (G,H)$

$v_i = H + V_i\ (G,H) + cV'_i\ (G,H)$

The function U and V may be either expressed in terms of developments in power series of G and H or by numerical tables.

The inverse field-to-grid transformation has a similar format

$$G = u_i + g_i\ (u_i,v_i) + cg'_i\ (u_i,v_i)$$
$$H = v_i + h_i\ (u_i,v_i) + ch'_i\ (u_i,v_i)$$

(5.3)

The transformation between the body $Oxyz$ and celestial RGC $OXYZ$ systems of coordinates is the attitude. To determine the attitude means to find the expressions of ψ, θ and ϕ in function of time.

5.2. _Equations for attitude determination_

Two measurements contribute to the attitude determination: gyroscope readings and timing of star transits through the star-mapper grids.

Each gyroscope measures the inertial rotation of the satellite about one of three non coplanar axes called gyroscope input axes. The gyroscope reading is done automatically by an electronic system that determines the torque needed to maintain the gyroscope spin axis parallel to the input axis. Using a torque model that includes the torques produced by solar radiation pressure, gravity gradient and gyroscope reaction, one can derive the time derivatives of the three angles ψ, θ and ϕ.

$$\left.\begin{array}{l} \dfrac{d\psi}{dt} = \Psi(t) \\[2mm] \dfrac{d\theta}{dt} = \Theta(t) \\[2mm] \dfrac{d\phi}{dt} = \Phi(t) \end{array}\right\}$$

(5.4)

However, these readings are not sufficiently accurate nor is the torque model sufficiently precise to allow integrating these differential equations.

The use of star-mapper observations provides a direct link between the attitude angles, time and the coordinates of the observed star. Let t be the time when a star image has been observed to be crossing a barycentric line of equation (5.1)

$$G = G_0 + \varepsilon H$$

The corresponding line on the sky is obtained by applying the field-to-grid transformation (5.3). One obtains a certain equation

$$F(u, v, g_i, cg'_i, h_i, ch'_i) = 0 \tag{5.5}$$

Assuming that the field-to-grid transformation is calibrated, this reduces to a certain non linear equation

$$f(u, v, \Delta u, \Delta v) = 0 \tag{5.6}$$

where Δu and Δv are the corrections for this transformation.

If λ and β are the coordinates of the star in the RGC system of coordinates, the relation between IGC and RGC coordinates are given by the transformation $\mathcal{F}(\psi, \theta, \phi)$ so that, expressed in terms of λ and ξ, one gets by elementary calculations

$$\left. \begin{array}{l} \xi = u \pm \gamma/2 = x(\lambda, \beta, \psi\beta(t), \theta(t), \phi(t)) \\[2mm] \eta = \quad v \quad = y(\lambda, \beta, \psi(t), \theta(t), \phi(t)) \end{array} \right\} \tag{5.7}$$

so that the condition expressing that the star is observed on a barycentric line of the star-mapper is of the form

$$Z(\lambda, \beta, \gamma, \psi(t), \theta(t), \phi(t), \Delta u, \Delta v) = 0 \tag{5.8}$$

Each observation of a star-mapper transit gives a Z equation in ψ, θ and ϕ for different times. The quantities Δu and Δv are computed using the calibrated grid-to-field transformation. How this calibration is done will be shown later. Finally λ and β are taken from the best available star catalogue (Input Catalogue at the first treatment, improved values later).

5.3. _Representation of the attitude_

Using equations (5.8) and if necessary equations (5.4) one must determine the three function of t. Since at a given instant, one may get only a single relation (5.8) between the three unknown functions of time, it is necessary to represent them by some analytical functions.

The difficulty is that there exists no simple analytical representation. Between gas jets, the functions are smooth, but the effect of a gas jet actuation is to modify the first derivative so that there is at that moment a discontinuity in the derivatives. But actually, gas jets have a duration between 0.05 and 0.5 seconds and the passage from the previous situation to the next is continuous.

The simplest solution to describe this is to use cubic splines which are third order polynomials with continuity conditions in the extremities of intervals. But generally, in this case, the observations are unsufficient in number and this solution is not acceptable. In one of the consortia (NDAC), numerical integration of the modelled torques improved by gyroscope data is used and then smoothly upgraded by the results of star-mapper observations. This numerical approach was found to give quite satisfactory results. In the other consortium (FAST) a fully analytical approach was adopted and applied in two different ways.

i) Long term representation - Outside eclipses, when the torques are essentially functions only of the rotation angle ψ, it has been analytically shown that the variation of attitude may be represented, in absence of gas jets, by a periodic function of ψ and that this function is insensitive to initial conditions within a certain vicinity of a given solution. If the time span is one satellite rotation, and the precision required is of the order of 0".05, these periodic functions may be developed in Fourier series of $\psi(t)$ up order 15 (or less) :

$$\psi = \psi_0 + \psi_n (t - t_n) + \sum_{j=1}^{15} P_j \cos j \ (\psi_n (t - t_n) + \psi_j) \qquad (5.9)$$

This development has a secular term, representing the rotation. There are two analoguous functions for θ and ϕ. When there is a gas jet actuation, the initial conditions are changed, but not in a manner that contradicts the above results so that one can keep for all the periods between gas jet actuation the same coefficients P_j of (5.9) and modidy only ψ_n. Using the simplified relation (5.9), one has to determine the P_j, ψ_j and ψ_n. However, one should add to it a dynamical description of the gas jet effects. If δ_n is the impulsive function of the gas jets and t_n, t'_n are the times of beginning and end of the actuation, the first part of (5.9) becomes

$$\psi = \psi_0 + \psi_1 (t - t_n) + \psi_2 (t - t_n)^2 + \int_{t_n}^{t'_n} \int_{t_n}^{t'_n} A_n \delta (t - t_n) \ dt + \dots \qquad (5.10)$$

Similar equations hold for the other two angles θ and ϕ. So, in summary, the integral is computed using a model and one has to determine 90 coefficients for the trigonometric series and about 80 coefficients for the second order representation of the secular terms. The number star transits and consequently of equations is of the order of 700. This is quite sufficient to determine the parameters by a least square procedure.

ii) Short term representation - In case of eclipses or penumbra, the main cause of the torque - the solar radiation pressure - disappears or is quickly modified and the preceding model does not apply. Then the adopted solution is to represent each angle by a Legendre polynomial of high order (between 5 and 15). In comparison with the first case, the proportion of observation equations to the number of parameters to be determined is smaller by a factor of about 2 and the continuity conditions are not well taken into account. This mode has to be used as rarely as possible, since it gives a precision which is about twice worst than the long term representation.

5.4. *Precision of attitude determination*

If one takes into account only the precisions obtained for the star transit estimates one may expect accuracies of the results to be of the order of 20, 35 and 80 mas respectively in ψ, θ, ϕ . In practice, the modeling error may be large, and biases are introduced in the short term representation. In particular, if observations are lacking at the beginning or the end of the period because, the model being only numerical, it is unable to perform extrapolations. In some instance errors of 1" or 2" have been shown to exist. The long term representation, fully dynamical, does not suffer this drawback. It is expected from simulations, that accuracies should be of the order of 30, 50 and 100 mas. In real situations, this could not yet • be checked, since the star positions used are given by the Input Catalogue whose r.m.s. error is of the order of 0".25 to 0".30 and this is reflected in the overall precision of the result. With these uncertainties, the actual precisions for the long term representation in the first treatment are found to be :

    35-55   mas for ψ
    60-90   mas for θ
    130-170 mas for ϕ

These numbers are to be multiplied by 2 for the short term polynomial representation.

6. Reduction on a great circle

The data acquired by the satellite is, as we have already stated, divided into data sets of a few hours (from 3 to 8). The reduction on a great circle is the first step to obtain astrometric information from the observations treated as shown in the preceding sections. The complete data so gathered is the following

- The attitude from the star-mapper.
- Grid coordinates deduced from the phases which give only the distance from a central line of a slit. The slit number is deduced from the position of the star as given by the Input Catalogue and transferred to grid coordinates using the field-to-grid transformation and the attitude by formulae that are inverse to those used to determine the attitude. Unfortunately, due to errors in the Input Catalogue, errors of one or more grid steps may occur for some stars.
- Field-to-grid transformation, basic angle and a few other calibrated quantities.

- The ordinates of the stars with respect to the RGC computed from the Input Catalogue positions and the attitude.
- The theoretical geometric position of each observed star at some mean instant t_0 of observation corrected for parallax and proper motion. This reference position denoted as α_c (computed abscissa) is to be improved by the great circle reduction. At each instant t when the star is observed, the difference between the instantaneous apparent position and the reference position is computed in such a way that any difference between the observed and computed apparent abscissa is interpreted as a correction to the reference position at time t_0.

6.1. *The equations*

Let λ and β be the coordinates of the star in the RGC reference frame and u, v, their local field coordinates, referred to the instantaneous great circle (IGC). With respect to the X axis of the satellite, the abscissa along the IGC is $u \pm \gamma/2$ (+ preceding, - following field of view). Let us take the ordinary Euler angles to represent the attitude (see figure 6)

$$\psi - \phi = \Omega N$$

$$\phi = N X$$

$$\theta = RGC, IGC$$

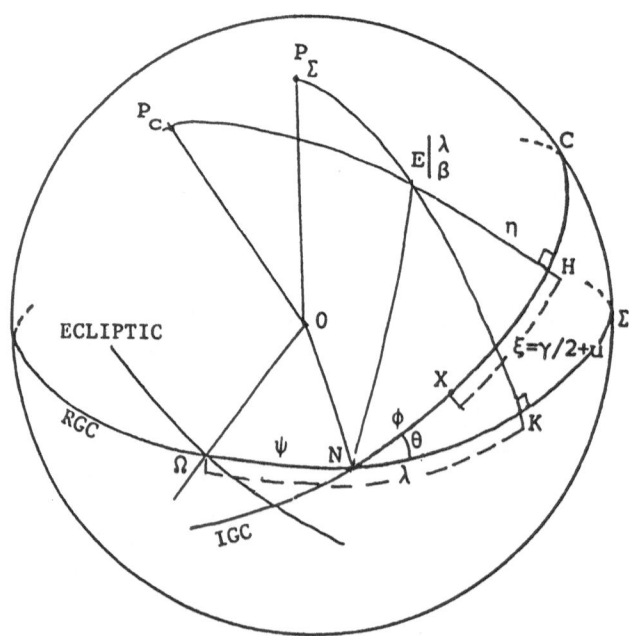

Figure 6 - Transformation of coordinates between the RGC and the IGC systems.

Let us compute NE in two different manners

$$\cos(NE) = \cos(NH)\cos(EH) = \cos(NK)\cos(EK)$$

$$= (\lambda - \Omega N)\cos\beta \qquad = \cos(\phi \pm \frac{\gamma}{2} + u)\cos\upsilon$$

$$\cos(\lambda - \psi + \phi) = \frac{\cos(\phi \pm \gamma/2 + u)\cos\upsilon}{\cos\beta}$$

u is determined from the observed grid coordinate, G and the other grid coordinate H by the grid-to-field transformation as given in section 5.1

$$u = G + U_i(G,H) + cU'_i(G,H)$$

where third order polynomials are used to describe U and U' ($i = 1$ or -1, preceding or following field). So finally,

$$u = G + \sum A_{ijk} G^J H^k + c \sum A'_{ijk} G^J H^k$$

Solving for λ, one gets an equation of the form

$$\lambda = F(\beta, u\,(i, A, A'), \psi(t), \theta(t), \phi(t), \gamma)$$

In the basic equation, we assume that
- γ is not exactly known,

$$\gamma = \gamma_0 + \Delta\gamma$$

- the attitude parameter $\psi(t)$ as determined by the attitude determination is not sufficiently accurate:

$$\psi(t) = \psi_0(t) + \Delta\psi(t)$$

where $\psi_0(t)$ is the along scan attitude determined by the star-mapper.
- parameters A_{ijk} and A'_{ijk} are also not well known:

$$A_{ijk} = A'^{\circ}_{ijk} + \Delta A_{ijk}$$

$$A'_{ijk} = A'^{\circ}_{ijk} + \Delta A'_{ijk}$$

These are the unknowns that are added to $\Delta\lambda$.

Let us compute F with the reference values and develop the difference. One obtains

$$\lambda_0\big(\beta, u\,(\gamma_0, A, A'^{\circ}), \psi_0(t), \theta(t), \theta(t), \phi(t)\big) =$$

$$= -\Delta\lambda + \frac{\partial F}{\partial \gamma_0}\Delta\gamma + \sum_{ijk}\frac{\partial F}{\partial A^{\circ}_{ijk}}\Delta A_{ijk} + \sum \frac{\partial F}{\partial A'^{\circ}_{Aijk}}\Delta A'_{ijk} + \frac{\partial F}{\partial \psi_0}\Delta\psi_0(t) + \eta + O(\varepsilon^2)$$

where η represents the photon noise effect on the grid coordinates and second order effects $O(\varepsilon^2)$ have been neglected.

6.2. The design matrix
The ensemble of these equations for all observed grid coordinates of a set of stars assumed to be well observed at each observation frame and

called active stars is a system that can be formally written in matrix form as

$$y = Ax + e$$

where e is a stochastic variable.

It can be also written as

$$y = A_a x_a + A_s x_s + A_i x_i + e$$

corresponding to attitude, star and instrumental unknowns. A_a and A_s are very sparse (one non zero element per row), A_i is completely filled.

In a typical data set of a little more than 7 hours of observations, one has about 50 000 equations (four per observing frame). There are typically 1000 active stars chosen for their good observations not double, not too faint and 25 instrumental unknowns. Two cases occur :

1. a single uncorrelated value of $\Delta\psi$ is computed for each of the 12 000 observation frames. It is called the *geometric solution* . It is simpler, but steps of 1".2 (grid width) may sometime happen and no attitude may be determined in observation frames without stars,
2. $\psi(t)$ is represented in a polynomial form (spline functions). Each function represent some 25 to 40 observation frames. Cubic splines are defined by four parameters with some conditions at the edges, which slightly increase the number of equations of conditions. This reduces the attitude unknowns to about 1200-1500. This is the *smoothed solution* , which is highly preferable but sometimes may fail if there is not enough observations for a given spline.

Even in the case of the geometric solution, there are significantly more equations than unknowns.

6.3. Geometric solution
It is the first that is computed. The computation goes as follows :
- Elimination of the attitude unknowns.
- Choleski factorization of the normal equations and solution for the instrumental parameters and star positions.
- Substitution of the solution in the observation equations and solving for the attitude parameters.

The equations have a rank deficiency of one allowing all λ be determined modulo a constant. It is resolved by the minimum norm method, forcing the sum of abscissae corrections be equal to 0.

In some cases, the reduced star abscissae cluster in batches separated by s=1".208 showing the existence of grid step errors in grid coordinates. One cluster is arbitrarily chosen correcting the other grid coordinates by an integer multiple of s.

6.4. Attitude smoothing
A second approach is then performed with attitude smoothing equations.

Instrumental parameters are not recomputed. The equation writes as

$$- A_l \, x_l + y = A_a \, Bx_b + A_s \, x_s + e_l$$

where $x_a = Bx_b$, B being the diagonal matrix of the B spline representation.

The computation goes as follows :
- star unknowns are eliminated,
- after reordering, the attitude unknowns are computed,
- they are substituted in the equation which are solved for star abscissae.

Therefore, two sets of attitude and star parameters are available at the end. They are compared using statistical tests. As a rule, whenever, the smoothed solution has good statistical tests, it is chosen. This happens in about 80% of the cases. Otherwise, the geometric solution is kept. The most frequent cause of failure of the smoothed solution is the absence of observations during a short B spline.

6.5. *Passive stars*
All stars that have not participated to the main reduction are called passive stars. Their abscissae are then determined. The instrumental and attitude parameters are substituted in the equations and star abscissae are computed. Similarly minor planet positions are determined: one per observation frame.

6.6. *Accuracy of the solution*
The precision of the determination of abscissae on the reference great circle depends upon the precision of the grid coordinates which improves as the number of photons received increases. Table 3 gives the rms currently obtained by the great circle reduction using the geometric and the smoothed solutions as a function of star magnitudes. This is compared to the mean rms of the grid coordinates over 25 observation frames, a typical number of observations for an actual data set treated by the great circle reduction

TABLE 3

Magnitude	Mean r.m.s geometric	Means r.m.s smoothed	r.m.s of mean grid coordinates*
< 6	3.0 mas	1.8 mas	0.6 mas
6 to 7	3.3 mas	2.2 mas	1.0 mas
7 to 8	3.5 mas	2.6 mas	1.5 mas
8 to 9	3.8 mas	3.1 mas	2.1 mas
9 to 10	4.5 mas	3.8 mas	2.8 mas
10 to 11	5.4 mas	4.7 mas	4.3 mas
> 11	7.0 mas	6.0 mas	6.4 mas

* computed assuming 25 independent observations (observation frames)

This table shows that there is still place for improvement. Indeed in the first treatments as those made until now, the star positions used for attitude determination have a mean error of about 0.3 arc second. This introduces a bias in the abscissa determination which depends upon the star-mapper attitude. The calibrations used are not yet the best possible. Possible grid step errors also shift the solution for some stars, increasing the apparent error. It is expected that all these causes of error do not show in the r.m.s. and that the actual accuracy is significantly worse than the numbers given in table 3. However, after several iterations in which improved star catalogues will be used, the final accuracy should be of the order of the r.m.s. for the smoothed solution and possibly better.

7. Computation of astrometric parameters

The next step of the reduction is a synthesis on the sphere. Starting from the projections on great circles, the problem is to find what is the track of the star on the sky. Each star will be observed on 30 to 50 reference great circles out of the 3000 that will be obtained during the full mission. In addition, because of the rank deficiency of the great circle solution, the provisional origins of the great circles have to be determined.

In this part of the reduction, the apparent motion must be exactly modelled. This excludes double and multiple stars whose determined centroid is direction dependent (see section 8) and those stars that may have grid step errors. Actually, in FAST, we have chosen *a priori* 40 000 stars that are known not being double, whose positions in the INCA catalogue have an r.m.s. better than 0".4 and are the brightest possible so that they are well observed. They are called primary stars on which a sphere reconstitution process is applied in order to determine the positions of all RGC origins. Then, the astrometric parameters can be determined for all simple stars.

7.1. *Sphere reconstitution*

Let us call λ°_0 and β°_0 the ecliptic longitude and latitude at time t as it is assumed to be known. Let us also call μ°_{λ}, μ°_{β} and ϖ_0 the reference components of the yearly proper motion and the reference parallax. The actual values of these parameters differ by unknown quantities $\Delta\lambda_0$, $\Delta\beta_0$, $\Delta\mu_{\lambda}$, $\Delta\mu_{\beta}$, $\Delta\varpi$ that have to be determined.

At a time t_i, the position of the star is given by

$$\lambda(t_i) = \lambda^{\circ}_0 + \Delta\lambda_0 + \left(\mu^{\circ}_{\lambda} + \Delta\mu_{\lambda}\right)(t_i - t_0) + \frac{(\varpi + \Delta\varpi)(1+f(t))}{\cos\beta} F(t,\lambda,\beta)$$

$$\beta(t_i) = \beta^{\circ}_0 + \Delta\beta_0 + \left(\mu^{\circ}_{\beta} + \Delta\mu_{\beta}\right)(t_i - t_0) + (\varpi_0 + \Delta\varpi)(1+f(t))G(t,\lambda,\beta)$$

$$(7.1)$$

where $1+f(t_i)$ is the distance between the observer and the barycenter of the solar system expressed in astronomical units. It is given by the ephemerides of the Earth motion as well as $L(t_i)$, the longitude of the barycentre. The functions F and G are the projection factors of the parallactic displacement on both axes.

Since the position of the RGC on the sky is known by the exact position of its pole, one can compute the reference abscissa a_0 on the RGC from its intersection with the ecliptic plane, the reference values $\lambda^{\circ}{}_0$, $\beta^{\circ}{}_0$ etc... and from (7.1). Linearizing the resulting expression, one gets

$$a = a_0 + A_i \left[\Delta\lambda_0 + \Delta\mu_\lambda \, (t_i - t_0) \right] + B_i \left[\Delta\beta_0 + \Delta\mu_\beta (t_i - t_0) \right] + C_i \, \Delta\varpi$$

This has to be equated to the abscissa a^* obtained by the great circle reduction. However a_0 is affected by the error $\Delta\Omega$ of the origin and also by other errors that are assumed to be essentially dependent upon thermal effects due to the solar illumination and therefore are modelled as periodic functions of the angular distance to the projection of the Sun position on the RGC. Taking into account these corrections, one gets general equations of the form

$$a^* - a_0 = A_y \, \Delta\lambda_i + B_y \, \Delta\beta_i + A'_y \Delta\mu_{\lambda_i} + B'_y \, \Delta\mu_{\lambda_i} + C_y \, \Delta\varpi_i + \Delta\Omega_j$$
$$\sum_{k=1}^{k} \left[\, p_k \cos k(a^* - s_j) + p'_k \sin k(a^* - s_j) \, \right] \tag{7.2}$$

where index i refers to stars and index j refers to RGC's; p_k and p'_k are instrumental unknowns and s_j is the abscissa of the Sun on the RGC.

With 3000 RGC's and about 400 primary stars on each RGC, one gets of the order of 1 200 000 equations with 200 000 unknown astrometric parameters, 3000 origins and a dozen of instrumental parameters.

Grouping the equations star by star (index i), one may eliminate the astrometric unknowns and be left with about one million equations that depend only on the origins and instrumental unknowns. This system of equations is used to form the least square equations of conditions that are solved by an iterative method of successive approximations. There has not been yet any significant solution made using actual data, but simulations have shown that origins should be determined to an accuracy of 0.4 mas even if a few grid step errors remain.

As in the case of reductions on a great circle, the sphere reconstitution equations have a rank deficiency. A fixed and a linearly time dependent rotations do not modify the solution. These can be fixed by six binding linear conditions which will have to be determined later.

7.2. _Astrometric parameter determination_

Once the origins and the instrumental parameters are determined, these quantities are substituted in equation 7.2 and each group of equations pertaining to a given star is solved for the astrometric parameters.

This is done not only for primary stars, but also for the others, with the exception of double and multiple stars. A specific algorithm has been devised in order to solve for the remaining grid step errors. This is not very difficult unless there is a large number of them.

The final precision expected from this reduction depends of course upon the duration of the mission. Because of the peculiarities of the scanning law, it depends also upon the latitude of the star. It is given in the case of the nominal mission in figure 7. The actual mission, if it lasts four years, should approach closely these expectations though with slightly worst results for parallaxes and slightly better for proper motions.

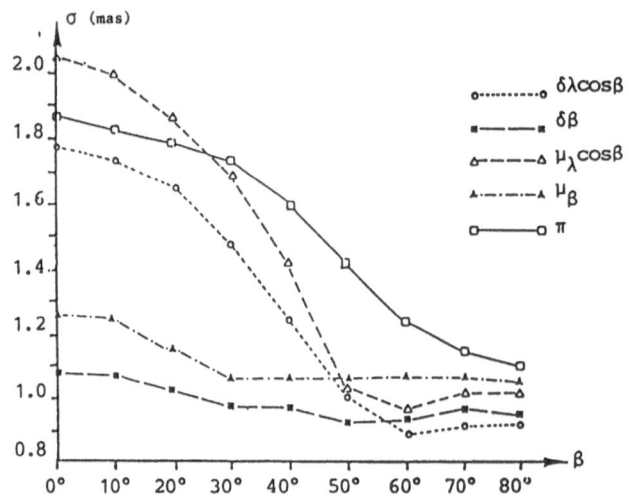

Figure 7 - Expected final precisions in the case of the nominal mission.

8. Double and multiple star treatment

Double and multiple stars do not give the same signal as single stars. It is the sum of the signals given independently by each of the components. If the intensity and the relative position of the components are perfectly known, it is not difficult to compute the five modulation parameters. But the inverse problem cannot be solved from the observation on a single RGC. It is therefore necessary to gather all observations over a number of RGC's whose directions in the sky are known and solve the inverse problem.

The main difficulty arises from the fact that although the relative angles of RGC's are known, the absolute abscissae of the signals are not known, so that there is no way to transform the phases into absolute position information, as it is done in VLBI for instance in the image retrieval by the CLEAN method.

285

8.1. First approximation, double stars

The only way is to start from approximate solution, and then vary the parameters. The basic equations are as follows where we assume the modulation coefficients M_1 and M_2 equal to a mean value.

Let us consider the case of a double star and neglect the background

$$A_1 = I_1 + I_1 M_1 \cos(\omega t + \phi_1) + I_1 M_2 \cos(2\omega t + 2\phi_1)$$

$$A_2 = I_2 + I_2 M_1 \cos(\omega t + \phi_2) + I_1 M_2 \cos(2\omega t + 2\phi_2)$$

The sum of the modulations is

$$B = J_0 + J_0 N_1 \cos(\omega t + \theta_1) + J_0 N_2 \cos(2\omega t + 2\theta_2)$$

with

$$J_0 = I_1 + I_2$$

$$I_0 N_1 \cos\omega t \cos\theta_1 - J_0 N_1 \sin\omega t \sin\theta_1 =$$

$$= I_1 M_1 \cos\omega t \cos\phi_1 - I_1 M_1 \sin\omega t \sin\phi_1 + I_2 M_1 \cos\omega t \cos\phi_2 - I_2 M_1 \sin\omega t \sin\phi_2$$

giving

$$J_0 N_1 \cos\theta_1 = I_1 M_1 \cos\phi_1 + I_2 M_1 \cos\phi_2$$

$$J_0 N_1 \sin\theta_1 = I_1 M_1 \sin\phi_1 + I_2 M_1 \sin\phi_2$$

and, similarly

$$J_0 N_2 \cos 2\theta_2 = I_1 M_2 \cos 2\phi_1 + I_2 M_2 \cos 2\phi_2$$

$$J_0 N_2 \sin 2\theta_2 = I_1 M_2 \sin 2\phi_1 + I_2 M_2 \sin 2\phi_2$$

After some transformations, one gets

$$\left(J_0 N_1\right)^2 = I_1^2 M_1^2 + I_2^2 M_1^2 + 2 I_1 I_2 M_1^2 \cos(\phi_1 - \phi_2)$$

$$\left(J_0 N_2\right)^2 = I_1^2 M_2^2 + I_2^2 M_2^2 + 2 I_1 I_2 M_2^2 \cos 2(\phi_1 - \phi_2)$$

The last two equations are modified putting

$$\left(J_0 N_1\right)^2 = \left(J_0 M_1\right)^2 - 2 I_1 I_2 M_1^2 \left(1 - \cos(\phi_1 - \phi_2)\right)$$

$$\left(J_0 N_2\right)^2 = \left(J_0 M_2\right)^2 - 2 I_1 I_2 M_2^2 \left(1 - \cos 2(\phi_1 - \phi_2)\right)$$

Let us call $x = (\phi_1 - \phi_2)/2$. One gets

$$I_1 I_2 \sin^2 x = \frac{J_0^2 (M_1^2 - N_1^2)}{4M_1^2}$$

$$I_1 I_2 \sin^2 x \cos^2 x = \frac{J_0^2 (M_2^2 - N_2^2)}{16M_2^2}$$

Solving in x, one gets the projected separation by $s(\phi_1 - \phi_2)/2\pi$ $+ks$ (k is known because largely separated double stars are sufficiently well known). From these, one then computes $I_1 I_2$ and since $I_1 + I_2 = J$, are the roots of

$$I^2 - J_0 I + (I_1 I_2) = 0.$$

Finally the first two equations can be modified and become

$$J_0 N_1 \cos(\theta_1 - \phi_1) = I_1 M_1 + I_2 M_1 \cos(\phi_1 - \phi_2)$$

$$J_0 N_1 \sin(\theta_1 - \phi_1) = \qquad + I_2 M_1 \sin(\phi_1 - \phi_2)$$

giving $\theta_1 - \phi_1$ and hence ϕ_1.

Similarly, the other two equations give ϕ_2 and $\theta_2 - \phi_2$.

Theoretically, from the knowledge of $\phi_1 - \phi_2$ over two RGC's, one may determine the separation ρ and the angle of position θ of the double star. This is actually computed from many RGC's.

8.2. Improvement of the solution

Once a first solution has been obtained, it is sufficiently close to the actual solution to permit an adjustment of all parameters. For a fixed double star, one would take

$$\rho = \rho_0 + \Delta\rho$$

$$\theta = \theta_0 + \Delta\theta$$

$$M_{11} = M_1° + \Delta M_{11}$$

$$M_{12} = M_2° + \Delta M_{12}$$

$$I_1 = I_1° + \Delta I_1$$

and similar unknowns for the second star.

One may also adjust linear a quadratic movements in ρ and θ, or some orbital parameters like Thiele-Innes coefficients.

Once the solution computed by a least square adjustment is known, one goes back to each observation and gets residuals and the smoothed values of θ_1-ϕ_1; θ_2-ϕ_1; θ_1-ϕ_2; θ_2-ϕ_2 and

$$\left(C_1\theta_1 + C_2\theta_2 \right) - \left(C_1\phi_1 + C_2\phi_2 \right)$$

The latter provides the value of the shift from the position observed to that of one of the components.

9. Iterations

The precision and the accuracy obtained at the great circle reduction depend upon a certain number of quantities that cannot, in a first run, be determined with a sufficient accuracy because they use the Input Catalogue in which most of the star positions have an r.m.s. error of about 0".3 and some stars have much worst positions. This is the case of the attitude in ϕ and θ which is not improved by the great circle reduction. This is also the case of grid coordinates which may have errors equal to an integer number of grid periods because the Input Catalogue is consulted in order to determine to which slit the observation relates. Finally some calibrations such as grid-to-field for star-mapper slits depend also on original star positions.

9.1. *Principle of iterations*
When a year or a year and a half of data will be collected and processed, it is expected that the results will be correct to 10 mas or possibly better. This is one and a half orders of magnitude better than the Input Catalogue. Therefore if the reduction is repeated using the new values of star positions, all errors just described will desappear. The iteration procedure is based upon this remark.

   Not all the computations have to be repeated. The analysis of photon counts on the star-mapper or on the main grid does not depend upon the star positions so that the times of transit through the star-mapper and the modulation curves by the main grid have not to be redetermined. The attitude is recomputed using not only the new star positions, but also the improved $\psi(t)$ as obtained by the preceding great circle reduction which gives results much more precise than with the star-mapper. Grid coordinates are recomputed in so far the grid numbers are redetermined. This gives new inputs to the great circle reduction which is repeated, providing new abscissae to the sphere reconstitution and astrometric parameter determination.

   This process may be reiterated if the results appear still not being satisfactory. Figure 8 presents the overall reduction scheme as adopted by FAST consortium, including the relations with other HIPPARCOS tasks.

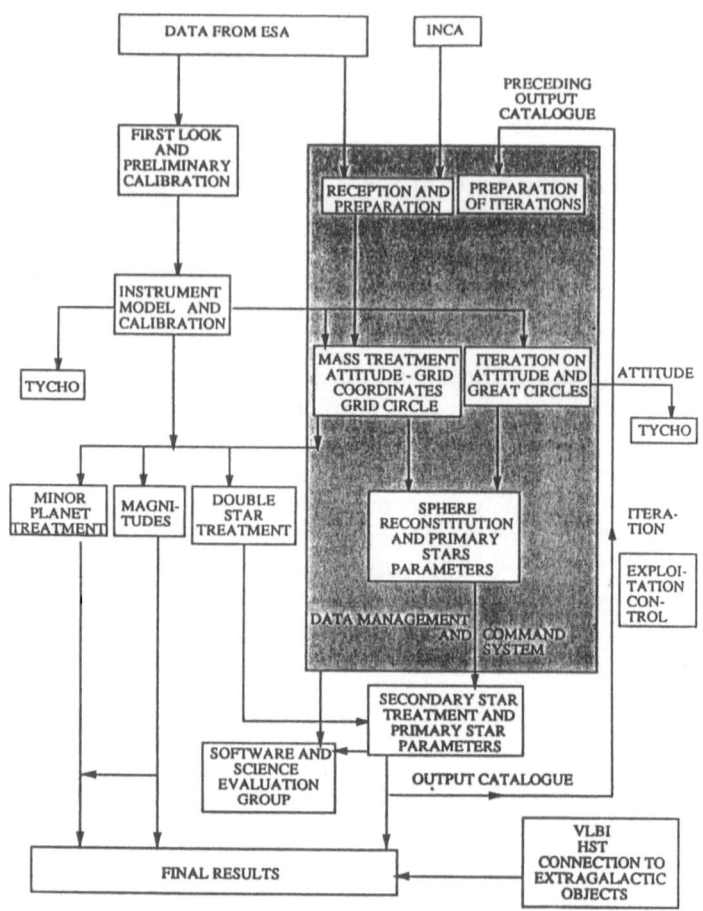

Figure 8 - Schematic descriptions of the FAST reduction procedures.

9.2. *Setting the sphere rotation*

Once the final sphere solution will have been obtained, the stars will be refered to a consistent HIPPARCOS reference frame, which will have to be compared with existing reference frames. It actually can be linked to any of them, using the possibility of a rotation and a time dependent rotation provided by the sphere reconstitution rank deficiency. If $\mathcal{R}$ and $\mathcal{R}'$ are matrices representing these rotations, any star direction $\boldsymbol{E}$ may be replaced by

$$\boldsymbol{E}'(t)=(\mathcal{R}+(t-t_0)\,\mathcal{R}')\boldsymbol{E}(t)$$

and proper motion vector $\boldsymbol{M}$ by

$$\boldsymbol{M}'(t)=\mathcal{R}'\boldsymbol{M}(t)\ .$$

These equations can be written for a certain number of stars whose position is known in both coordinate frames and solved for $\mathcal{R}$ and $\mathcal{R}'$.

It would be possible to refer HIPPARCOS to the FK5 system, since all FK5 stars are in the HIPPARCOS observation program. But it is better to link it to a more stable and better defined reference frame, the one constructed by VLBI and based upon the positions of quasars whose proper motions are assumed to be very small.

The problem is that quasars are too faint to be observable by HIPPARCOS. However, there are some radio stars which are part of the HIPPARCOS catalogue and are sufficiently good radio-emitters to be observable by VLBI, at least in its relative mode (phase locked to a near-by quasar). The accuracy with which proper motion of such star may be (and in two cases have already been) obtained is a fraction of a mas/year. These are mostly RS CVn stars which are very close binaries (separation a few mas) and are radio flare stars. It is not clear where the emitting point is fixed with respect to the photocenter observed by HIPPARCOS, but the important information is proper motion which is in the mean the same.

9.3. *Double and multiple star inclusion*

Another difficult point in the iteration procedure is the inclusion of double stars in the results. We have seen that the point giving the phase is RGC dependent and therefore is not a fixed point in the system of two or more stars. An additional task of the double and multiple star reduction is to determine for each RGC the position of this point with respect to the components as shown in section 8.2. In treating the abscissae for astrometric parameter determination, one has to correct abscissae in order to refer to one of the components and compute proper motion and parallax for this component. If orbital or linear motion has been found for the couple, it has also to be corrected for. Only then can one get the astrometric parameters of physical double or multiple stars.

9.4. *Expected final accuracies*

Clearly, they depend upon the duration of the mission. Very precise prediction were made for the nominal $2^{1/2}$ year mission. They were the following (Perryman et al., 1989)

$$\sigma_{pos} = 1.1 \text{ mas} \quad \left\{ \begin{array}{c} \sigma_\lambda \cos\beta = 1.3 \text{ mas} \\ \\ \sigma_\beta = 1.0 \text{ mas} \end{array} \right.$$

$$\sigma_\varpi = 1.5 \text{ mas}$$

$$\sigma_\mu = 1.6 \text{ mas yr}^{-1} \quad \left\{ \begin{array}{c} \sigma_{\mu\lambda} \cos\beta = 1.8 \text{ mas yr}^{-1} \\ \\ \sigma_{\mu\beta} = 1.4 \text{ mas yr}^{-1}. \end{array} \right.$$

They were based on an average abscissa standard error of 3.5 mas for a star of magnitude 9.

The conditions are now modified in the following manner :
- Only 65% of the data is useful instead of 100%.
- The abscissa determination seems to be about 10% better than expected (see table 3), but one has to take into account the degradation of the transmission which lowers the photon counts with time. This degradation is modelled by $1+(P-1)/24$, P being number of years of mission.
- The standard errors of proper motion improve as $P^{-3/2}$.
- The standard error of positions and parallaxes improve like $P^{-1/2}$.

From these, it results that the present expectations, as function of the duration P of the mission are given in table 4.

TABLE 4

P	σ_{pos}	σ_x	σ_μ
1.5 yr	1.9 mas	2.5 mas	4.3 mas yr^{-1}
2.0 yr	1.6 mas	2.2 mas	2.8 mas yr^{-1}
2.5 yr	1.5 mas	2.0 mas	2.1 mas yr^{-1}
3.0 yr	1.4 mas	1.9 mas	1.6 mas yr^{-1}
4.0 yr	1.3 mas	1.7 mas	1.1 mas yr^{-1}
5.0 yr	1.2 mas	1.5 mas	0.8 mas yr^{-1}

So, even with a five year mission one would get hardly the accuracy predicted for the nominal mission in positions and parallaxes, but a significant improvement (factor 1.6) in proper motions would be achieved. However, this is by all means better than the nominal 2 mas announced as the expected objective in 1980.

10. TYCHO Program

Looking back to the basic equation (5.8) that links star positions λ and β with respect to a celestial reference frame and the attitude angles ψ, θ, ϕ, provided that the calibrated Δu, Δv of the grid-to-field transformation are known, one sees that the definition of the celestial reference plays no role. Thus, it is not necessary to restrict it to the RGC reference frame, but it can as well be written in equatorial coordinates as in any other. Also, if in the attitude determination, λ and β are assumed known, ψ, θ and ϕ being unknown, the reverse could just as well be true. This is the idea behind what is called the TYCHO program or experiment which uses the star-mapper observations of stars while the attitude is provided by the HIPPARCOS main mission, the objective being to determine positions and whenever possible, proper motions and parallaxes of all observable stars crossing the star-mapper.

10.1. *Preparation of the reduction*

In contrast with what is provided by ESA for the main mission reduction, all the data continuously recorded by the star-mapper is analysed. This analysis cannot be made without some preparation so that one knows where to look for significant star signatures in the data. This is done using a TYCHO Input Catalogue which consists of about 3 million brightest stars on the sky. It has been obtained by merging the Hubble Space Telescope "Guide Star Catalog" which ranges between magnitudes 8-9 to 16-17 and the data stored in the Centre de Données Stellaires that is principally the HIPPARCOS Input Catalogue data base. The accuracy of the positions in the TYCHO Input Catalogue is of the order of 1" to 2". The limiting magnitudes are $B = 12.8$ and $V = 12.1$. It is believed that only 30 to 40% of these stars are bright enough to be recognized. However, this limit ensures that wrong magnitude estimates would not significantly bias the completeness of the survey.

This catalogue is being used together with the attitude data to predict the times of crossing the barycenter of each group of slits of the star-mapper. For this particular task, it is not necessary to wait for final accurate HIPPARCOS attitude and the on-board attitude is sufficiently good, since its precision matches the precision of the TYCHO Input Catalogue. Finally the prediction task provides also a *"predicted group crossing interval"* , interval of time centered on the predicted crossing time.

10.2. *Detection of stars*

A star appears in the star-mapper recordings as four peaks already described in section 4 superposed to a certain background. Since TYCHO program is aimed at measuring as many stars as possible, it is necessary to lower as much as possible the signal to noise ratio. For this reason, it is important to measure the background and then to subtract it from the signal.

Background is determined in building the distribution function of star counts over a *"frame period"* of five main mission observationperiods and comprising 6400 samples in each channel. The distribution is composed of a Poisson distribution in the lower part due to the background and a more irregular distribution due to bright stars. Both distributions are easily separated, and the value of the background is determined from the median value of the lower end of the distribution. Normally, the background is of the order of 3 counts per sampling period of 1/600s, but it increases largely when the satellite is in the outer Van Allen belts where it may be anything above 40 counts per sampling period, a limit at which even bright stars of magnitude 7 are no more detectable so that no attitude becomes available and, even if the data is recovered, no reduction is possible.

From time to time there is also another kind of perturbation called spikes. They are high counts during one or two sample periods, uncorrelated between the two channels. The origin is probably fluorescence due to high energy protons. Using a non linear filter, it is possible to suppress the most conspicuous spikes.

The detection of transits is done in a more involved manner than for the main mission. The reason is that one needs to get as many stars as possible, as close as possible to the background level, while for attitude, only well observed stars are useful. For this reason, several successive filtering are applied to the data and the counts in the four slits are folded together.

10.3. Identification
The preceding algorithm is continuously performed over all the available data. The detected signals may or may not correspond to real stars. It is necessary to recognize them from spurious transits.

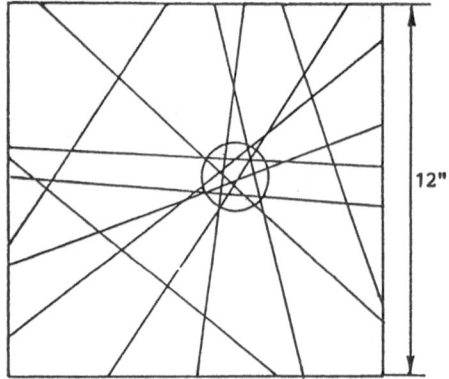

Figure 9 - Identification of star observations.

Let us represent a map of the sky around the predicted position of the star. The star observed at a certain time *t* is on the projection of the mean slit on the sky at this very instant. Its locus is therefore, a straight line on the map. If one draws these lines for all the observations of this star, one should obtain lines that converge to a single point - the actual position of the star - within the observing errors. Figure 9 shows an exemple of such a 12"x12" map. The circle represents the expected errors. One can see that three observations do not correspond to the star.

10.4. Astrometry
For the last part of the TYCHO data reduction, only recognized observations are kept. The equations of condition are very similar to those written in the sphere reconstitution. But the attitude being provided by the

HIPPARCOS main mission, the positions of the slits in the sky are exactly known. So, only the part relative to the star is kept, together with some instrumental parameters representing the grid to field transformation for each group of slits of the star-mapper.

The expected precision of the astrometric reduction of TYCHO data is essentially derived from the vertical slit observations, which are almost twice as precise as the inclined slit observations (see figure 4). The expected positional errors for a four year mission is :

8 mas for magnitude 9

15 mas for magnitude 10

30 mas for magnitude 11

These numbers are also representative of the expected errors in parallaxes and annual proper motions. This is less than for HIPPARCOS main mission, but, of course, the real interest of TYCHO is the fact that such results will be obtained for about a million stars.

REFERENCES

Green, R.M. (1985), *Spherical astronomy* (Cambridge University Press, Cambridge).

Kovalevsky, J. (1990), *Astrométrie Moderne* (Springer-Verlag, Heidelberg).

Perryman, M.A.C., Hassan, H. and 22 other authors, *The HIPPARCOS Mission, Pre-Launch Status* , Vol. 1, *The HIPPARCOS Satellite* , ESA SP-1111 (European Space Agency, Paris).

Perryman, M.A.C., Turon, C. and 21 other authors, Vol. 2, *The INPUT Catalogue* , ESA SP-1111 (European Space Agency, Paris).

Perryman, M.A.C., Lindegren, L., Murray, C.A., Hoeg, E., Kovalevsky, J. and 49 other authors, *The HIPPARCOS Mission, Pre-Launch Status* , Vol. 3, *The HIPPARCOS Satellite* , ESA SP-1111 (European Space Agency, Paris).

Shao, M., Colavita, M.M., Hines, B.E., Staelin, D.H., Hutter, D.J., Johnson, K.J., Mozurkewitch, D., Simon, R.S., Hershey, J.L., Hughes, J.A. and Kaplan, G.H. (1988), *The Mark III stellar interferometer* , Astronomy and Astrophysics, Vol. 193, p. 357-371.